The Hazardous Waste Q&A

Revised Edition

Other Books by Travis P. Wagner:

The Complete Guide to the Hazardous Waste Regulations
1991, Published by Van Nostrand Reinhold

Hazardous Waste Identification and Classification Manual
1990, Published by Van Nostrand Reinhold

The Hazardous Waste Q&A

Revised Edition

An In-depth Guide To
The Resource Conservation and Recovery Act
and
The Hazardous Materials Transportation Act

Travis P. Wagner

VAN NOSTRAND REINHOLD
New York

Library of Congress Catalog Card Number
ISBN 0-442-01331-0

Printed in the United States of America

Van Nostrand Reinhold
115 Fifth Avenue
New York, New York 10003

Chapman and Hall
2-6 Boundary Row
London, SE1 8HN, England

Thomas Nelson Australia
102 Dodds Street
South Melbourne 3205
Victoria, Australia

Nelson Canada
1120 Birchmount Road
Scarborough, Ontario MIK 5G4, Canada

16 15 14 13 12 11 10 9 8 7 6 5 4 3 2

Library of Congress Cataloging-in-Publication Data

Wagner, Travis.
The hazardous waste Q & A : an in-depth guide to the Resource Conservation and Recovery Act and the Hazardous Materials Transportation Act / Travis P. Wagner. — Rev ed.
p. cm.
Includes index.
ISBN 0-442-01331-0
1. Hazardous wastes—Law and legislation—United States--Miscellanea. 2. Hazardous substances—Transportation—Law and legislation—United States—Miscellanea. I. Title. II. Title: Hazardous waste Q and A.
KF3946.Z9W34 1992
344.73'04622—dc20
[347.3044622] 92-15755
CIP

*To Roger and Sharon Meunier,
alias Mom and Dad, for
their continuous support
and inspiration.*

About the Author

Travis P. Wagner is one of the nation's leading experts in hazardous waste regulatory compliance. He is an environmental scientist for Science Applications International Corporation in Falls Church, Virginia. He assists various federal agencies in developing technical guidance and regulations for hazardous waste management. Mr. Wagner, a noted speaker and trainer, teaches Environmental Policy at the USDA Graudate School in Washington, D.C. Much of the extensive regulatory expertise on RCRA in *The Hazardous Q&A* came from the author's experience as a regulatory information specialist with the U.S.EPA's RCRA/Superfund Hotline, where he advised federal agencies, industry, congressional staff, lawyers, scientists, and engineers on thousands of regulatory and policy issues and problems. Mr. Wagner also is the author of *The Complete Guide to the Hazardous Waste Regulations* and the *Hazardous Waste Identification and Classification Manual*. He has also written for *Pollution Engineering, Hazmat World,* and *Environmental Waste Management*. Mr. Wagner received his degree in Environmental Science from Unity College in Maine.

Contents

Preface

This book focuses on the regulation of hazardous waste under the Resource Conservation and Recovery Act and the Hazardous Materials Transportation Act. The regulations spawned by these acts have evolved from a few simple mandates to a myriad of regulations that attempt to protect the environment and humans without unreasonably burdening the economy. The result is one of the most complex set of requirements known to our society. Unfortunately, because the responsible agencies (EPA and DOT) have limited budgets, regulatory compliance assistance is relatively nonexistent, even though potential public relations, potential fines, liability, citizen suits, and financial institutions' requirements for "clean" sites for financial transactions are making compliance increasingly important.

I wrote *The Hazardous Waste Q&A* to help those who have to deal with the regulations. I have presented the material in a question and answer format primarily to facilitate learning of this complicated subject.

Acknowledgments

The following people provided technical review and comment for which I am most grateful.

Kim Gotwals, SAIC
Charlotte Mooney, U.S. Environmental Protection Agency*
Ingrid Rosencrantz, SAIC
Hilary Sommer, Duke Power Co.
Andrew Teplitzky, U.S. Environmental Protection Agency*
Lance Traves, Clayton Environmental Services
Jim Wilson, Perry-Wagner, Inc.
Denise Wright, U.S. Environmental Protection Agency*

Special thanks to Kathie Gabriel for patiently word processing the entire book, Karen Burchard for editorial assistance, and Rebecca Dils for assisting with the figures and tables.

*The review and comment offered by this reviewer in no way signifies endorsement or approval by the U.S. Environmental Protection Agency of the contents of this book.

A Guide for Readers

This book is intended to assist anyone with an interest in or need to understand hazardous waste regulations. The book is based on the federal RCRA and HMTA regulations. Many states are authorized to operate part or all of the RCRA program in lieu of the federal program. Such state programs must be at least equivalent to and consistent with the federal program. Although many state programs are a "carbon copy" of the federal program, some states have adopted more stringent regulations. Thus, it is important to consult with the appropriate state agency to determine if there are additional requirements.

The book is generally patterned after Titles 40 and 49 of the Code of Federal Regulations (CFR). Because the CFR tends to be confusing and illogical in its presentation and arrangement of material, *THE HAZARDOUS WASTE Q&A* may deviate slightly from the CFR in its presentation. The sole purpose of this deviation is to facilitate ease in the understanding of the regulations.

To understand the hazardous waste regulations, you must become familiar with the definitions of terms used for this is where most regulatory confusion occurs. The terms are clearly marked and are defined in Appendix A. It cannot be stressed enough how important it is to be very careful when reading a regulation; check the definitions of terms.

The information contained in this book is based on the regulations, *Federal Register* preambles, OSWER Directives, Regulatory Interpretation Letters (RILs), Program Implementation Guidance (PIGs), EPA policy memos, EPA guidance manuals, DOT guidance manuals, and RCRA/Superfund Hotline Monthly Reports (HMR). I have tried to be as accurate and objective as possible and have not included material that judges the need, value, or practicality of the implementation of the hazardous waste regulations.

Some of the questions contained in this volume concern the application of regulations to specific situations. Because the regulations are complex, their applicability can depend heavily upon individual circumstances. The reader should, therefore, avoid reading more into the responses than has been provided. However, the questions do provide the information needed to understand typical regulatory situations and thus, the necessary tools to make a prudent decision. Contact the appropriate state or EPA Regional Office for additional information. The addresses and telephone numbers of the EPA Regional Offices and states are listed in Appendix G of this book.

You should note that compliance with applicable federal requirements does not necessarily relieve an individual from compliance with applicable state requirements.

Chapter 1

General Provisions

The Solid Waste Disposal Act (SWDA), enacted in 1965, was the first federal legislation that addressed the nation's solid waste management practices. This act was subsequently amended by the Resource Conservation and Recovery Act (RCRA) in 1976 and the Hazardous and Solid Waste Amendments of 1984 (HSWA). *Hazardous waste* was first addressed in 1976 under RCRA. These three acts together are commonly referred to as RCRA, pronounced wreckra."

The three major programs of RCRA are Subtitles C, D, and I. Subtitle C regulates hazardous waste; Subtitle D regulates solid waste (nonhazardous waste); and Subtitle I regulates underground storage tanks that hold petroleum products and hazardous substances, but does not include wastes. Table 1 shows the RCRA table of contents.

Title I of the Hazardous Materials Transportation Act (HMTA), which is administered by the Department of Transportation, regulates the transportation of hazardous waste in conjunction with RCRA.

Because the focus of this book is hazardous waste management, only Subtitle C is addressed. Subtitle C, hazardous waste management, incorporates 13 sections that serve as the basis for the development of the hazardous waste regulations that have been promulgated by the EPA. Subtitle C states what EPA must do to govern hazardous waste handling and disposal and provides EPA with the authority to carry out the provisions of the act. Specifically, the various sections of Subtitle C of RCRA state that "the Administrator [EPA] . . . shall promulgate regulations. . . ." Thus, the legislative branch passed a law requiring the execu-

TABLE 1 The RCRA table of contents.

THE SOLID WASTE DISPOSAL ACT

As amended by the Resource Conservation and Recovery Act of 1976, the Used Oil Recycling Act of 1980, the Solid Waste Disposal Act Amendments of 1980, and the Hazardous and Solid Waste Amendments of 1984.

Subtitle A—General Provisions

Subtitle B—Office of Solid Waste; Authorities of the Administrator

Subtitle C—Hazardous Waste Management

TABLE 1 *(cont.)*

Sec. 3016	Inventory of Federal Agency hazardous waste facilities
Sec. 3017	Export of hazardous waste
Sec. 3018	Domestic sewage
Sec. 3019	Exposure information and health assessment
Subtitle D—State or Regional Solid Waste Plans	
Sec. 4001	Objectives of subtitle
Sec. 4002	Federal guidelines for plans
Sec. 4003	Minimum requirements for approval of plans
Sec. 4004	Criteria for sanitary landfills; sanitary landfills required for all disposal
Sec. 4005	Upgrading of open dumps
Sec. 4006	Procedure for development and implementation of State plan
Sec. 4007	Approval of State plan; Federal assistance
Sec. 4008	Federal assistance
Sec. 4009	Rural communities assistance
Sec. 4010	Adequacy of certain guidelines and criteria
Subtitle E—Duties of the Secretary of Commerce in Resource and Recovery	
Sec. 5001	Functions
Sec. 5002	Development of specifications for secondary materials
Sec. 5003	Development of market for recovered materials
Sec. 5004	Technology promotion
Sec. 5005	Nondiscrimination requirement
Sec. 5006	Authorization of appropriations
Subtitle F—Federal Responsibilities	
Sec. 6001	Application of Federal, State, and local law to Federal facilities
Sec. 6002	Federal procurement
Sec. 6003	Cooperation with Environmental Protection Agency
Sec. 6004	Applicability of solid waste disposal guidelines to executive agencies
Subtitle G—Miscellaneous Provisions	
Sec. 7001	Employee protection
Sec. 7002	Citizen suits
Sec. 7003	Imminent hazard
Sec. 7004	Petition for regulations; public participation
Sec. 7005	Separability
Sec. 7006	Judicial review
Sec. 7007	Grants or contracts for training project
Sec. 7008	Payments
Sec. 7009	Labor standards

TABLE 1 *(cont.)*

Sec. 7010	Interim control of hazardous waste injection
Sec. 7012	Law enforcement authority
Subtitle H—Research, Development, Demonstration, and Information	
Sec. 8001	Research, demonstrations, training, and other activities
Sec. 8002	Special studies; plans for research, development, and demonstrations
Sec. 8003	Coordination, collection, and dissemination of information
Sec. 8004	Full-scale demonstration facilities
Sec. 8005	Special study and demonstration projects on recovery of useful energy and materials
Sec. 8006	Grants for resource recovery systems and improved solid waste disposal facilities
Sec. 8007	Authorization of appropriations
Subtitle I—Regulation of Underground Storage Tanks	
Sec. 9001	Definitions
Sec. 9002	Notification
Sec. 9003	Release detection, prevention, and correction regulations
Sec. 9004	Approval of State programs
Sec. 9005	Inspections, monitoring, and testing
Sec. 9006	Federal enforcement
Sec. 9007	Federal facilities
Sec. 9008	State authority
Sec. 9009	Study of underground storage tanks
Sec. 9010	Authorization of appropriations

tive branch to develop and implement regulations governing hazardous waste.

APPLICABILITY AND IMPLEMENTATION OF RCRA

Applicability

Q: Who is subject to Subtitle C of RCRA?

A: Any person who generates, transports, or manages a *hazardous waste* is subject to Subtitle C of RCRA. Thus, determining what is a hazardous waste is the key question, because only those wastes that meet the definition of hazardous waste are subject to Subtitle C.

Q: Under RCRA, is there a difference between hazardous waste and toxic waste?

A: Yes; although the terms *hazardous waste* and *toxic waste* are often used interchangeably; in regard to RCRA regulations there is an important distinction between the two. Hazardous waste denotes a regulated waste; only certain waste streams are designated as hazardous. This designation is not based solely on toxicity; it also includes other physical characteristics that present an environmental or health threat, as well as the quantity generated, damage case history, and environmental fate. Toxic waste is a hazardous waste that is specifically regulated because of its human toxicity. Every waste stream is toxic to some degree, but it is not necessarily hazardous. Thus, those wastes that pose a serious threat when mismanaged are differentiated.

Q: How long have hazardous wastes been regulated under RCRA?

A: RCRA was enacted in 1976, but it wasn't until May 19, 1980, that EPA completed the first phase of the RCRA hazardous waste regulatory program. The May 19th rule was not effective until November 19, 1980, when hazardous waste management facilities had to either close or comply with RCRA.

Q: Does RCRA regulate hazardous wastes disposed of before November 19, 1980?

A: Generally no; except for a few provisions addressed in Chapter 10 (i.e., corrective action), hazardous wastes disposed of before November 19, 1980, are not regulated under RCRA, but are subject to the Comprehensive Environmental Response, Compensation, and Liability Act (CERCLA or Superfund).

RCRA is written in the present tense and its regulatory scheme is prospective. EPA believes that Congress intended the hazardous regulatory program under Subtitle C of RCRA to control hazardous waste management activities that take place after the effective date of the Phase I regulations (November 19, 1980). Thus, the Subtitle C regulations did not by their terms apply to inactive (either closed or abandoned) disposal facilities. This statement can be found in the May 19, 1980, *Federal Register* (45 *FR* 33170). However, hazardous waste placed in a surface impoundment, tank, or drum before November 19, 1980 and remaining in storage after that date is subject to RCRA, because this is considered active storage. Active storage, on or after November 19, 1980, is subject to RCRA requirements based on *Environmental Defense Fund v. J. Lamphier,* 714 F.2N.D. (1983).

Rule making

Q: How do provisions of a law become a regulation?

A: The executive branch (e.g., EPA, DOT) implements the intentions of laws passed by Congress by developing and enforcing rules or regulations.

RCRA specifically requires EPA to promulgate regulations only after public notice and the opportunity for public comment and hearings in accordance with the Administrative Procedures Act. EPA meets the public notification requirements by publishing notices of its intentions to promulgate regulations in the *Federal Register.*

Q: What is meant by promulgation?

A: A regulation is *promulgated,* made known to the public, by being signed by an agency head and then published as a final rule in the *Federal Register.*

***Q: What is the* Federal Register?**

A: The *Federal Register* is a compendium of notices, announcements, and descriptions of the activities of the federal government. It is published every business day by the federal government (see Appendix G for further information).

Q: How does the general rule-making procedure work?

A: The general rule-making procedure is as follows. First, a proposed rule is developed by an agency and published in the *Federal Register.* The proposed rule establishes a public comment period, usually 60 days, and a public hearing if requested by at least five people. After the public comment period closes, the proposed rule is revised as appropriate in light of the comments. A final rule is subsequently promulgated.

Q: If an agency publishes a proposed rule, is it obligated to promulgate a final rule?

A: No; when a notice of proposed rule making is published by an agency, there is no obligation to promulgate a final rule unless Congress has mandated otherwise. The agency may formally withdraw the proposal and close the administrative docket through a *Federal Register* notice.

Q: Can an agency promulgate a final rule that is significantly different from the proposed rule?

A: No; only modest adjustment or deletion of portions of the rule is generally allowed. If there is a need to make a rule significantly more stringent or broader in scope than originally proposed, the agency must publish another notice of proposed rule making describing the revised action and soliciting additional public comments.

Q: Can an agency later change a promulgated final rule?

A: Yes; an agency can change a promulgated final rule by publishing proposed amendments, amendments to a final rule, interim final amendments, technical corrections, technical amendments, and clarification notices.

Q: What is the format of a published rule?

A: A proposed, final, or amended rule consists of three major parts: the heading, preamble, and text.

Q: What is contained in the heading?

A: The heading of the notice supplies identifying and descriptive information. This information includes who published the notice (e.g., EPA, DOT), the regulations affected, and a descriptive title of the program addressed (e.g., Hazardous Waste Management System; Waste Identification).

Any phrase used in the heading of a notice that contains the words *rule* or *amendment* and does not include the word *proposed,* means that the notice is promulgating a new, or changed regulation. This is important because everyone affected by the regulation will have a finite period to comply, that period often beginning on the date the notice is published in the *Federal Register*. Proposed rules and amendments are published to inform the public of their content, and these types of notices usually establish a time span during which the public is invited to comment on the proposed rules and amendments.

Q: What is contained in the preamble of a rule?

A: The preamble contains a summary of the action being taken, effective dates, addresses, and contacts for additional information.

The major portion of the preamble provides information explaining the purpose and intent of the regulations, why they are written as they are, who they are intended to affect, and other information used in support of the regulations. The preamble is written in "nonregulatory" language and, most importantly, although it explains the intent of the regulatory language, nothing in the preamble carries the force of the law.

Q: What is contained in the text of a rule?

A: The preamble is followed by the text of the actual regulations as either proposed or promulgated. This part of the notice contains the regulations as they will read when they become effective after promulgation as a final rule. When promulgated, they become part of the Code of Federal Regulations (CFR) and carry the full force of the law.

Q: How is the Federal Register *referenced?*

A: Each volume (year) of the *Federal Register* is consecutively page-numbered and each daily issue receives a consecutive number. The following heading of a typical notice serves as an example.

"226091/Federal Register/Vol. 57, No 245/Thursday, December 21, 1992/ Final Rule"

> The above heading indicates that this issue of the *Federal Register* was published on Thursday, December 21, 1992; it is the 245th daily issue published in 1992; it is part of Volume 57 (the 1992 volume); it is from the Final Rules section of that day's issue; and it is page 226091. The accepted notation for identifying this page for reference purposes is 57 *FR* 226091. However, since that notation does not identify which daily issue that page is contained in, the date is often included (i.e., December 21, 1992).

Q: Beyond statutory language and promulgated regulations, is there other information available to assist the regulated community?

A: Yes; further clarification of regulations is provided through the issuance of guidance documents and policy directives.

Q: What are guidance documents and policy directives?

A: Guidance documents are issued primarily to elaborate and provide direction on the implementation of a regulation. They essentially explain how to do something. In contrast, policy directives specify procedures that must be followed pertaining to a regulation. (See Appendix G for information on obtaining these materials.)

Code of Federal Regulations

***Q: What is the* Code of Federal Regulations?**

A: The Code of Federal Regulations, often denoted by the acronym CFR, is the compilation of all final Federal regulations in effect in the United States. The full text of all final regulations (not including the preamble) promulgated by all Federal government agencies is included in the CFR (see Appendix G).

Q: What is the format of the CFR?

A: The regulations are grouped under "Titles" in the CFR, and each Title is divided into "Chapters." The hazardous waste regulations can be found in "Title 40—Protection of Environment" and "Title 49—Transportation." Title 40 and 49 are generally identified or annotated as "40 CFR" and "49 CFR."

Each Chapter of a Title is divided into numerous "Parts." Each Part is further divided into "Subparts." The Subparts are comprise "Sections." Sections are sometimes expressed with the symbol "§"

Q: What are the codified Parts of Subtitle C of RCRA?

A: The codified Parts of Subtitle C of RCRA, contained in Title 40 CFR, are:

Part 124—Procedures for decisionmaking (permits)
Part 260—Definitions, petitions for rulemaking changes, delisting procedures
Part 261—Hazardous waste identification
Part 262—Generator standards
Part 263—Transporter standards
Part 264—Treatment, storage, and disposal facilities: final operating standards
Part 265—Treatment, storage, and disposal facilities: interim status standards
Part 266—Hazardous waste fuel burned for energy recovery, used oil fuel, and specific recycling activities
Part 268—Land disposal restrictions

Part 270—Permits and operation during interim status
Part 271—State programs

Q: What are the codified Parts of HMTA?

A: The codified Parts of HMTA, contained in Title 49 CFR, are:

Part 106—General procedures, and petitions for rulemaking, hearings, and reconsideration
Part 107—Program procedures and exception procedures
Part 171—Definitions
Part 172—Hazardous Materials Table, shipping papers, shipper's certification, package markings, labeling, and placarding
Part 173—Shipper and shipping requirements
Part 177—Highway shipments
Part 178—Packaging requirements

Petitions

Q: Is there a procedure under RCRA for a person to have a regulation modified or revoked?

A: Yes; RCRA contains provisions (Subpart C of Part 260) that allow anyone to petition the EPA Administrator to modify or revoke any regulation promulgated under Parts 260 through 265.

Q: What is required in a petition?

A: The general information required in a petition includes:

- Petitioner's name and address
- A statement of the petitioner's interest in the proposed action
- A description of the proposed action, including (where appropriate) suggested regulatory language
- A statement of the need and justification for the proposed action, including any supporting tests, studies, or other relevant information

Q: Is this the only information required for a petition?

A: No; other information is required depending on the particular petition. The specific information required (and available petitions) is contained in the following regulatory sections:

Section 260.20—General requirement to revoke or modify a regulatory provision
Section 260.21—Petitions for equivalent testing or analytical methods
Section 260.22—Petitions to amend Part 261 to exclude a waste produced at a particular facility ("delisting")
Sections 260.30-31—Variances from classification as a solid waste
Sections 260.32-33—Variances to be classified as a boiler
Sections 260.40-41—Variances from regulations of certain hazardous waste recycling activities

Q: What is the process for submitting a petition?

A: There is no specified form or application for submitting a petition, but the regulations specify in detail the information that must be presented. Petitions should be sent by certified mail to:

The Administrator (A-100)
U.S. Environmental Protection Agency
401 M Street, SW
Washington, DC 20460

Q: How does EPA respond to a petition?

A: Pursuant to Section 260.20, the EPA Administrator must make a tentative decision to grant or deny a petition and publish a notice of the tentative decision, either in the form of an advanced notice of proposed rule making (ANPR), a proposed rule, or a tentative determination to deny the petition in the *Federal Register*. An informal public hearing may be held if the Administrator decides that one is warranted or if one is requested.

After evaluating public comments, the Administrator must make a final decision by publishing in the *Federal Register* a regulatory amendment or a denial of the petition.

Q: A generator submitted a petition under Section 260.22 to amend Part 261 to exclude a hazardous waste produced at a particular facility, but EPA's final decision was to deny the petition. What options does the generator have for appealing EPA's decision to deny the petition?

A: A generator who has had a petition denied by EPA may appeal to the U.S. Court of Appeals for the District of Columbia Circuit, pursuant to Section 7006(a)(1) of RCRA.

The generator may also wish to petition EPA for reconsideration of the decision. A petition for reconsideration does *not* extend the time to file suit in court, because denial of a petition is a final agency action.

Chapter **2**

Waste Identification and Classification

Determining that a waste is a RCRA hazardous waste is perhaps the most important and by far the most complex step in ascertaining one's responsibility under RCRA. Careful attention should be paid to each step in this determination. Although the regulations use a step-by-step identification process, many factors can affect the classification of a waste, and they may not necessarily appear in a logical order. This is due mainly to EPA's redefinition of solid waste, and because certain hazardous wastes are subject to special considerations under various sections of RCRA.

HAZARDOUS WASTE IDENTIFICATION

Solid Waste

Q: What is the protocol in determining if a waste is hazardous?

A: There is a basic step-by-step process in determining if a waste is hazardous.

Q: What is this step-by-step process in determining if a waste is hazardous?

A: The waste identification process is as follows:

1. Determine if the material is defined as a solid waste.
2. Determine if the solid waste is excluded from regulation.

3. If the waste is not excluded, determine if the waste is specifically listed as a hazardous waste.
4. Regardless of whether the waste is listed, determine if the waste exhibits a characteristic of a hazardous waste (ignitable, corrosive, reactive, or toxic).
5. If the waste is neither listed nor exhibits an above mentioned characteristic, the waste is not regulated under Subtitle C of RCRA. If the waste is hazardous, determine if it will be recycled and how.

Q: What is a solid waste?

A: A solid waste is any "discarded material" that is not excluded by Section 261.4(a) or that has been delisted. (Delisting is the process for a generator to have a listed hazardous waste classified to a nonhazardous waste through a petition.)

The term *solid waste* does not refer to its physical state; it is a regulatory term only. Thus, industrial wastewater may be a solid waste.

Q: What is the definition of a discarded material?

A: A *discarded material* is any material that is disposed, stored, or treated before its disposal; that is burned as a fuel, treated, recycled, abandoned, considered inherently wastelike (e.g., certain dioxin wastes); or that is stored or accumulated before recycling. There are some exceptions to the definition of solid waste regarding specific recycling activities [Section 261.6(a)(2)].

Q: If the waste is considered a solid waste, what is the next step?

A: Assuming a waste meets the definition of solid waste, the second step is to determine if the waste is a listed hazardous waste, meaning, is the waste or the process that generated the waste specifically listed in 40 CFR Part 261? Assuming that it is not, the third and final step is to determine if the waste exhibits a characteristic of hazardous waste. The four characteristics are to be applied to each waste stream that does not meet a listing.

Excluded Wastes

Q: What wastes are excluded?

A: The wastes listed below are not subject to Subtitle C requirements, because they are either excluded from the statutory definition of solid waste (Section 3001 of RCRA) based on EPA's interpretation, not intended by Congress to be regulated under Subtitle C based on RCRA's legislative history indicating Congressional intent, or subject to regulation under other EPA statutes.

The excluded wastes under Section 261.4 are:

- Domestic sewage
- Any mixture of domestic sewage and any other waste that passes through a sewer system to a publicly owned treatment works (POTW)
- Irrigation return flows
- Source, special nuclear, or by-product material as defined by the Atomic Energy Act (AEA)
- Materials subjected to in situ mining techniques that are not removed from the ground during extraction
- Certain pulping liquors used in the production of paper
- Spent sulfuric acid used to produce virgin sulfuric acid
- All household wastes and resource recovery facilities that burn only household waste. (Hotel, motel, septic sewage, and campground waste are all considered household waste.)
- Materials returned to the soil as fertilizers, such as manure and crops
- Mining overburden returned to the mine site
- Fly ash waste, bottom ash waste, slag waste, and flue gas emission control waste generated primarily from the combustion of coal or other fossil fuels (the "utility waste exemption")
- Drilling fluids, produced waters, and other wastes associated with the exploration, development, or production of crude oil, natural gas, or geothermal energy
- Specific wastes from the tannery industry containing primarily trivalent chromium instead of hexavalent chromium
- Solid waste from the extraction and beneficiation of ores and minerals, including phosphate rock overburden from uranium mining (the "mining waste exclusion," or the Bevill Amendment)
- Cement kiln dust
- Discarded wood that fails only the Characteristic Toxicity Test (a test to determine if a waste exhibits a hazardous characteristic) for arsenic as a result of being treated with arsenical compounds

***Q:** Concerning the domestic sewage exemption, can a person legally discharge hazardous waste into a sewer system?*

A: A person can dispose of hazardous waste into a POTW system under RCRA, however, storage or treatment prior to discharging into a POTW is not excluded from RCRA and is subject to applicable storage regulations. In addition, it is important to note that these facilities are subject to applicable pretreatment standards under the Clean Water Act (CWA). POTWs have local jurisdiction. Thus, they can legally prohibit discharges that interfere with the operation of their system or that may cause them to violate their National Pollution Discharge Elimination System (NPDES) permit.

***Q:** Are there exclusions available under RCRA for hazardous waste?*

A: Yes; there are conditional exclusions for wastes generated in a product-storage tank and for laboratory samples.

Product-Storage Tanks

***Q:** What is the exclusion concerning product-storage waste?*

A: Any waste generated in a product-storage tank, raw material storage tank, product-transport vessel or vehicle, or manufacturing process unit is not subject to regulation until it is removed from that unit or if the waste remains in the unit 90 days after the unit ceases operation [Section 261.4(c)].

Samples

***Q:** What is the exclusion for laboratory samples?*

A: Under Sections 261.4(d) and (e), there are exclusions for two types of samples: samples being analyzed for the purpose of hazardous waste identification and samples being used for treatability studies.

***Q:** What is the exclusion concerning samples sent for identification purposes?*

A: For a sample to be excluded under Section 261.4(d), it must be in the process of being analyzed for the sole purpose of hazardous waste identification, in which case the sample would be excluded from RCRA during its storage and transportation. However, there may be applicable Department of Transportation (DOT) or U.S. Postal Service requirements.

Q: What is the status of the sample after it is analyzed?

A: Once the sample has been analyzed, it must be sent immediately back to the sample collector. If the laboratory keeps the sample or does not send it back to the original sample collector, the sample is no longer excluded and is subject to applicable regulations. If the sample is sent back to the collector, the collector becomes the generator. If the laboratory keeps the sample, it would be considered the generator.

Q: Is there a specified size limit for these samples?

A: Yes; there is an approximate size limit of 1 gallon for the sample exclusion (46 *FR* 47427, September 25, 1981).

Q: What is the exclusion for treatability samples?

A: People who generate or collect samples for the purpose of conducting treatability studies are not subject to Parts 261 through 263 nor are the samples included in the quantity determinations for generator categorization [Sections 261.4(e) and (f)].

Q: What is a treatability study?

A: As defined in Section 260.10, "a *treatability study* means a study in which a hazardous waste is subjected to a treatment process to determine: (1) whether the waste is amenable to the treatment process, (2) what pretreatment (if any) is required, (3) the optimal process conditions needed to achieve the desired treatment, (4) the efficiency of a treatment process for a specific waste or wastes, or (5) the characteristics and volumes of residuals from a particular treatment process." A treatability study is not a means to commercially treat or dispose of hazardous waste.

Q: What treatability studies are covered under the exemption?

A: EPA has listed (53 *FR* 27293, July 19, 1988) several types of studies that are exempted:

- Liner compatibility studies
- Corrosion studies
- Toxicological and health effects studies
- Material compatibility studies relating to:
  - Leachate collection systems
  - Geotextile materials
  - Pumps
  - Personal protective equipment

Q: When is a treatability sample exempt from the regulations?

A: The regulations [Section 261.4(e)] exempt waste samples from Subtitle C when:

- The sample is being transported to the laboratory for testing or is being transported back to the sample collector after testing
- The sample is being stored by the sample collector or laboratory before testing or after testing prior to its return to the generator
- The sample is being analyzed to determine its characteristics or composition
- The sample is being stored at the laboratory for a specific purpose, such as a court case or enforcement action

Samples subject to the exemption must still comply with DOT and U.S. Postal Service (USPS) shipping requirements.

Q: What if a sample is not subject to DOT or USPS regulations?

A: If DOT, USPS, or other shipping requirements do not apply, the following information must accompany the sample:

- The name, mailing address, and telephone number of the originator of the sample

- The name, address, and telephone number of the facility that will perform the treatability study
- The quantity of the sample
- The date of shipment
- A description of the sample, including its EPA hazardous waste number

Q: What are the quantity limits?

A: For each process being evaluated for each generated waste stream, a generator or sample collector is limited to no more than 1000 kg for a nonacute hazardous waste; 1 kg of acute hazardous waste; or 250 kg of soil, water, or debris contaminated with acute hazardous waste.

Q: What are the requirements for facilities conducting treatability requirements?

A: Facilities conducting treatability testing are excluded from permit requirements provided they:

- Notify EPA at least 45 days before conducting a treatability study
- Obtain an EPA identification number
- Test no more than 250 kg of hazardous waste per day
- Maintain records that document compliance with the treatment rate limits, storage time, and quantity limits
- Maintain on-site all treatability contracts and shipping papers for at least 3 years
- Submit an annual report to EPA, by March 15, that estimates the number of studies and the amount of waste expected to be used in treatability studies during the current year. The report covering the previous year must include information on:
  - The name, address, and EPA identification number of the facility conducting the treatability studies
  - The types (by process) of treatability studies conducted
  - The names and addresses of those for whom studies have been conducted (including their EPA identification numbers)
  - The total quantity of waste in storage each day
  - The quantity and types of waste subjected to treatability studies
  - When each treatability study was conducted
  - The final disposition of residues and unused samples from each treatability study
- Notify EPA by letter when a facility is no longer planning to conduct treatability studies

Q: How are mobile treatment units (MTUs) classified concerning treatability studies?

A: EPA has determined that MTUs conducting treatability studies can qualify for this exemption (53 *FR* 27297, July 19, 1988).

Q: What are the requirements for MTUs?

A: Each MTU or group of MTUs operating at the same location is subject to the treatment rate, storage, and time limitations and the notification, record-keeping, and reporting requirements that are applicable to stationary laboratories or testing facilities conducting treatability studies. That is, a group of MTUs operating at one location will be treated as one MTU facility for purposes of Sections 261.4(e) and (f). Furthermore, these requirements apply to each location where an MTU will conduct treatability studies.

Listed Hazardous Wastes

Q: What is a "listed" hazardous waste?

A: EPA has specifically listed wastes in Sections 261.30 through 261.33 that are presumed to be hazardous. If a waste meets the listing description it is presumed to be hazardous regardless of its concentration. However, generators have the opportunity to demonstrate that a particular listed waste is not hazardous by petitioning to delist a waste at a particular generation site based on specified criteria.

The listed hazardous wastes consist of wastes from nonspecific sources (F code), wastes from specific sources (K code), and commercial products (U and P code).

The RCRA-listed hazardous wastes can be found in 40 CFR Sections 261.31 through 261.33, as well as in Appendix C of this book.

Q: What criteria does EPA use to list wastes as hazardous?

A: EPA has listed wastes based on their toxicity, reactivity, corrosivity, and ignitability. For hazardous wastes listed because they meet the criteria of toxicity, EPA's principal focus is on the identification and concentration of the waste's constituents and the nature of the toxicity presented by the constituents. If a waste contains significant concentrations of hazardous waste constituents, EPA is likely to list the waste as hazardous unless it is evident that the waste constituents would not be mobile or persistent should they migrate (Section 261.11).

Q: Is EPA's listing information documented and available to the public?

A: Yes; a detailed explanation for each hazardous waste listing is contained in EPA's *Listing Background Documents*. The listing documents are organized into the following sequence: (1) the EPA basis for listing the waste or waste stream; (2) a brief description of the industries generating the listed waste stream; (3) a description of the manufacturing process or other activity that generates the waste and identification of waste composition, constituent concentrations, and annual quantity generated; (4) a summary of the adverse health effects of each of the waste constituents of concern; and (5) a summary of damage case histories involving the waste.

Q: What is the difference between acutely and nonacutely hazardous waste?

A: Any waste that is designated as an acutely hazardous waste (H) is subject to reduced weight limits regarding generator categories and more stringent requirements concerning the determination of empty containers. All P-code wastes and F020, F021, F022, F023, F026, F027, and F028 are all designated as acutely hazardous waste.

Wastes from Nonspecific Sources

Q: What are the wastes from nonspecific sources?

A: The first category of listed hazardous wastes is generally material-specific wastes generated by a variety of processes. These wastes are further broken down into solvent wastes, electroplating wastes, and dioxin wastes.

Q: What is a solvent waste?

A: Solvent wastes are designated as wastes F001 through F005. For a waste to be classified as a solvent waste, the purpose of the material must have been to mobilize or solubilize a constituent. Thus, if a material was used solely as a reactant or a feedstock, it is not classified as a solvent waste (OSWER Directive No. 9444.08).

Q: If a waste stream contains various solvent wastes (F001 through F005), is the entire waste stream considered a hazardous waste?

A: The mere presence of any of the components listed in F001 through F005 in a waste does not necessarily make it a listed hazardous waste. For

F001, F002, F004, and F005, it will meet the listing criteria if it contains 10 percent or more of a given solvent that is listed. It can also meet the listing criteria if it is a mixture of 10 percent or more of the listed solvents. For F003, the mixture rule applies in a modified form. According to the rule, mixtures containing F003 solvents are covered under the listings only under two conditions: (1) the mixture contains only F003 constituents, or (2) the mixture contains one or more F003 constituents and 10 percent or more of the other listed solvents prior to use (50 *FR* 53315, December 31, 1985).

Q: How are the following mixtures classified under RCRA?

A: *1. Solvent containing 15 percent xylene (F003), 15 percent toluene (F005), and 70 percent Water.*
2. Solvent containing 80 percent xylene (F003), 5 percent methylene chloride (F001), and 15 percent water.
3. Solvent containing 80 percent xylene (F003) and 20 percent water.

A: 1. The first mixture when spent would be a listed hazardous waste under RCRA, designated as an F005/F003 waste.
2. The second mixture is not a listed waste because the methylene chloride (F001) concentration prior to use is less than 10 percent and it contains constituents other than F003. The mixture, however, will probably be ignitable and therefore classified as D001.
3. The third mixture is a listed waste if it is considered to be a commercial or technical grade xylene solution.

Q: Many dry cleaning facilities use carbon-filter cartridges to filter the solvent perchloroethylene (F002). When the cartridges are removed for disposal, are they classified as a hazardous waste?

A: Yes; these filters are classified as F002 hazardous waste because they contain a spent listed solvent. The weight of the entire filter is counted to determine the appropriate generator category.

Q: What are the electroplating wastes?

A: The wastes F006 through F012 and F017 are electroplating wastes. EPA defines electroplating as anodizing and chemical etching and milling (51 *FR* 43351, December 2, 1986). Further clarification of the individual processes is included in EPA's listing background documents.

Q: What are the dioxin wastes?

A: The wastes F020 through F023 and F026 through F028 are the dioxin wastes classified as acutely hazardous. These wastes are hazardous regardless of whether dioxin is present or not. Dioxin itself is not a listed hazardous waste, but is a contaminant in certain organic processes.

Wastes from Specific Sources

Q: What are the wastes from specific sources?

A: The second category of listed hazardous wastes contains those generated from specific sources. These listings, under the designation of K codes, refer to hazardous wastes from specific industrial processes, such as *untreated process wastewater from the production of toxaphene* (K098).

Commercial Chemical Product Wastes

Q: What are the commercial chemical product wastes?

A: The third category of listed hazardous wastes is commercial chemical products, designated by either a U or P code. The P-code wastes are considered acutely hazardous and are subject to more stringent requirements concerning empty containers and generator weight limits. For a waste to be categorized as a U or P waste, it must be a commercial chemical product in an unused form. The definition of commercial chemical products includes technical grades, pure forms, off-specification products, sole-active-ingredient products, or spill or container residues of these products.

If a commercial chemical is used and subsequently becomes spent, it does not meet any of the U or P listings, but may meet one of the other F or K listings or exhibit a characteristic (45 *FR* 78540, November 25, 1980). For example, if a person were to dispose of unused methylene chloride, it would be considered a U080 hazardous waste. However, if that person used the methylene chloride as a solvent and wanted to dispose of it, it would be classified as a F001 hazardous waste.

Q: If an unused chemical product contains more than one listed U-code commercial chemical, is it a U-code waste when disposed?

A: No; if a material contains more than one active ingredient that is a listed U- or P-code material, it does not receive the U or P listing and can become hazardous only if it exhibits a hazardous characteristic.

Q: When does a commercial chemical product become a hazardous waste?

A: A commercial chemical product is not considered a hazardous waste until it is intended to be discarded or if it is spilled, in which case the spill cleanup residue attains the appropriate U- or P-code listing. Thus, it can be stored indefinitely without RCRA restraints if the intent is to use it or to have it recycled.

Q: A state health department deemed a house unfit for habitation due to excessive chlordane levels in the soil around the house. The soil contaminated with chlordane was removed and placed in drums. The removal resulted in 45 drums of contaminated soil with an average chlordane concentration of 50 ppm. Is this contaminated soil a hazardous waste?

A: The contaminated soil is not a listed hazardous waste. Chlordane could be considered a listed hazardous waste only if it were discarded prior to use or were a spill residue. If chlordane met any one of these criteria, it would be a listed hazardous waste (U036). In this situation, however, none of these criteria were met, because chlordane was applied as a commercial chemical product. The generator must still determine if the contaminated soil exhibits a characteristic of hazardous waste (ignitability, corrosivity, reactivity, or toxicity). If the soil does not exhibit a characteristic, then it is not a RCRA hazardous waste.

Q: A commercial chemical product contains two active ingredients, methylene chloride and toluene. This unused product is spilled on the ground. How are the spill and spill residue classified?

A: If a commercial chemical product with more than one active ingredient is spilled, it will not be classified as U or P spill residue because, in the above example, it contains more than one active ingredient. However, if the contaminated soil exhibits a characteristic of hazardous waste, then the soil mixture is a hazardous waste.

Q: A facility generates beryllium dust from finishing operations. Is this waste beryllium considered a listed waste under P015?

A: No; the listing of beryllium dust, like all other substances listed in Section 261.33, includes only unused commercial chemical products. Beryllium dust (P015) is a raw material and should be labeled as such.

Q: Is asbestos a hazardous waste?

A: No; asbestos was listed as a hazardous waste (U013) in the May 19, 1980 *Federal Register* (45 *FR* 33126). However, on November 25, 1980 (45 *FR* 78532), asbestos was removed from the list due to overlapping regulation for the disposal of asbestos under the Clean Air Act's National Emission Standards for Hazardous Air Pollutants (NESHAPS).

Q: Are polychlorinated biphenyls (PCBs) a hazardous waste?

A: No; PCBs are not a listed hazardous waste under RCRA. PCBs themselves will not exhibit a characteristic of hazardous waste. Currently, they are regulated under the Toxic Substances Control Act.

Q: Pursuant to TSCA, a PCB transformer must be decontaminated using an appropriate solvent. If toluene (>10 percent) is used, is the resultant PCB/toluene mixture regulated under RCRA (as F005) or TSCA?

A: If a waste mixture contains a listed RCRA hazardous waste and PCBs (greater than 50 ppm), the waste must be disposed of in accordance with TSCA [Section 268.42] and managed in compliance with RCRA. [50 *FR* 49261, November 29, 1985 and Section 761.1(e)]. This means that the mixed waste could only be stored by the generator for only 90 days, and the waste must be disposed of by a TSCA incinerator with a 99.9999 percent destruction and removal efficiency for wastes containing >500 ppm PCB.

Empty Containers

Q: What is the status of hazardous waste residues remaining in an empty container?

A: Any hazardous waste remaining in a container that is considered *empty* under Section 261.7 is no longer a hazardous waste.

There are definitions for nonacutely and acutely hazardous waste containers, paper bags, compressed gas, and tanks.

Q: What is the definition of an empty nonacutely hazardous waste container?

A: A nonacutely hazardous waste (e.g., most F, K, and U wastes) container is considered *empty* if it is thoroughly emptied using common industry practices and contains less than 1 inch of residue on the bottom or less than 3 percent by weight for containers less than 110 gallons or less than 0.3 percent by weight for containers greater than 110 gallons (47 *FR* 36098, August 18, 1982).

Q: How is the one inch measured if the container is not flat?

A: If a container bottom is not flat (e.g., rounded or conical) the 1 inch is to be measured at the deepest point (47 *FR* 36093, August 18, 1982).

Q: A generator has several drums that contain less than 0.5 inch acetone on the bottom. Because there is less than 1 inch of hazardous waste residue remaining in the drum, is this drum considered an empty container and, thus, not subject to further regulation?

A: No; Section 261.7(b)(1)(i) requires that all wastes must be removed using common industry practice, such as pouring and pumping. In this case, a drum contains liquid acetone that probably could have been removed further by common industry practices. Thus, these containers would not be considered empty.

Q: What is the definition of an empty acutely hazardous waste container?

A: A container or liner holding an acutely hazardous waste (P-code and certain F-code wastes) is considered *empty* if it is triple-rinsed with an appropriate solvent, rinsed using another method shown to be equivalent, or has the liner removed (47 *FR* 36094-5, August 18, 1982).

Q: What is the definition of an empty paper bag?

A: A paper bag can meet the definition of an empty container by repeatedly beating an inverted bag to ensure thorough emptiness (OSWER Directive No. 94411.15).

Q: What is the definition of an empty compressed gas container?

A: A container holding a hazardous waste that is a compressed gas must be emptied until the pressure of the material inside the container is equal to atmospheric pressure. When the pressures are equal, the container is considered *empty* (47 *FR* 36094, August 18, 1982).

Q: What is the definition of an empty tank?

A: There is no definition for an empty tank. Although the empty container definition is commonly applied, the tank owner should check with the state or appropriate EPA regional office for a conclusive determination.

Q: Are there special requirements for drum reconditioners?

A: Yes; a drum reconditioner who recycles empty hazardous waste containers and generates residue as a result of reconditioning the containers must analyze the residues for the characteristics of a hazardous waste. Although the containers were empty, a new waste (the residue) has been generated, and it must be tested for the characteristics (47 *FR* 36096, August 18, 1982).

Delisting

Q: What is delisting?

A: Delisting is a request for formal rule making (by petition) for a generator to reclassify a listed hazardous waste as a nonhazardous waste at a specific site. Section 261.22 sets forth the requirements to be followed in submitting a delisting petition.

Q: What is required to delist a hazardous waste?

A: A waste is listed if it contains hazardous constituents (Part 261, Appendix VIII) or if it exhibits a characteristic of hazardous waste (ignitable, corrosive, reactive, or toxic). Part 261, Appendix VII, categorizes each listed waste and lists the hazardous constituents (i.e., Appendix VIII constituents) for which the waste is listed. A petitioner must demonstrate that the Appendix VIII hazardous constituents for which the waste was listed are not present in the waste and that the waste does not fail any of the ICRT characteristics. As required by the Hazardous and Solid Waste Amendments (HSWA), a facility must also demonstrate that the waste will not be hazardous for other reasons. This includes other Appendix VIII hazardous constituents.

Q: What additional information is required in a delisting petition?

A: EPA has outlined the additional information that a petitioner should include to satisfy the HSWA requirements (OSWER Directive No. 9433.07).

- A list of raw materials and a description and schematic diagram of manufacturing processes that may contribute waste, wastewater, or rinse water to the waste stream
- An evaluation of representative samples for the characteristics of hazardous waste, including an analysis of the potential for sulfide and cyanide gas generation
- The total oil and grease and total organic carbon (TOC) content of the waste
- A statement indicating that samples analyzed and reported in the petition are considered representative of any potential variation
- The average and maximum quantities of waste generated per month and per year
- The quality assurance procedures followed during sampling and analysis
- An explanation of why planned changes of equipment, feedstock materials, and manufacturing processes will not alter the chemical makeup of the waste stream

Characteristics of Hazardous Wastes

Q: What are the hazardous waste characteristics?

A: Section 3001 of RCRA requires EPA to develop and promulgate criteria for identifying characteristics of hazardous waste that are separate from the listed wastes. The primary responsibility for determining whether a waste exhibits a characteristic rests with the generator. Characteristics were to be selected that were measurable by standard, available testing protocols. Thus, EPA established that ignitability, corrosivity, reactivity, and toxicity are the characteristics of a hazardous waste.

Q: When does a generator test for the characteristics of a hazardous waste?

A: To identify a hazardous waste, the protocol is to first determine if a solid waste is a RCRA-listed hazardous waste. If it is not listed, a generator must then determine if the waste fails any of the characteristics of hazardous waste: ignitability, corrosivity, reactivity, and/or toxicity (as determined by the toxicity characteristic leaching procedure test). If a listed hazardous waste also exhibits a characteristic, both the listing code and the characteristic code need to be identified.

Q: Are there specified test methods?

A: Yes; the regulations outline the specified test methods. Additionally, the EPA manual, *Test Methods for Evaluating Solid Waste, Physical/ Chemical Methods* (more commonly known as SW-846) provides further information on sampling and analyzing procedures for complying with RCRA. It is available from the U.S. Government Printing Office (GPO).

Ignitability

Q: What constitutes an ignitable hazardous waste?

A: A waste is an ignitable waste (designated as D001) if it meets *any* of the following conditions (Section 261.21):

1. A liquid that has a flash point of less than 140°F as determined by either a Pensky-Martens or a Setaflash closed-cup test.
2. Solids that can spontaneously combust through friction, absorption, or loss of moisture. Currently, there is no standardized test to determine this.
3. It is an ignitable compressed gas as defined by the Department of Transportation (DOT) in 49 CFR 173.300.
4. It is an oxidizer as defined by DOT in 49 CFR 173.151.

Q: How is a liquid defined?

A: *Liquids* are determined by using the paint filter test (PFT), EPA test method no. 9095 (contained in SW-846). A sample of the waste is placed on a paint filter (400-micron or no. 60 mesh). If any liquid seeps through the filter within 5 minutes, it is considered a liquid.

Q: Section 261.21(a)(1) excludes ignitable wastes with less than 24 percent alcohol. What is included in this exclusion?

A: There is an exclusion from the ignitable characteristic for aqueous solutions that fail the flash point test and contain less than 24 percent alcohol.

Originally, this exclusion was intended for alcoholic beverages such as wine (45 *FR* 33108, May 19, 1980). However, the regulatory language is ambiguous regarding the extent of this exclusion. OSWER Directive No. 9443.02 clarified this exclusion by stating, "while the Agency's intent was that this exemption apply to potable beverages only, because the term alcohol was used instead of ethanol, all aqueous wastes which are ignitable only because they contain alcohols (here using the term alcohol to mean any chemical containing the hydroxl [-OH] functional group) are excluded from regulation.'' The directive also defines the term *aqueous solution* by stating, ''with respect to what constitutes an aqueous solution, such a solution is one in which water is the primary component. This means that water constitutes at least 50% by weight of the sample.''

Q: What is the definition of an ignitable compressed gas?

A: An *ignitable compressed gas* must first meet the definition of a *compressed gas,* which is ''any material or mixture having in the container an absolute pressure exceeding 40 p.s.i. at 70°F or, regardless of the pressure at 70°F, having an absolute pressure exceeding 104 p.s.i. at 130°F; or any liquid flammable mixture having a vapor pressure exceeding 40 p.s.i. absolute at 100°F.'' A material meeting the definition of a compressed gas is ignitable if it is either a mixture of 13 percent or less (by volume) with air, forms a flammable mixture, or the flammable range with air is wider than 12 percent regardless of the lower limit; the flame projects more than 18 inches beyond the ignition source with valve opened fully, or, the flame flashes back and burns at the valve with any degree of valve opening; or there is any significant propagation of flame away from the ignition source.

Q: Section 261.21(a)(3) states that a waste is ignitable if it is an "ignitable compressed gas" as defined by DOT's regulations (49 CFR 173.300). The DOT has a definition for* flammable compressed gas, *but none for* ignitable compressed gas. *What is the correct definition?

A: Although there is a discrepancy in term, a flammable compressed gas is the same as an ignitable compressed gas and, thus, classified as an ignitable hazardous waste under RCRA (HMR).

Q: What is an oxidizer?

A: An oxidizer is a substance that yields oxygen readily when involved in a fire, thereby accelerating and intensifying the combustion of organic material. There is no prescribed test for determining this classification. DOT does give examples, in 49 CFR Section 173.151, of oxidizers, such as chlorate, permanganate, inorganic peroxide, or a nitrate. As a general rule of thumb, an oxidizer has a prefix of ''per'' or ends with ''ate.''

Q: A generator produces a semiliquid waste that is not a listed waste and does not exhibit a characteristic of hazardous waste; however, if the waste loses its moisture, it will spontaneously combust. Because the waste is not a hazardous waste when it is first generated, can its status change?

A: Yes; even if a waste does not meet the definition of a hazardous waste when it is first generated, it can become a hazardous waste, and thus regulated, at any time.

Corrosivity

Q: What constitutes corrosivity?

A: A waste is a corrosive waste (designated as D002) if it meets *any* of the following criteria (Section 261.22):

1. It is *aqueous* and has a pH of 2 or less or 12.5 or more.
2. It is a *liquid* and corrodes steel at a rate of 6.35 mm or more per year as determined by the National Association of Corrosion Engineers.

Q: A generator's caustic waste (pH 13) does not fail the paint filter test and thus is considered a solid (nonliquid) waste. The waste is not a listed waste. Is it a corrosive hazardous waste because of the characteristic?

A: No; a waste that is considered a ''solid'' waste by the virtue of the paint filter test is not considered a corrosive waste, regardless of its corrosive properties (45 *FR* 33109, May 19, 1980).

Reactivity

Q: What constitutes a reactive waste?

A: The reactivity characteristic (designated D003) requires a determination based more on subjective judgment than a standardized quantitative determination, except for explosives (class A or B) and toxic (sulfide and cyanide) gas generation (Section 261.23).

If a material can cause the generation of harmful vapors or fumes that can present a danger to human health or the environment, reacts violently with water, or has the ability to explode without detonation, it is considered a reactive hazardous waste.

Q: Are there quantitative limits for toxic gas generation?

A: Yes; for sulfide and cyanide gas generation, using the prescribed test method found in Chapter 7.3.3 of SW-846, if free (not total) cyanide is generated at 250 mg/kg or more, or free sulfide is generated at 500 mg/kg or more, it is considered a reactive hazardous waste.

Toxicity

Q: What is the toxicity characteristic?

A: The toxicity characteristic, (TC), is designated as D004 through D043. The TC is designed to identify wastes that are likely to leach hazardous constituents into groundwater under improper management conditions. EPA established a testing procedure, known as the toxicity characteristic leaching procedure (TCLP), that extracts constituents from a solid waste in a manner that simulates the leaching action that can occur in a landfill. EPA has made the assumption that industrial waste would be codisposed of with nonindustrial waste in an actively decomposing municipal landfill situated over an aquifer (Section 261.24).

Q: What if a generator does not intend to land-dispose waste?

A: For the purposes of determining whether a waste is hazardous according to the toxicity characteristic, it is irrelevant that a generator does not, in fact, codispose of hazardous waste in a municipal landfill or in any type of landfill.

Q: What is the TCLP?

A: The TCLP toxicity test, described in Part 261, Appendix II, requires a representative sample of the waste to be subjected to an acidic solution. After 24 hours, the resulting extract is then tested to determine if it meets *or* exceeds any of the limits of any of the contaminants identified in Table 2. The contaminants include volatiles, semi-volatiles, metals, pesticides, and herbicides.

Q: What if the waste is a liquid?

A: If the waste is a liquid, it does not need to be subjected to the acidic solution. The liquid will be considered the extract and analyzed for the contaminants accordingly.

Q: A generator produces waste from a process that uses organic materials only. If the generator is confident that metals or pesticides would not be present in the waste, must the entire TCLP be conducted?

A: No; Section 262.11(c) allows a generator to apply his knowledge of the waste when determining if it is hazardous. Thus, a generator can have a partial TCLP (e.g., volatiles and semi-volatiles) conducted if he has sufficient knowledge to insure the accurate classification of the waste.

TABLE 2 Toxicity characteristic levels.

EPA HW No.	Contaminant	Limit (mg/L)
D004	Arsenic	5.000
D005	Barium	100.0
D018	Benzene	0.5
D006	Cadmium	1.0
D019	Carbon tetrachloride	0.5
D020	Chlordane	0.03
D021	Chlorobenzene	100.0
D022	Chloroform	6.0
D007	Chromium	5.0
D023	o-Cresol	200.0
D024	m-Cresol	200.0
D025	p-Cresol	200.0
D016	2,4-D	10.0
D026	Cresol (total)	200.0
D027	1,4-Dichlorobenzene	7.5
D028	1,2-Dichloroethane	0.5
D029	1,1-Dichloroethylene	0.7
D030	2,4-Dinitrotoluene	0.13
D012	Endrin	0.02
D031	Heptachlor	0.008
D032	Hexachlorobenzene	0.13
D033	Hexachlorobutadiene	0.5
D034	Hexachloroethane	3.0
D038	Isobutanol	1000.0
D008	Lead	5.0
D013	Lindane	0.4
D009	Mercury	0.2
D014	Methoxychlor	10.0
D035	Methyl ethyl ketone	200.0
D036	Nitrobenzene	2.0
D037	Pentachlorophenol	100.0
D038	Pyridine	5.0
D010	Selenium	1.0
D011	Silver	5.0
D039	Tetrachloroethylene	0.7
D015	Toxaphene	0.5
D040	Trichloroethylene	0.5
D041	2,4,5-Trichlorophenol	400.0
D042	2,4,6-Trichlorophenol	2.0
D017	2,4,5-TP Silvex	1.0
D043	Vinyl Chloride	0.2

Source: 40 CFR Section 261.24

Special Categories of Wastes

Mixture Rule

Q: What happens if a generator mixes hazardous waste with a nonhazardous waste?

A: A mixture of any amount of a listed hazardous waste and a solid (nonhazardous) waste is considered a hazardous waste [Section 261.3(a)(2)(iii)].

There is a limited exclusion for *de minimis* mixtures in wastewater treatment systems meeting certain conditions. Wastewater-treatment systems subject to either an NPDES permit or pretreatment standards have specific exclusions for the effluent from the mixture rule under Section 261.3(a)(2)(iv). These exclusions include wastewater mixed with specified spent solvents, de minimis commercial chemical product losses, and laboratory wastes. All these exclusions are subject to certain conditions.

Q: What is the difference if the mixture is with a listed waste versus a characteristic waste?

A: Yes; if the mixture is hazardous solely because of an I, C, or R characteristic and the resultant mixture no longer retains that characteristic, it is not considered a hazardous waste. An example is F003 (a listed waste), which is listed solely because of the ignitability characteristic. Hence, a mixture of F003 and a nonignitable, nonhazardous waste would become nonhazardous, provided the mixture no longer exhibited the ignitability characteristic (46 *FR* 56588, November 17, 1981).

Q: If waste is generated from treating hazardous waste (e.g., incineration) does it remain hazardous?

A: Yes; waste generated from the treatment, storage, or disposal of a hazardous waste, including any sludge (pollution control residue), spill residue, ash, leachate, or emission control dust, will remain a hazardous waste unless it is delisted, or, in the case of a characteristic waste, it no longer exhibits the characteristic [Section 261.3(c)].

Q: Wasn't there a court decision that overturned the mixture and derived-from rules?

A: On December 6, 1991, the U.S. Court of Appeals for the District of Columbia Circuit ruled that EPA had failed to give sufficient notice and opportunity for public comment in promulgating the "mixture" and "derived-from" rules. The court therefore vacated the rules and remanded them to EPA (see *Shell Oil Co. v. EPA,* no. 80-1532 *et al.*, D.C. Cir., December 6, 1991).

Because of the potential chaos that this action could cause, EPA issued an interim final rule on March 3, 1992 (57 *FR* 7628), reinstating the rules on an interim basis until April 28, 1993, pending full notice and comment in accordance with the Administrative Procedures Act. EPA published a proposal to modify these rules, including the establishment of a concentration-based hazardous waste determination (May 20, 1992, 57 FR 21450). Therefore, until final action is taken, the mixture and derived-from rules remain in full effect.

Contained-In Policy

Q: What is the contained-in policy?

A: Because the mixture rule requires that a hazardous waste be mixed with a *solid waste* for it to be regulated under that rule, EPA developed the contained-in policy. Material that is not necessarily a waste that contains a listed hazardous waste is managed as if it were a waste. An example is groundwater contaminated with a listed hazardous waste. Although the groundwater is not a hazardous waste, the listed hazardous waste component is subject to regulation, thereby requiring the entire mixture to be managed as hazardous (OSWER Directive 9481.00-6).

Low-Level Radioactive Mixed Waste

Q: What is a low-level radioactive waste?

A: A low-level radioactive waste (LLW) is radioactive material that (a) is not high-level radioactive waste, spent nuclear fuel, or by-product material as defined in Section 11(e)(2) of the Atomic Energy Act (AEA) (i.e., uranium or thorium mill tailings) and (b) is classified by the Nuclear Regulatory Commission (NRC) as a low-level radioactive material.

Q: What happens if a low-level radioactive waste is mixed with a RCRA hazardous waste?

A: If a low-level radioactive waste contains a listed RCRA hazardous waste, or the LLW exhibits a characteristic of hazardous waste, the material is classified as a *mixed low-level waste* and must, therefore, be managed and disposed of in compliance with EPA regulations, Title 40 CFR Parts 124 and 260 through 280 and NRC regulations, Title 10, Parts 20, 30, 40, 61, and 70.

The management and disposal of mixed low-level radioactive wastes must also be in compliance with state requirements, in states that are EPA-authorized for the hazardous components of the waste, and with the NRC agreement state radiation-control programs for the low-level radioactive portion of the waste (OSWER Directive No. 9440.00-1).

Infectious Wastes

Q: What is infectious waste?

A: Infectious waste is waste capable of producing an infectious disease. There are factors necessary for induction of disease. These factors include:

- Presence of a pathogen of sufficient virulence
- Dose
- Portal of entry
- Resistance of host

Therefore, for a waste to be infectious, it must contain pathogens with sufficient virulence and quantity so that exposure to the waste by a susceptible host could result in an infectious disease (OSWER Directive No. 9401.00-2).

Q: What are some examples of infectious wastes?

A: EPA has listed (OSWER Directive No. 9401.00-2) examples of wastes that are considered infectious:

- Isolation wastes
- Cultures and stocks of infectious agents and associated biologicals
- Human blood and products
- Pathological wastes
- Contaminated sharps (e.g., needles)
- Contaminated animal carcasses, body parts, and bedding

Q: What is the regulatory status of infectious waste?

A: Currently, EPA does not regulate infectious waste. Although EPA has clear statutory authority under Section 1004(5) to regulate infectious waste, it has not yet listed any waste for its infectious potential, nor has EPA established infectious characteristics to be used to identify a waste as hazardous.

Congress has, however, authorized a pilot program (Medical Waste Tracking Act, PL-100-582) for medical wastes that include some infectious wastes. This act is strictly a demonstration program to track the disposition and transportation of medical wastes.

Q: Does EPA provide any guidance on how to properly manage infectious waste?

A: Yes; EPA has published a *Guide to Infectious Waste Management* (order No. PB86 199130) that is available from the National Technical Information Service (see Appendix G).

Hazardous Constituents

Q: What are the hazardous constituents?

A: The hazardous constituents are designated substances used as parameters for various purposes. There are: Appendix VII, the *hazardous waste constituents;* Appendix VIII, the *hazardous constituents;* and Appendix IX, the *groundwater monitoring constituents*.

Q: What criteria are used to designate a substance as a hazardous constituent?

A: The hazardous constituents are substances that have been shown in reputable scientific studies to have toxic, carcinogenic, mutagenic, or teratogenic effects on humans or other life forms and include such substances as those identified by EPA's Carcinogen Assessment Group.

Q: What are the Appendix VII constituents?

A: The Appendix VII constituents, found in Appendix VII of Part 261, are also known as the *hazardous waste constituents*. The Appendix VII table identifies each listed hazardous waste and lists the Appendix VIII hazardous constituents for which the waste was listed.

Q: What are the Appendix VIII constituents?

A: The Appendix VIII constituents, found in Appendix VIII of Part 261 and Appendix D of this book, are also known as the *hazardous constituents*. These constituents are used as a basis for the EPA Administrator to list a waste as hazardous. They are also used for delisting wastes and as a trigger for corrective action.

Q: If a generator analyzes waste, and it contains Appendix VIII constituents, is that waste hazardous?

A: No; if a generator tests a waste and finds Appendix VIII constituents, this does not render the waste hazardous; the generator must use the criteria established in Section 261.3 (i.e., listing and characteristics) for this determination. The Appendix VIII constituents are justification for EPA to list a hazardous waste. Only the EPA Administrator can designate a listed hazardous waste through a formal rule-making procedure.

Q: What are the Appendix IX constituents?

A: The Appendix IX constituents, found in Appendix IX of Part 264 and

Appendix E of this book, are substances specifically designated for groundwater monitoring purposes at permitted facilities. Previously, the entire Appendix VIII list was required for monitoring purposes. However, a number of these Appendix VIII constituents were impossible to test for in water, unstable in water, or inappropriate for groundwater characterization. Thus, the Appendix IX list was developed to select constituents that are better suited to characterize the underlying groundwater (52 *FR* 25942, July 9, 1987).

RECYCLING HAZARDOUS WASTES

Applicability

Q: How is material classified concerning recycling activities?

A: Under Subtitle C of RCRA, EPA has the authority to regulate hazardous wastes. Hazardous wastes, however, are defined in the statute as a subset of solid wastes. Thus, it is necessary to define what a solid waste is to determine the extent of EPA's jurisdiction under Subtitle C. Pertaining to recycling activities, EPA must determine whether a recyclable secondary material is a solid waste before it can assert Subtitle C jurisdiction over management of that waste. EPA considers a secondary material to be a material that potentially can be a solid and hazardous waste when recycled.

Q: What are secondary materials?

A: The recycling regulations refer to the following types of secondary materials: spent materials, sludges, by-products, scrap metal, and commercial chemical products recycled in ways that differ from their normal use. The term *secondary material* is used primarily for enhancing the understanding of the recycling regulations, but the term is not actually used in the regulatory text.

Q: Specifically, how is recycling regulated?

A: The actual recycling process is unregulated (except waste burned as fuel and use constituting disposal). For example, a generator can distill solvents on-site without the distillation unit being regulated (OSWER Directive No. 9441.24). However, the generation, transportation, and storage prior to recycling are regulated unless the specific waste is excluded. Thus, a facility that distills solvents from off-site sources must have interim status or a permit for the storage of the waste. A generator may recycle and/or store wastes prior to recycling them without interim status or a permit, provided the waste is generated on-site and the accumulation is done in accordance with Section 262.34 (i.e., accumulate in tanks or containers for fewer than 90 days).

Q: Wastes generated from the treatment of hazardous waste remain a hazardous waste. However, if a person recycles a hazardous waste and generates a usable material, is that material a hazardous waste?

A: No; there is an exclusion from the derived-from rule for products derived from hazardous waste. For example, if a person places spent solvent in a distillation unit, the material distilled would no longer be regulated if it is considered a product. Only the still bottoms remain regulated, because they were derived from hazardous waste and are themselves waste [Section 261.3(c)(2)(i)].

Q: Are there certain recyclable materials excluded from regulation?

A: Yes; Sections 261.2(e) and 261.6(a)(3) exclude certain recyclable materials from regulation under Parts 262 through 266. These materials are:

- Used oil
- Industrial ethyl alcohol
- Scrap metal
- Used batteries returned for regeneration (does not include reclamation)
- Materials used or reused as ingredients to make a product, provided they are not reclaimed before use
- Wastes used or reused as effective substitutes for commercial products without prior reclamation
- Wastes returned to the original process from which they were first generated without first being reclaimed

Q: If an excluded material is to be recycled, how is it regulated?

A: Most recycled materials are considered solid wastes by EPA, depending on both the recycling activity itself and the nature of the recycled material. The following four types of recycling activities are potentially subject to RCRA regulation:

- Uses that actually constitute ultimate disposal such as placing waste directly on the ground unless that was their intended use (e.g., pesticides)
- Burning waste or waste fuels for energy recovery or using wastes to produce a fuel

- Reclamation, which is the regeneration of wastes or the recovery of material from wastes
- Speculative accumulation, which is either accumulating wastes that are potentially recyclable, but for which no recycling (or no feasible recycling) market currently exists, or accumulating wastes before recycling unless 75 percent of the accumulated material is recycled during a 1-year period

The definition of solid waste under Section 261.2 also distinguishes among five types of secondary materials: spent materials, sludges, by-products, commercial chemical products, and scrap metals. The regulatory status of each secondary material is outlined in Table 3.

Q: What are spent materials?

A: *Spent materials* are materials that have been used and as a result of such use have become contaminated by physical or chemical impurities, such that they can no longer serve the purpose for which they were produced without regeneration. The following materials are considered spent after use: wastewater, solvents, catalysts, acids, pickle liquor, foundry

TABLE 3 **Classification of wastes when recycled.**

Type of waste	Reclamation	Speculative accumulation	Burned as fuel	Use constituting disposal
Spent materials listed or exhibiting an HW characteristic	yes*	yes	yes	yes
Sludges that are a listed HW	yes	yes	yes	yes
Sludges exhibiting an HW characteristic	no	yes	yes	yes
By-products that are a listed HW	yes	yes	yes	yes
By-products exhibiting an HW characteristic	no	yes	yes	yes
Commercial chemical products listed or exhibiting an HW characteristic	no	no	yes	yes

*Yes = material is defined as a solid waste when recycled; no = material is not a solid waste when recycled.
Source: 40 CFR Section 261.2(c)

sands, lead-acid batteries, and activated carbon. Spent activated carbon can also be classified as a sludge if it has been used as a pollution control device [Section 261.1(c)(1)].

Q: What is a sludge?

A: *Sludges* are residues from pollution control technology. Examples include bag house dusts, flue dusts, wastewater treatment sludges, and filter cakes (Section 260.10).

Q: What are by-products?

A: *By-products* include residual materials resulting from industrial, commercial, or agricultural operations that are not primary products, are not produced separately, are not fit for a desired end use without substantial further processing, and are not spent materials, sludges, commercial chemical products, or scrap metals. Examples of by-products include mining slags, drosses, and distillation column bottoms [Section 261.1(c)(3)].

Q: What are commercial chemical products?

A: *Commercial chemical products* include commercial chemical products and intermediates, off-specification variants, spill residues, and container residues that are listed in Section 261.33 or that exhibit an ICRT characteristic.

Q: What is scrap metal?

A: *Scrap metal* includes bits and pieces of metal parts that are generated by metal-processing operations or result from consumer use. Examples of scrap metal include bars, turnings, rods, sheets, wire, radiators, scrap automobiles, and railroad boxcars.

Specific Recycling Activities

Q: What are the specific recycling activities?

A: The specific recycling activities, addressed under Section 261.2 and Part 266, are speculative accumulation, use constituting disposal, reclamation, precious metal recovery, lead-acid battery reclamation, and burning and blending of waste fuels.

Q: What is speculative accumulation?

A: A material is accumulated speculatively if it is accumulated before being recycled. A material is *not* accumulated speculatively, however, if the person accumulating it can show that the material is potentially recyclable and has a feasible means of being recycled and at least 75 percent of the accumulated material is recycled or transferred to a different site for recycling in a calendar year. The 75 percent requirement may be calculated on the basis of either volume or weight and applies to waste accumulated during a calendar year beginning with the first day of January [Section 261.1(c)(8)].

The only exceptions to this rule are hazardous commercial chemical products (listed or characteristic) because they are not considered wastes when stored prior to recycling.

Q: What is used to calculate the 75 percent recovery?

A: In calculating this percentage, the 75 percent requirement is to be applied to each material of the same type (e.g., slags from a single smelting process) that is recycled in the same way (i.e., from which the same material is recovered or that is used in the same way) [Section 261.1(c)(8)].

Q: What is use constituting disposal?

A: *Use constituting disposal* is defined as either:

- Applying materials to the land or placing them on the land in a manner constituting disposal
- Applying materials contained in a product to the land or placing them on the land in a manner constituting disposal

Examples of such use include use as a fill, cover material, fertilizer, soil conditioner, or dust suppressant and use in asphalt or building foundation materials [Section 261.1(c)(1)].

Products that include listed hazardous wastes as ingredients are classified as solid and hazardous wastes when placed directly on the land for beneficial use, unless and until the product is formally delisted. Products that include characteristic hazardous wastes as ingredients are classified as solid and hazardous wastes only if the product itself exhibits hazardous waste characteristics.

Q: What is the status of a material used in this manner?

A: Hazardous secondary materials are considered solid wastes when applied to the land in these ways, except for listed commercial chemical products whose ordinary use involves application to the land. Consequently, the use of hazardous-waste-derived fertilizer and hazardous-waste-derived products where the components are physically inseparable is not regulated as land disposal. Therefore, hazardous waste generator, transporter, and storage requirements apply prior to use, and applicable land disposal requirements under Parts 264 and 265 apply to the activity itself.

Q: What is reclamation?

A: Reclamation is defined as the regeneration of waste materials or the recovery of material with value from wastes. Reclamation includes such activities as dewatering, ion exchange, distillation, and smelting. However, simple collection, such as collection of solvent vapors, is not considered reclamation. Use of materials as feedstocks or ingredients, such as the use of a material as a reactant in the production of a new product, also is not considered reclamation [Section 261.1(c)(4)].

Q: Are precious metals regulated when reclaimed?

A: Yes; hazardous wastes that contain precious metals are subject to reduced requirements when being reclaimed to recover economically significant amounts of gold, silver, platinum, paladium, irridium, osmium, rhodium, and ruthenium, or any combination of these [Section 266.70(a)].

Q: What are these reduced requirements?

A: These reduced requirements are that:

- Generators and transporters must have an EPA identification number and must use a manifest.
- Storage facilities must have an EPA identification number, comply with manifest requirements, and keep records to demonstrate that 75 percent of all received precious metal wastes are being reclaimed per calendar year to satisfy the speculative accumulation provision.

Q: Is a reclaimer of precious metals from hazardous waste required to obtain a permit?

A: No; the reclaimer is not required to obtain a permit or comply with the standards of Parts 264 or 265 for precious-metal-recovery operations. A reclaimer can be considered a "designated facility" for the purposes of the manifest (see definition) if a Part A permit application is filed with an attached statement explaining that only precious metal reclamation will be conducted. However, the facility will not receive interim status for the precious-metal-recovery operations unless requested. The Part B permit application is not required for these operations.

Q: How are people reclaiming lead-acid batteries regulated?

A: People who generate, transport, collect, or store spent lead-acid batteries, but do not reclaim them, are not subject to regulation under Parts 262 through 270.

Those who store spent batteries that they reclaim must notify EPA and obtain an EPA identification number and comply with the storage facility requirements of Parts 264 and 265 Subparts A through L, with the exception of waste characterization (Sections 264.14 and 265.23) and manifest-related requirements (Sections 264.71-72 and 265.71-72).

Burning and Blending of Waste Fuels

Q: How are burning and blending of waste fuels regulated?

A: Another form of recycling is the burning of hazardous waste and used oil as fuel for legitimate energy recovery. Legitimate energy recovery is roughly defined as a material with a heating value of 5,000 to 8,000 BTU/lb (48 *FR* 11159, March 16, 1983) in an industrial furnace or boiler (see definitions).

There are two categories of waste fuel: hazardous-waste fuel and used oil-fuel. Burning hazardous-waste fuel in a unit other than an industrial furnace or boiler or burning fuel with fewer than 5,000 to 8,000 BTU/lb must comply with the incinerator standards under Part 264 and 265, as well as the requirement to obtain a permit or interim status. Burning hazardous waste fuel in an industrial boiler or furnace is subject to special management standards under Part 266.

Hazardous-Waste Fuel

Q: What is the definition of hazardous waste fuel?

A: *Hazardous-waste fuel* is any hazardous waste (listed or exhibiting a characteristic) that is being burned for legitimate energy recovery. Any used oil that is mixed with any hazardous waste, including small-quantity generator waste, is considered a hazardous-waste fuel when being burned [Section 266.40(c)].

Q: How is hazardous waste as fuel regulated?

A: The hazardous-waste-as-fuel activities are divided into two phases, Phase I and Phase II.

The Phase I regulations were promulgated (50 *FR* 49164, November 29, 1985) to institute administrative controls for marketers and burners of hazardous-waste fuels. The rule applies to the generation, transportation, and storage of all hazardous-waste used as fuels or used to produce a fuel. The rule also prohibits the burning of hazardous waste fuels in non-industrial boilers, unless the boiler complies with the incinerator standards under Parts 264 and 265.

On February 21, 1991 (56 *FR* 7134), EPA published the final Phase II rule for industrial furnaces and boilers (known as the BIF rule). The regulations establish emission limits on toxic organic compounds, toxic metals, hydrogen chloride, chlorine gas, and particulate matter. In addition, the rule subjects owners and operators of these devices to the general facility standards for hazardous waste management facilities, including the requirement to obtain a RCRA permit or interim status.

Q: What are the prohibitions of burning hazardous waste?

A: A person may market a hazardous-waste fuel only to other marketers who have notified EPA of their hazardous-waste-fuel activities, have obtained an EPA identification number, and use burners that have an industrial furnace or boiler. The waste-as-fuel regulations require all marketers and burners to notify EPA of this activity, even if they have previously notified and received an EPA identification number for other activities (Section 266.31).

No hazardous-waste fuel may be burned in a cement kiln that is located within the incorporated boundaries of a city with a population of 500,000 or greater [Section 266.31(c)].

Q: What is a marketer?

A: A *marketer* is any person who sends hazardous-waste fuel directly to a burner; produces, processes, or blends hazardous-waste fuel; or distributes hazardous-waste fuel, but does not process, produce, or blend the fuel. The term *marketer* has no association with monetary transactions. Thus, a person who distributes hazardous-waste fuel for no monetary profit would still be considered a marketer (Section 266.34).

Q: What are the requirements for marketers?

A: Pursuant to Section 266.34, marketers of hazardous-waste fuel must:

- Notify EPA of regulated waste-as-fuel activities.
- Comply with the appropriate accumulation requirements (e.g., 90 days for on-site generators under Section 262.34) or a permit or interim status for nongenerators.
- Use a manifest for all off-site shipments.
- Obtain a one-time certification from the burner or another marketer before initiating a shipment. This certification must state that the burner or marketer is in compliance with all applicable requirements of the waste-as-fuel regulations. A copy of this certification must be kept for at least 3 years.

Q: What are the requirements for generators of hazardous waste fuels?

A: Generators of hazardous-waste fuels are subject to the same requirements as other hazardous-waste generators under Part 262. However, if a generator sends hazardous-waste fuel directly to a burner, the generator would become a marketer and would also have to comply with the marketer requirements (Section 266.32).

Q: What are the requirements for transporters of hazardous-waste fuel?

A: Transporters of hazardous-waste fuel are subject to the Part 263 requirements for transporters, as well as applicable Department of Transportation (DOT) requirements (see Chapter 4).

Q: What are the requirements for burners of hazardous waste fuels?

A: Pursuant to Section 266.35, burners of hazardous-waste fuel must:

- Notify EPA of the hazardous-waste-fuel-burning activities.
- Retain a copy of each certification that was sent to a marketer for at least 3 years.
- Comply with the applicable storage requirements. This means either storing 90 days under Section 262.34 if the burner is burning only hazardous-waste fuel generated on-site or, if the burner is burning hazardous-waste fuel generated off-site, filing a Part A permit application as a storage facility to obtain interim status to store waste before burning.

Used-Oil Fuel

Q: What is the definition of used oil?

A: *Used oil* means oil that is refined from crude oil, used, and as a result of that use, contaminated by physical or chemical means [Section 266.40(b)].

Used oil mixed with any hazardous waste, including small-quantity-generator waste, is subject to regulation under the hazardous-waste-as-fuel requirements discussed in the previous section [Section 266.40(c)].

Q: Does used oil meet the definition of solid waste?

A: *Used oil* does not need to be a solid waste (per Section 261.2) to be regulated under RCRA Subtitle C, because the authority to regulate used oil'' is found in Section 3014 of RCRA.

Q: Is used oil a hazardous waste?

A: Used oil is not currently a listed hazardous waste, but it can be hazardous if it exhibits a characteristic. To dispose of used oil, the generator must determine if the oil is hazardous. The following regulations govern only the burning of used oil. Other used-oil recycling activities are not currently regulated.

Q: If waste oil contains hazardous components, how is it regulated?

A: Any used oil that contains more than 1,000 ppm of total halogens (fluorine, chlorine, bromine, iodine, and astatine) is presumed to have been mixed with a hazardous-waste and would be regulated as a hazardous-waste fuel. However, provided that it can be documented that mixing has not occurred, the oil may be handled under the used-oil rules. For example, commercial cutting oils commonly contain chlorinated substances as additives [Section 266.40(c)].

Thus, used cutting oil containing chlorine at a concentration higher than 1,000 ppm may not have been mixed with a hazardous waste. Used-oil fuel not mixed with hazardous waste is considered either specification or off-specification oil fuel.

Q: What is specification oil?

A: *Specification oil* means any used oil that does not exceed any of the specification levels in Table 4. However, all used oil being burned for energy recovery, unless it is a hazardous-waste fuel, is assumed to be off-specification oil unless demonstrated otherwise. This demonstration can be accomplished by a laboratory analysis [Section 266.40(e)].

TABLE 4 **Specification levels for used oil fuels.**

Constituent/property	Allowable level
Arsenic	5 ppm maximum
Cadmium	2 ppm maximum
Chromium	10 ppm maximum
Lead	100 ppm maximum
Flash point	1000°F minimum
Total Halogens	4,000 ppm maximum

Source: 40 CFR Section 266.40

Q: Table 4 states that used oil can contain up to 4,000 ppm of halogen, yet the level for presumption that the oil was mixed with hazardous waste is 1,000 ppm. Is this a discrepancy?

A: No; it can be demonstrated that used oil with more than 1,000 ppm of total halogens was not mixed with hazardous wastes. If this is demonstrated, and the total halogen level is below 4,000 ppm, it is considered specification oil. If it can be demonstrated that used oil was not mixed with hazardous waste, and the level is above 4,000 ppm, it is considered off-specification oil. Otherwise, the oil must be treated as a hazardous-waste fuel [Section 266.40(e)].

Q: What is used for this demonstration?

A: This demonstration can be accomplished by a laboratory analysis. A total chlorine test may be used instead of a total halogen test to satisfy this requirement.

Q: What is off-specification oil?

A: *Off-specification oil* means any used oil that exceeds any specification level identified in Table 4. However, off-specification oil may be blended to meet the specification without a permit or interim status. Blending is considered treatment prior to recycling and, thus, is not regulated.

Q: What are the requirements for managing used-oil fuel?

A: Once it is documented that used-oil fuel is specification oil, no further regulations apply. If the used-oil fuel is off-specification oil, the management is subject to the regulations under Part 266. The initial party to claim that a fuel is specification oil must retain all appropriate paperwork substantiating the claim for at least 3 years, except for burners who burn specification oil that they generate on-site [Section 266.43(b)(6)].

Off-specification oil may be marketed only to burners or other marketers who have notified EPA of their used-oil-management activities and have an EPA identification number. Generators of off-specification oil are subject to regulation only if they burn the oil on-site or market the oil directly to a burner. Transporters are not subject to these regulation unless they are also marketers [Section 266.41(a)].

Q: What is a marketer of used-oil fuel?

A: Marketers of used-oil fuel are those who generate and market fuel directly to a burner; receive used oil from other generators and produce, process, or blend the used-oil fuel (including those sending blended or processed used oil to brokers or other intermediaries); and distribute but do not process or blend used- oil fuel [Section 266.43(a)].

Q: What are the requirements for used-oil marketers?

A: Marketers of used-oil fuel must notify EPA of used-oil-fuel activities with EPA Form 8700-12. The notification is required even if the marketer has previously notified EPA of other hazardous waste activities.

A marketer must conduct and maintain records of analyses, unless the marketer handles the oil as off-specification oil fuel.

The marketer must obtain a one-time certification from the burner or marketer before initiating a shipment. The certification must state that the burner or marketer has notified EPA of its used-oil-fuel activities and that, if the receiving facility is a burner, the facility will burn the off-specification oil only in an industrial furnace or boiler.

Q: What is used for documenting the movement of used-oil fuel?

A: Because used-oil fuel is not a hazardous waste (unless it exhibits a characteristic), an invoice system is required instead of a manifest. A copy of each invoice received or sent must be retained for at least 3 years [Section 266.43(b)(4)].

Q: What information is required to be in the invoice?

A: The invoice must contain:

- An invoice number
- Marketer's EPA identification number
- Receiving facility's EPA identification number

- Names and addresses of both the receiving and shipping facility
- Quantity of off-specification-oil fuel
- The date of shipment or delivery
- The statement: *This used oil is subject to EPA regulation under 40 CFR Part 266.*

*Q: **How long must the invoice be kept?***

A: A burner must maintain a copy of any invoice it has received or any waste analysis conducted on the used oil for at least 3 years.

*Q: **What units can a burner use to burn off-specification oil?***

A: A burner may burn off-specification oil only in an industrial furnace, boiler, or a used-oil-fired space heater (provided that the used oil for the space heater is generated on-site or is from do-it-yourself oil changers) [Section 266.44(b)].

*Q: **Must a burner notify EPA of used-oil activities?***

A: Yes; a burner of either specification or off-specification used-oil fuel must notify EPA of its used-oil-fuel activities (except for those using a used-oil-fired space heater). This notification must state the location and a general description of the used-oil-management activities [Section 266.44(c)].

*Q: **Must the burner notify the shipper?***

A: Yes; before accepting the first off-site shipment of off-specification oil, the burner must certify to the shipper that EPA has been notified of its used-oil-fuel activities and that burning will be done only in an industrial furnace or boiler [Section 266.44(c)].

Chapter 3

Generators

RCRA, HMTA, and Superfund all place the most accountability and liability for the environmentally sound disposition of hazardous wastes on the generator. Generators of hazardous waste must comply with the regulations contained in 40 CFR Part 262. The standards are designed to ensure, among other things, proper record keeping and reporting, and proper accumulation (treatment and storage); the use of the Uniform Hazardous Waste Manifest system to track shipments of hazardous waste; the use of proper labels and containers; and the delivery of the waste to a permitted treatment, storage, or disposal facility. In addition, any generator who ships waste off-site to a waste-management facility is considered a shipper subject to the applicable HMTA regulations administered by DOT.

GENERAL REQUIREMENTS

Applicability

Q: What is a generator?

A: A generator is any person, by site, whose act or process produces hazardous waste or whose act first causes a hazardous waste to become subject to regulation (Section 260.10).

Q: If a site involves a RCRA corrective action, what is considered generation?

A: At a site involving RCRA corrective action, the physical removal of material from the ground constitutes generation. Thus, any person removing material (waste) from a site is the generator and must determine if it is a hazardous waste in accordance with Section 262.11.

Q: If an owner of a tank hires a contractor to remove waste from a raw-material-storage tank, who is considered the generator?

A: Both the tank cleaner and the owner of the tank are considered generators based on the definition of generator in Section 260.10.

In this scenario, EPA would like the parties to select a person to accept the duties of the generator, although in this example EPA would normally look at the owner of the tank as the generator (45 *FR* 72026, October 30, 1980). However, EPA may apply the definition of generator to each of these parties, because it is the *act or process* of each of these parties that produces the hazardous waste. EPA has stated that it will hold the persons that *generated* the waste jointly and severally liable even though that person may not be the owner or operator. Thus, persons are potentially liable as generators even though they may not have accepted the duties of the generator (OSWER Directive No. 9451.01).

Q: While on maneuvers, a U.S. naval vessel generates various hazardous wastes on board. These wastes are placed in containers while still on the vessel. The vessel docks at a shipyard and the wastes are unloaded from the ship and placed on the pier. The owner or operator of the shipyard stores the waste for shipment off-site. Who is the generator of the waste?

A: The naval vessel is considered the site where the waste is generated. Language in the October 30, 1980, *Federal Register* (45 *FR* 72024) states that in certain cases a waste is not generated until it is removed from product or raw material transport vessel. This naval vessel is not a product- or raw-material-transport vessel; it is the site where a process produces a hazardous waste and is the generator according to the definition in Section 260.10.

Q: Is a generator whose waste is sent to a hazardous waste facility in another state responsible for complying with that state's requirements?

A: Yes; a generator who sends waste out of state must comply with that state's hazardous waste regulations, laws, and program. If the state program is not authorized by EPA, the generator must comply with the RCRA regulations and any applicable state requirements. Where an unauthorized state's requirements are less stringent than the federal requirements, the RCRA regulations apply and vice versa.

Liability

Q: Can generators be held responsible, in the legal sense, for damages resulting after hazardous waste leaves their control?

A: Yes; generators may be liable for damages caused by hazardous wastes that have left their physical control. Such liability may be imposed by Section 107 of Superfund (CERCLA) or by application of negligence, nuisance, trespass, contract law, or other legal doctrines.

Q: Is the generator liable for any problems that occur before ultimate disposal of a hazardous waste, even if the generator complies with all the applicable regulations?

A: Compliance with the regulations merely ensures that EPA will not bring an enforcement action for violations under Subtitle C of RCRA. Compliance does not necessarily insulate a generator from liability under common and statutory law and, more importantly, Section 107 of Superfund.

Determining Generator Categories

Q: What regulations are applicable to a generator of any amount of hazardous waste?

A: A generator of *any* amount of waste must determine if the waste is hazardous (Section 262.11).

Q: Must the generator perform tests for identifying a hazardous waste?

A: No; the regulations require only that the generator *determine* if a waste is a hazardous waste; the waste does not necessarily have to be analyzed to make this determination. Section 262.11(c) allows the generator to apply knowledge of the hazardous characteristic of the waste in light of the materials or the processes used in determining if a waste is hazardous.

Q: What if a generator unknowingly misidentifies a waste?

A: EPA does not recognize a defense for a good-faith mistake in identifying one's waste and, thus, a generator is fully responsible for the proper identification/classification of a hazardous waste (45 *FR* 12727, February 26, 1980).

Q: Once a waste is identified as hazardous, what is the next step?

A: The next step is to determine the appropriate generator category and the requirements that must be met. To accomplish this, the quantity of hazardous wastes generated by a facility must be counted. In general, a generator must tally the weight of all hazardous wastes generated in a calendar month. The total weight of *countable* hazardous wastes determines the appropriate generator category and applicable requirements, as outlined in Table 5.

TABLE 5 Requirements for hazardous-waste generators.

	Requirements		
	Small-quantity generators	Medium-quantity generators	Large-quantity generators
Quantity limits	<100 kg/mo	Between 100–1000 kg/mo	1,000 kg/mo or greater
Management of waste	State-approved or RCRA permitted	RCRA-permitted facility	RCRA-permitted facility
Manifest	Not required*	Required	Required
Exception report	Not required	Required after 60 days	Required after 45 days
Biennial report	Not required	Not required	Required
Personnel training	Not required	Basic training required	Required
Contingency plan	Not required	Basic plan required	Full plan required
EPA ID number	Not required*	Required	Required
On-site storage limits	May accumulate up to 999 kg	May accumulate up to 6,000 kg for up to 180 days or 270 days if waste is to be transported over 200 miles	May accumulate any quantity up to 90 days
Storage requirements	None	Basic requirements with the technical standards under Part 265 for containers or tanks	Full compliance with technical standards under Part 265 for containers or tanks

*Although not legally required under RCRA, many transporters will not handle hazardous waste without these items.

Q: What are the countable **hazardous wastes?**

A: Hazardous wastes must be counted if they are:

- Accumulated on-site for any length of time prior to their subsequent management
- Packaged and transported off-site
- Placed directly into an on-site, Subtitle C-regulated treatment, storage, or disposal unit
- Generated from a product-storage tank or manufacturing process unit

Q: Are there wastes that do not have to be counted?

A: Yes; hazardous wastes do *not* have to be counted if they are:

- Specifically excluded from regulation, including spent lead-acid batteries, used oil, and commercial chemical products sent off-site for reclamation.
- Nonacutely hazardous waste remaining in an *empty container,* as defined in Section 261.7.
- Managed in an elementary neutralization unit, totally enclosed treatment unit, or wastewater treatment unit as these units are defined in Section 261.10.
- Discharged directly to a publicly owned treatment works (POTW) without being accumulated or treated prior to discharge.
- Produced from the on-site treatment of previously counted hazardous waste. For example, a facility generates 995 lb of methyl isobutyl ketone (MIK), then distills the spent MIK in the facility's on-site distillation unit. The distillation unit, after the distillation process, generates 15 lb of waste MIK still bottoms. However, because the facility already counted the spent MIK before it was treated, the still bottoms do not have to be counted because, technically, it is still the same waste.
- Continuously reclaimed without storing the waste prior to reclamation, such as dry-cleaning solvents, although a generator must count any hazardous waste residue removed from the unit, as well as spent cartridge filters derived from listed dry-cleaning solvents, such as perchloroethylene.

Q: After the hazardous waste has been counted, what is the next step?

A: The generator next determines the appropriate category based on the amount of *countable* hazardous waste generated per calendar month, as outlined in Table 6.

TABLE 6 Categories of hazardous waste generators.

Waste generated	Category
<100 kg/mo =	Small-quantity generator
100 to 1,000 kg/mo =	Medium-quantity generator
1,000 or more kg/mo =	Large-quantity generator

(1 kilogram equals approximately 2.2 pounds)

Q: Previously, weren't there just two categories of generators?

A: Yes; before the Hazardous and Solid Waste Amendments of 1984, there were two categories of hazardous waste generators: those generating less than 1,000 kg/mo and those generating 1,000 kg/mo or more. The amendments reorganized the categories of generators into three tiers based on the generation of nonacutely hazardous waste per month: less than 100 kg, from 100 to 1,000 kg, and 1,000 kg or more. Although the categories were restructured, they have not been given standardized names. Various publications, as well as EPA itself, have created confusion among the regulated community by using differing terms that are nondescriptive, such as "conditionally exempt small-quantity generator." For the purposes of better understanding the generator requirements, this book uses the following terms: small-quantity generators (<100 kg/mo), medium-quantity generators (between 100 and 1,000 kg/mo), and large-quantity generators (1,000 kg/mo or more).

Q: Can a generator change categories?

A: Yes; because a generator is regulated corresponding to the amount of hazardous waste generated per calendar month, it is possible to change categories monthly. For example, a generator normally produces less than 100 kg/mo; however, once a year, the generator conducts a facility-wide product storage and manufacturing process unit cleanout. This annual operation generates more than 1,000 kg during that month. The generator is considered a large-quantity generator for that month only and must manage the waste in accordance with Part 262. After that month, the generator, assuming less than 100 kg of hazardous waste is generated, can revert to the small-quantity generator category.

SMALL-QUANTITY GENERATORS

Applicability

Q: What is a small-quantity generator?

A: A small-quantity generator (SQG) is a generator who, in a calendar month, generates less than 100 kg of nonacutely hazardous waste or less than 1 kg of an acutely hazardous waste.

Q: What are the requirements for small-quantity generators?

A: A small-quantity generator is excluded from Parts 262 through 270 of RCRA provided (1) the waste is identified (Section 262.11); (2) the generator does not accumulate at any one time hazardous wastes in quantities of 1,000 kg or more; and (3) the waste is either treated or disposed of on-site or the generator ensures delivery to an off-site storage, treatment, or disposal facility, any of which [Section 261.5(f)(3)]:

- Is permitted under Part 270 of RCRA
- Has interim status under Part 270 of RCRA
- Is authorized to manage hazardous waste by a state with an authorized program under Part 271 of RCRA
- Is permitted, licensed, or registered by a state to manage municipal or industrial hazardous waste
- Is a facility that beneficially uses, reuses, or legitimately recycles or reclaims the hazardous waste
- Is a facility that treats the waste prior to beneficial use or reuse or to legitimate recycling or reclamation

Q: Are small-quantity generators not liable if their waste causes environmental damage?

A: Regardless of the regulatory status of a generator's hazardous waste under RCRA, Superfund (Section 107) does not exclude a generator's hazardous waste based on the amount or management of the waste. Thus, it is in the generator's best interest to ensure that his hazardous waste is managed in the most secure manner possible. If a generator's hazardous waste is ever involved in a release, the generator is liable for all or part of the cleanup costs.

Q: If one-third of the hazardous waste in an on-site storage tank is emptied every week and the waste is hauled off-site, could the generator qualify for a small-generator limit as long as the amount remaining on-site does not exceed 100 kg?

A: The generator can qualify as a small-quantity generator only if the *total* amount of hazardous waste generated is less than 100 kg per calendar month (Section 261.5).

Q: Section 261.5(d)(3) states that an SQG need not include spent materials that have been reclaimed and subsequently reused on-site in the quantity determination, provided they have already been counted once. The regulation does not specify, however, whether this allowance applies only within a month or applies to all waste counting. For example, if an SQG counts and reclaims a solvent on-site in October and uses it again in November, must the SQG include the spent solvent in the quantity determination for November?

A: Yes; the SQG must include the reused material in the quantity determination for the subsequent month, assuming that it becomes a spent material and, hence, a hazardous waste again in November. All counting occurs on a month-to-month basis, so the "multiple counting" exemption applies only within one month. Therefore, an SQG would count a material just once only if he reclaimed and reused it more than once within one month.

Q: A small-quantity generator who always generates less than 100 kg of hazardous waste per month exceeds the 100-kg accumulation limit. During the accumulation period, additional small quantities of waste are generated and placed into partially full drums. The original 100 kg of regulated waste is shipped off-site, while the partially full drums remain. Is the waste remaining in the partially full drums excluded under Section 261.5?

A: No; all waste generated and accumulated after the SQG's accumulation limit is exceeded is also subject to the Part 262 regulations (generator requirements) because the total accumulation is still above the SQG limit for that month. Even when the original regulated waste is shipped off-site, the remaining partially full drums are still regulated and must be removed within their respective time limits, per Section 261.5(f).

Q: A dry cleaner generates 200 kg/mo of hazardous waste (listed spent solvent) from a dry-cleaning process. At the facility the owner distills these solvents and 90 kg/mo of still bottoms is produced. How much waste has been generated at this facility?

A: This facility generated 200 kg. In determining the quantity of hazardous waste generated, a generator need not include hazardous waste produced by on-site treatment of hazardous waste [Section 261.5(d)(2)].

Q: What is required if a small-quantity generator accumulates on-site more than 1,000 kg of waste?

A: If a small-quantity generator accumulates 1,000 kg or more of hazardous waste on-site, the generator loses the SQG exclusion, and all the accumulated waste is subject to full regulation under Part 262 generator standards and must be sent to a designated facility within 180 days.

***Q:* What are the requirements for waste accumulation units used by small-quantity generators?**

A: There are no technical standards specified for accumulation units used by SQGs.

***Q:* A small-quantity generator sends hazardous waste to a facility that stores the hazardous waste in an underground tank before it is sent to a recycler. The facility is not treating the waste prior to recycling. Is the facility required to have a RCRA storage permit?**

A: Section 261.5(g)(3)(v) allows small-quantity generator waste to go to a facility that recycles waste or treats the waste prior to recycling. This exemption does not include storage prior to recycling. However, Section 261.5(g)(3)(iv) allows the waste to go to a facility that is registered, licensed, or certified by the state to handle solid waste.

***Q:* A small-quantity generator sends waste to a facility that is registered by the state to manage solid wastes. This facility accepts wastes from other small-quantity generators and, after collecting enough waste for a bulk shipment (over 1,000 kg), sends it to a facility for disposal. Is manifesting required in this scenario?**

A: No; manifesting is not required in this scenario. Even if the storage facility collects 1,000 kg and sends the waste off-site, no manifesting is required because the waste generated came from SQGs of less than 100 kg. SQG wastes are exempt from Parts 262 through 270, including manifesting requirements.

***Q:* A small-quantity generator mixes F001 (spent solvent) waste with a D001 (ignitable) waste and then sends this mixture to be recycled. Does this mixing activity require a treatment permit?**

A: No; Section 261.5(g)(3)(v)(B) allows a small-quantity generator to treat waste prior to recycling without having to obtain a permit.

***Q:* If an acutely hazardous waste is spilled, what SQG quantity limits for the spill residue apply?**

A: If 1 kg of acutely hazardous waste is generated or accumulated prior to a spill, then all the spill residue is subject to regulation (even if under

100 kg) because the waste when spilled was subject to full regulation per Section 261.5(i). However, if the SQG had not exceeded the 1-kg limit of acutely hazardous waste in generation or accumulation, and less than 1 kg of acutely hazardous waste was spilled, then the 100-kg limit for spill residue would apply, per Section 261.5(e)(2).

Q: If a small-quantity generator periodically becomes subject to the Section 262.34 generator regulations, e.g., only once every 5 years, must the facility personnel training be updated annually per Section 265.16(c)?

A: No; the SQG must comply with the Section 265.16(c) requirements only when subject to all of Section 262.34, which, in this case, is once every 5 years.

Q: A generator has two accumulation areas on-site. In one area, the generator stores waste identified as small-quantity generator waste. In another area at the same site, the generator stores large-quantity generator (LQG) waste, which was produced during the months the generator exceeded the 1,000-kg per-month limit. Must the generator count waste stored in his LQG accumulation area when determining if the 6,000-kg accumulation limit is exceeded in Section 262.34(d)(1).

A: The 6,000-kg cap for the SQG waste stream applies to all waste accumulated on-site. *On-site* means all contiguous property (Section 260.10). The definition does not refer to *units* or *accumulation areas*. Therefore, the generator must count all waste, including both SQG and LQG waste that is on-site, to determine compliance with Section 262.34(d)(1).

REQUIREMENTS FOR MEDIUM- AND LARGE-QUANTITY GENERATORS

Q: What are the requirements for medium- and large-quantity generators?

A: In addition to determining if their waste is hazardous and determining the appropriate generator category, both medium- and large-quantity generators must:

- Minimize their hazardous waste
- Comply with the land disposal ban
- Notify EPA and obtain an EPA identification number
- Comply with the manifest procedures
- Follow DOT's pretransport requirements

In addition to these requirements, there are additional requirements specifically for medium- and large-quantity generators discussed later in this chapter.

Waste Minimization

Q: What is waste minimization?

A: *Waste minimization* has been defined differently by different organizations. EPA in its October 1986 report to Congress on the minimization of hazardous waste, defined waste minimization as:

> The reduction, to the extent feasible, of hazardous waste that is generated or subsequently treated, stored, or disposed of. It includes any source reduction or recycling activity undertaken by a generator that results in either: (1) the reduction of total volume or quantity of hazardous waste or (2) the reduction of toxicity of hazardous waste, or both, so long as the reduction is consistent with the goal of minimizing present and future threats to human health and the environment. Waste minimization does not include treatment of hazardous waste.

Q: What is required for a generator regarding waste minimization?

A: When generators submit the biennial report (EPA Form 8700-13), it must contain information that outlines: (1) efforts undertaken during the year to reduce the volume and toxicity of waste generated, and (2) the changes in volume and toxicity of waste actually achieved during the year in comparison with previous years [Section 262.41(a)(6) and (7)].

In addition, generators must use a manifest containing a certification (Item No. 16) that the generator has a program to reduce the volume or quantity and toxicity of hazardous waste to the degree determined by the generator to be economically practicable. The program must include a practicable method currently available to the generator that minimizes the present and future threat to human health and the environment.

Land Disposal Ban

Q: Are generators required to determine if their waste is subject to the land disposal restrictions?

A: Yes; generators must determine if their waste is subject to the land disposal prohibitions [Sections 262.11(d) and 268.7(a)].

Q: What are the land disposal prohibitions?

A: There are land disposal prohibitions on certain hazardous wastes, as described in Chapter 8. Although the ultimate responsibility for compliance with the land disposal prohibitions is placed on the land-disposal facilities, generators remain responsible for determining whether their wastes are subject to the prohibitions [Sections 262.11(d) and 268.7(a)].

To determine the applicability of a waste, the generator can either base the determination on knowledge of the waste, testing, or both.

Q: What is required if a generator determines that a waste to be shipped off-site is a restricted waste and exceeds the applicable treatment standards (i.e., prohibited from land disposal)?

A: If a generator determines that the waste does not meet the specified treatment levels (and is thus prohibited from land disposal), the generator must notify the designated facility in writing.

Q: What information is required in the notification?

A: According to Section 268.7(a)(1), the notice must contain the following:

- EPA hazardous waste numbers
- The corresponding treatment standard
- Waste analysis data (where applicable)
- The manifest number associated with the shipment of waste

Q: What is required if the waste meets the treatment standards and is thus allowed to be land disposed?

A: If a generator determines that a restricted waste can be land-disposed of without further treatment, each shipment of that waste to a designated facility must have a notice and certification stating that the waste meets all applicable treatment standards. The signed certification must state the following:

> *I certify under penalty of law that I have personally examined the waste through analysis and testing or through knowledge of the waste to support this certification that the waste complies with the treatment standards specified in 40 CFR 268.32 or RCRA Section 3004(d). I believe that the information submitted is true, accurate, and complete. I am aware that there are significant penalties for submitting a false certification, including the possibility of a fine or imprisonment.*

Notification Requirements

Q: What are the notification requirements for generators?

A: Section 3010 of RCRA requires any person generating, transporting, or owning or operating a treatment, storage, or disposal facility involving hazardous waste to notify EPA of the regulated activity using EPA Form 8700-12, *Notification of Hazardous Waste Activity*. This procedure requires notification of a facility's location, a general description of the regulated activity, and identification of the types and estimates of the quantities of hazardous wastes.

Q: If a generator notified of activities in 1982 and subsequently generated a new waste, is renotification required?

A: No; a new form is not required if a facility generates or transports hazardous wastes not previously identified on the facility's original form, except for waste-as-fuel activities (45 *FR* 12747, October 30, 1980). It is important to note that the mere filing of a notification form does not automatically subject a person to RCRA regulation. A facility will become subject to Subtitle C only when that facility generates, transports, or manages a hazardous waste.

Q: If EPA lists a new hazardous waste, must a generator file a subsequent notification for generating or handling that new waste?

A: Pursuant to Section 3010, a generator must renotify for newly listed wastes only if the Administrator specifically requires renotification. The *Federal Register* notice for any newly listed waste will explicitly state if renotification is required.

Q: What does the waste code designation D000 on the notification form refer to?

A: The waste code designation D000 is not a valid designation. This number is used solely as a sample EPA toxicity code designation. Although this number has been used as a catchall category, it has no regulatory significance.

Q: The notification form has a box for infectious waste. Is infectious waste regulated?

A: No; EPA has not yet listed wastes because of their infectious potential nor established characteristics to be used by a generator to identify a waste as infectious. EPA has originally proposed to regulate wastes because of their infectious potential, but never promulgated the rule or changed the notification form.

Q: Can large corporations with several facilities notify for all facilities in a single notification?

A: No; separate notification must be submitted for each facility that is subject to the regulations.

Q: Is all information in the notification form available to the public, or can a generator request confidentiality on, for example, a portion of the waste stream?

A: All information that EPA receives from notification will be made available to the public. However, notifiers may request that the information they submit be kept confidential. EPA will consider such requests only if the owner or operator submits a written substantiation of the claim with the notification, in accordance with 40 CFR Part 2. A written substantiation does not guarantee that EPA will ultimately determine that the information submitted warrants confidential treatment.

EPA Identification Number

Q: What is an EPA identification number?

A: Upon receipt of a completed notification form, EPA issues an appropriate identification number [Section 262.12(a)]. These 12-digit identification numbers are obtained from a state or EPA regional office by submitting EPA Form 8700-12, *Notification of Hazardous Waste Activity*. The identification number is given to each generation site. Thus, if the facility relocates, a new identification number must be obtained.

Q: Is a new identification number required if a facility obtains a new owner?

A: If the ownership or operational control changes for the facility, a new identification number is not required, although it is recommended that a new number be obtained.

Q: A university has many research facilities on campus that generate hazardous waste. What should the university do concerning EPA identification numbers?

A: Several basic configurations exist for college campuses. A rural campus might have several buildings on one contiguous piece of property. This would be considered a single, or individual, generation site, even though one or more hazardous wastes are generated from one or more sources. One EPA identification number would be assigned, and small-quantity generator status would be determined by looking at the total hazardous waste generated or accumulated on the site.

However, many university campuses are divided by public roads or other rights-of-way not under univeristy control. In this case, each generation site (i.e., each section of the campus bisected by a public road) would be a generator (or small-quantity generator) and assigned its own EPA identification number.

Q: Hazardous waste is to be received in a U.S. port and then shipped to England. Must the receiving shipping company have an EPA identification number?

A: Yes; the receiving facility must have an EPA identification number to receive the waste, per Section 263.11. If it stores the waste, it would need a storage permit unless it is a *transfer facility*.

Q: Would a generator who also transports or stores hazardous waste have a single identification number for all RCRA transactions?

A: Yes; if the generator also stores on-site and/or transports hazardous waste, the generator notifies of these other activities and receives, or modifies, one EPA identification number.

Q: Corporation A, a manufacturing firm, owns a large site. Corporation B, a wholly owned subsidiary of Corporation A, has a permitted treatment unit on the site. Corporation B has an identification number associated with this waste-management activity. Corporation C, another wholly owned subsidiary of Corporation A, is also located on this site and will be generating hazardous waste in the near future. Should Corporation C use the identification number associated with the site, although a different corporation, or is Corporation C required to obtain its own identification number?

A: Section 262.12 requires a generator to have an EPA identification number before treating, storing, disposing of, transporting, or offering for transportation, hazardous waste. The definition of generator, in Section 260.10 is keyed to both person and site: "any person by site whose act or process produces hazardous waste. . . ." The definition of person in Section 260.10 is "an individual, trust, firm, joint stock company, Federal agency, corporation (including a government corporation), partnership, association, State, municipality, commission, political subdivision of a State, or any interstate body." The definition of individual generation site in Section 260.10 is "the contiguous site at or on which one or more hazardous wastes are generated." An individual may have one or more sources of hazardous waste, but is considered a single or individual generation site, if the site or property is contiguous.

In this situation Corporation B and Corporation C are two distinct entities (i.e., persons). They must each apply for a separate EPA identification number. Even though identification numbers are usually site-specific, where different people conduct different regulated activities on a site, a person conducting each regulated activity must obtain an EPA identification number. This does not preclude an EPA regional office or a state from issuing the same number to two persons.

Q: If a generator disposes of hazardous waste down a sewer to a POTW, does the generator need to obtain an identification number?

A: Section 262.12(a) states that a generator must not treat, store, dispose, transport, or offer for transport hazardous waste without having received an EPA identification number. If the generator is not doing any of these activities, an identification number is not needed.

Manifest Requirements

Q: What is the manifest program?

A: The Uniform Hazardous Waste Manifest is a component of a joint program between the Department of Transportation (DOT) and EPA. A generator who transports or offers for transportation hazardous waste for off-site treatment, storage, or disposal must prepare a manifest to accompany that shipment [Section 262.20(a)].

Q: What is the manifest?

A: The manifest, a one-page form with multiple carbon copies for the participants in the shipment, must identify the type and quantity of waste, the generator, the transporter, and the facility to which the waste is being shipped. The standard federal manifest has white numbered boxes and shaded lettered boxes. The shaded lettered boxes are for optional information that a state may require, and the white numbered boxes are required to be filled out by all users of a manifest.

Q: Why are there multiple copies of the manifest?

A: The manifest form is designed so that shipments of hazardous waste can be tracked from their point of generation to their final destination, the so-called cradle-to-grave system. The manifest is produced in a multicopy format so that each step in the shipment process can be documented. Thus, the generator of the hazardous waste completes the manifest and turns the shipment and remaining copies of the manifest over to the transporter. The transporter, or carrier, signs the manifest, returns one copy to the generator, and turns the shipment and manifest over to the designated facility (a facility with interim status or a permit). The facility signs the manifest, gives one copy back to the transporter, keeps a copy, and sends the remaining copy back to the generator, so the generator knows that the hazardous waste has reached its intended destination.

Some states require that a copy of the manifest also be sent to the state hazardous waste agency each time a shipment changes hands.

Q: What are the manifest requirements for designated facilities?

A: The designated facility that receives and signs the manifest has accepted the responsibility of that shipment and cannot ship the waste back to the generator or any other facility unless the generator or facility is classified as a designated facility, i.e., the generator must have a permit or interim status to accept the waste. If the original designated facility never actually signed the manifest, the waste can be shipped back to the generator. However, the waste would still be subject to the storage time limits.

Q: Is a generator required to keep copies of manifests on-site?

A: Section 262.40 does not specify that a generator must keep copies of manifests on-site. Copies can be kept at corporate headquarters. It must be noted, however, that Section 3007(a) of RCRA states that a generator must be able to provide to EPA or duly designated personnel information on or access to records regarding waste management.

Q: Where is the top copy ("original") of the manifest filed?

A: The regulations require that the generator receive a copy of the manifest with a handwritten signature of the facility owner or operator. The generator is required to keep a copy of the manifest signed by the first transporter until the generator receives the copy signed by the facility owner or operator. The owner or operator of the facility that receives the waste also must retain a copy of the manifest (signed by a facility representative and the transporter delivering the waste).

Q: A generator manifested 23 drums of hazardous waste to a facility. The designated facility decided to temporarily stop accepting all waste. (a) Can the

generator take the waste back without having a interim status or a permit? (b) Can the generator direct the waste to another facility?

A: (a) Yes; the generator can accept the waste if the transporter is unable to deliver the hazardous waste to the designated or alternate facilities [Section 262.20(b)]. The 90-day accumulation period still applies for the generator. For example, if the generator had stored the waste for 30 days before the initial shipment, then the generator has 60 days left (minus the days in transit) after accepting the waste back.

(b) Yes; the generator can direct the waste to another designated facility as long as the facility has the storage capacity and waste code authorization to accept the waste. Wastes that can be accepted are identified in either the facility's Part A or permit (see Chapter 9).

Q: Where does a generator obtain a manifest?

A: The acquisition protocol for the manifest form requires the generator to obtain the manifest from the consignment (disposal) state. However, if that state does not supply or require the use of a particular form, the manifest must be obtained from the generator's state. If the generator's state does not supply or require the use of a particular form, the generator may obtain the manifest from any source (49 *FR* 10496, March 20, 1984).

Q: Can states or private firms printing their own forms put their names, addresses, logos, and/or emergency response numbers on the form?

A: Yes; both entities can place this information in the margin or on the back of the form.

Q: Can a state preprint instructions or other applicable requirements on the manifest?

A: Yes; a state can preprint instructions or regulations in the margin or on the back of the form as long as the instructions or regulations do not require reporting additional information [Section 271.10(h)(1)(iv)].

Q: When shipping hazardous wastes by drum, should the weight or volume be entered in Item 14?

A: Either unit of measure is acceptable.

Q: A designated facility may require that such information as waste concentration and pH be included on the manifest. Where would this information go?

A: This type of information would go in Item 15 - "Special Handling Instructions and Additional Information."

Q: How are lab packs designated on the manifest?

A: DOT requires the outside container to show the proper shipping name and be labeled as required for each hazardous material in overpacks [49 CFR 173.25(a)].

The EPA lab pack, or DOT overpack, is required to be itemized according to the inner containers. The manifest will show each type of DOT shipping name represented by the contents, plus how many containers of each name.

Q: Is a generator allowed to place both hazardous and nonhazardous wastes on a manifest?

A: RCRA does not prohibit manifesting hazardous and nonhazardous waste on the same form. If the manifest is also the DOT shipping paper, DOT sets forth the way to designate the hazardous materials in 49 CFR 172.201(a)(1): enter the hazardous materials first. Color highlight the hazardous materials or place an "X" or "RQ" in Items 11 and 28.

Q: A generator loads waste directly into a bulk tank railcar. The rail transporter will then distribute the load among three trucks. How would the manifesting requirements be handled?

A: The generator would cut three manifests and, on each, would indicate that the total volume of waste transported will be split into thirds. Each manifest would require original signatures of the generator and the rail transporter. The signed manifests would be mailed to the trucking company, which would then give each of the three truck drivers one of the manifests.

Exception Reports

Q: Does the date of the generator's certification on a manifest designate the date of shipment?

A: The date of the generator's certification on a manifest does not necessarily have to be the date of shipment. The date of shipment is determined by the date of the initial transporter's signature. The time span for consideration of exception reporting is based on the date of the transporter's signature, not the generator's certification [Section 262.42(a)].

Q: What if a generator does not receive a copy of the manifest with the handwritten signature of the owner or operator of the designated facility within 35 days of the date the waste was accepted by the original transporter?

A: The generator must contact the transporter and/or the owner or operator of the designated facility to determine the status of the hazardous waste. If a copy has not been received within 10 days (45 days from the date the waste was accepted by the initial transporter—60 days for MQGs), the generator must submit an exception report to EPA (Section 262.42).

Q: What information is required in an exception report?

A: The report consists of a copy of the original manifest and a cover letter explaining the efforts taken to locate the waste and the results of those efforts.

Pretransport Requirements

Q: What are the pretransport requirements?

A: The pretransport requirements contained in Sections 262.30 through 262.33 require a generator to comply with the DOT regulations for shippers.

See Chapter 4 for a detailed explanation of the requirements for generators (shippers).

Q: Are there any additional requirements not identified in DOT's regulations?

A: Yes; Section 262.32(b) requires that generators shipping wastes in containers of 110 gallons or less must display the following words and information:

HAZARDOUS WASTE-Federal Law Prohibits Improper Disposal. If found, contact the nearest police or public safety authority or the U.S. Environmental Protection Agency.

Generator's Names and Address ____________________.

Manifest Document Number ____________________.

Because DOT requires additional information for containers, it is suggested that a generator/shipper use the standard yellow "Hazardous Waste Labels." These labels combine necessary DOT information, such as proper shipping names, and the United Nations/North America identification number, as well as required and important EPA information, such as the generator's name, address, EPA identification number, manifest document number, EPA waste code number, the accumulation start date, and the hazardous waste warning.

MEDIUM-QUANTITY GENERATORS

Applicability

Q: What is a medium-quantity generator?

A: A medium-quantity generator (MQG) is someone who generates between 100 to 1,000 kg of hazardous waste in a calendar month.

Q: Wasn't a MQG previously regulated as a large generator?

A: Yes; however, on March 24, 1986 (51 *FR* 10146), EPA removed medium-quantity generators from the existing small-quantity exclusionary provision in Section 261.5, established special requirements for hazardous waste generated by these generators, and promulgated final regulations that placed such generators under Part 262 (the standards applicable to generators of hazardous waste). EPA also promulgated specific amendments to Part 262 to exempt these generators from some of the standards for generators of more than 1,000 kg/mo.

Q: What are the general requirements for medium-quantity generators?

A: Medium-quantity generators must:

- Determine if their wastes are hazardous
- Obtain an EPA identification number
- Offer their wastes only to a transporter or hazardous waste management facility that has an EPA identification number

In addition to these requirements, a medium-quantity generator must comply with the requirements outlined below (see also Table 6).

On-Site Accumulation

Q: What are the on-site accumulation requirements?

A: A medium-quantity generator may accumulate hazardous waste onsite without a permit or interim status for up to 180 days (or 270 days if the waste must be transported over 200 miles) if (1) the generator does not accumulate 6,000 kg or more of hazardous waste; (2) waste is accumulated in either containers or tanks only; and (3) the generator complies with the requirements for personnel training, emergency procedures, preparedness and prevention, and the technical standards for accumulation units.

Q: Is it permissible for a MQG to ship waste to a facility more than 200 miles away, even though the generator could send the shipment to a facility less than 200 miles away?

A: Yes; there are no regulations addressing when a generator is permitted to ship waste in excess of 200 miles, and thus receive an extra 90 days storage time.

Personnel Training

Q: What is required for personnel training?

A: Medium-quantity generators must ensure that all employees are thoroughly familiar with proper waste handling and emergency procedures relevant to their responsibilities during normal facility operations and emergencies.

Emergency Procedures

Q: What are the requirements concerning emergency procedures?

A: At all times there must be at least one employee either on the premises or on call (i.e., available to respond to an emergency by reaching the facility within a short time) with the responsibility for coordinating all emergency response measures. This employee is deemed the *emergency coordinator* [Section 262.34(d)(5)].

Q: Does the emergency coordinator have any specific responsibilities?

A: Yes; according to Section 262.34(d)(5)(iv), the emergency coordinator must respond to any emergencies that may arise and, if appropriate, institute the following emergency measures:

- Contact the fire department and/or attempt to extinguish the fire.
- For any spill, contain the flow and commence cleanup wherever possible.

- For any fire, explosion, or release that meets a Superfund reportable quantity or a release that threatens human health or the environment, notify the National Response Center (1-800-424-8802).

Q: Are there additional requirements for MQGs concerning emergency procedures?

A: Yes; the MQG must post the following emergency information next to facility telephones [Section 262.34(d)(5)(ii)]:

- The name and telephone number of the designated emergency coordinator
- The telephone number of the fire department and appropriate emergency-response organizations
- The location of fire extinguishers, spill control equipment, and fire alarms

Preparedness and Prevention

Q: What are the requirements for preparedness and prevention?

A: A facility must be operated and maintained in a manner that will minimize the possibility of any fire, explosion, or an unplanned sudden or nonsudden release [Sections 262.34(d)(4) and 265.31].

Q: Are there other requirements for preparedness and prevention?

A: Yes; a medium-quantity generator must have emergency equipment, maintain adequate aisle space, and make prior arrangements with local emergency organizations.

Q: What is the required equipment?

A: Required equipment includes an alarm system, a communication device to contact emergency personnel (e.g., telephone), portable fire extinguishers, fire-control equipment, and an adequate fire-fighting water-supply system in the form of hoses or an automatic sprinkler system. This equipment must be routinely tested and maintained in proper working order (Section 265.32).

Q: What is required for adequate aisle space?

A: There must be adequate aisle space to allow the unobstructed movement of emergency equipment to any area of the facility. The regulations do not specify the aisle space; this should be determined upon consultation with local emergency organizations (Section 265.35).

Q: What are the prior arrangements that must be made?

A: Facilities must make prior arrangements with local emergency organizations and personnel for an emergency response. The arrangements should include notification of the types of waste handled, a detailed layout of the facility, a listing of facility contacts, and specific agreements with various state and local emergency response organizations. Refusal of any state or local authorities to enter into such arrangements must be documented and maintained in the facility's files (Section 265.37).

Accumulation Units

Q: What requirements must MQGs comply with concerning accumulation in containers?

A: Medium-quantity generators accumulating hazardous waste in containers must comply with Subpart I of Part 265, except for Section 265.176; this section requires ignitable and reactive wastes to be placed at least 50 feet from the facility's property boundary. The Subpart I requirements are outlined in Chapter 7 of this book.

Q: What are the requirements for accumulation in tanks?

A: Medium-quantity generators accumulating hazardous waste in a tank must comply with the following requirements pertaining to tank systems (Section 265.201):

- Treatment must not generate any extreme heat, explosions, fire, fumes, mists, dusts, or gases; damage the tank's structural integrity; or threaten human health or the environment in any way.
- Hazardous wastes or reagents that may cause corrosion, erosion, or structural failure must not be placed in the tank.
- At least 2 feet of freeboard (the distance between the top of the tank and the surface of the contents) must be maintained in an uncovered tank unless sufficient overfill-containment capacity is supplied.
- Continuously fed tanks must have a waste-feed cutoff or by-pass system.
- Ignitable, reactive, or incompatible wastes must not be placed in a tank unless these wastes are rendered nonignitable, nonreactive, or compatible.

Q: Are there inspection requirements for MQG tank systems?

A: Yes; at least once each operating day, the waste-feed cutoff and by-pass systems, monitoring equipment data, and waste level must be inspected. At least weekly, the construction materials and the surrounding area of the tank system must be inspected for visible signs of erosion or leakage [Section 265.201(c)].

LARGE-QUANTITY GENERATORS

On-Site Accumulation

Q: Is a large-quantity generator (LQG) allowed to accumulate hazardous waste?

A: Yes; a generator may treat or store waste on-site for up to 90 days without a permit or interim status, provided the following conditions, found in Section 262.34 and outlined below, are met:

- Storage or treatment occurs only in tanks or containers (no impoundments).
- The tanks or containers comply with Part 265 Subpart I, Standards for Containers, and Subpart J, Standards for Tanks.
- The generator does not accept shipments of hazardous waste generated from off-site sources.
- The waste is sent to a designated facility within 90 days, unless the waste is treated and rendered nonhazardous (by definition) within 90 days.
- The generator complies with the requirements for personnel training, preparedness and prevention, and contingency plan and emergency procedures of Part 265.

Q: A facility generates a waste that is in powder form and exhibits the characteristic of toxicity. The waste is stored in a tank pursuant to the standards specified in Section 262.34. When the tank is partially full, the generator pours in sand and mixes the contents of the tank until a homogeneous mixture is formed. The sand dilutes the original waste. The resulting mixture no longer exhibits a characteristic of a hazardous waste. How is the generator regulated under RCRA?

A: If the facility did not accumulate the waste for longer than the applicable period specified in 262.34 (90 days), then the facility would have to comply only with the applicable provisions of Section 262.34. Rendering a characteristic hazardous waste nonhazardous by dilution is treatment; however, such treatment does not require a permit if Section 262.34 is followed.

EPA clarified this interpretation in the March 24, 1986 *Federal Register* (51 *FR* 10168), which states, "Of course, no permitting would be required if a generator chooses to treat their hazardous waste in the generator's accumulation tanks or containers in conformance with the requirements of 262.34 and Subparts J or I of Part 265. Nothing in 262.34 precludes a generator from treating waste when it is in an accumulation tank or container covered by that provision. Under the existing Subtitle C system, EPA has established standards for tanks and containers which apply to both the storage and treatment of hazardous waste . . . the Agency believes that treatment in accumulation tanks or containers is permissible under the existing rules, provided the tanks or containers are operated strictly in compliance with all applicable standards."

Q: The generator standards in Section 262.34 state that accumulated hazardous waste should be placed in containers or tanks. What happens if the waste is accumulated in a surface impoundment?

A: The accumulation rule (Section 262.34) does not apply to surface impoundments. It applies only to waste that is to be shipped off-site in 90 days or less and is stored in approved containers or tanks. Hazardous waste stored in impoundments prior to shipping off-site must meet all interim status standards for impoundments, and the generator must obtain a permit or interim status.

Q: Will generators who store hazardous waste until they have accumulated a sufficient amount for hauling off-site be required to apply for a permit as a storage facility?

A: If the waste is stored in containers or tanks for less than 90 days before shipping off-site, no storage permit is required (Section 262.34). If the waste is stored for more than 90 days, however, a storage permit is required.

The 90-day rule merely exempts the generator from the requirement to obtain a storage permit. It does not change the technical requirements for storage. For example, generators must accumulate their hazardous waste either in shipping containers approved by DOT and in compliance with Parts 265, Subpart I, or in tanks that comply with Part 265, Subpart J.

Q: An industrial firm generates hazardous waste from a number of process units. Does each of these process units constitute a 90-day accumulation unit?

A: No; a generator may accumulate up to 55 gallons of hazardous waste or 1 quart of acutely hazardous waste in containers at or near any initial generation point. These initial generation points are considered *satellite accumulation areas*. As soon as the 55-gallon or 1-quart limit is attained, the generator has up to 3 days to move that container to the regular storage area. As soon as the container is at the regular storage area, the 90-day time limit starts. Satellite accumulation containers must be marked with the words *Hazardous Waste* [Section 262.34(c)(1)].

Q: When does the 90-day period begin?

A: The 90-day period starts as soon as the first drop hits the tank or drum (47 *FR* 1250, January 11, 1982).

Q: How is the period documented?

A: The date on which accumulation began must be marked on individual containers. The dating of tanks was accidentally omitted from the regulations, but some type of documentation should be kept to demonstrate compliance with the 90-day time limit.

Q: Are there marking requirements for tanks or containers?

A: Yes; in addition to the date on which accumulation began, each container and tank must have the words *Hazardous Waste* clearly marked them.

Personnel Training

Q: What is the purpose of the personnel training requirements?

A: The purpose of the training requirements is to reduce the potential for errors and violations that might threaten human health or the environment by ensuring that facility personnel acquire expertise in the areas to which they are assigned.

Each generator must establish a training program for appropriate facility personnel. The program must also include training for personnel to ensure facility compliance with all applicable regulations.

Q: What are the requirements of the personnel training program?

A: The program must contain an initial training program, as well as annual updates. The contents and format for the program are unspecified in the regulations. EPA accepts the use of on-the-job training as a substitute for or supplement to formal classroom instruction (45 *FR* 33182, May 19, 1980). However, the content, schedule and techniques used for on-the-job training must be described in training records maintained at the facility (see Chapter 5).

Q: A laboratory has 65 lab technicians who may handle hazardous wastes. Must all these lab technicians be trained to handle hazardous waste?

A: The lab technicians must have training to the extent necessary to ensure safe handling of the wastes. Per Section 262.34(a)(4), the generator must comply with Section 265.16 on training of personnel handling hazardous waste. Section 265.16(d) requires that training records be kept at the facility.

Preparedness and Prevention

Q: What are the requirements for preparedness and prevention?

A: The requirements for this section are contained in Part 265 and require that each facility be operated and maintained in a manner that will minimize the possibility of any fire, explosion, or an unplanned sudden or nonsudden release (Section 265.31).

Q: Are there other requirements?

A: Yes; a facility is required to have equipment, including an alarm system, a communication device to contact emergency personnel (e.g., telephone), portable fire extinguishers, fire-control equipment, and an adequate fire-fighting water-supply system in the form of hoses or an automatic sprinkler system. This equipment must be routinely tested and maintained in proper working order (Section 265.32).

There must be adequate aisle space to allow the unobstructed movement of emergency equipment to any area of the facility. The regulations do not specify the aisle space; this should be determined upon consultation with local emergency organizations (Section 265.35).

In addition, the facility must make prior arrangements with local emergency organizations and personnel for an emergency response. The arrangements should include notification of the types of waste handled, a detailed layout of the facility, a list of facility contacts, and specific agreements with various state and local emergency response organizations. Refusal of any state or local authorities to enter into such arrangements must be documented and maintained in the facility's files (Section 265.37).

Contingency Plan and Emergency Procedures

Q: What are the requirements of a contingency plan?

A: Each facility must have a contingency plan, as required in Part 265, that is designed to minimize hazards in the case of a suddent or nonsudden release, fire, explosion, or similar emergency. The plan must include a description of actions that will be undertaken by facility personnel, a detailed list and location of emergency equipment, and evacuation procedures (Section 265.52).

Q: Is a new plan required if a generator has previously prepared an SPCC plan?

A: No; if a generator has previously prepared a Spill Prevention, Control, and Countermeasure Plan (SPCC) in accordance with either 40 CFR Parts 112 or 300, or prepared some other emergency or contingency plan, he need only amend that plan to incorporate hazardous waste management provisions that are sufficient to comply with these provisions [Section 265.52(b)].

Q: What are the requirements for the contingency plan concerning emergency procedures?

A: The provisions of the plan must be carried out immediately whenever there has been a fire, explosion, or release of hazardous waste or constituents [Section 265.51(b)].

Q: What information is required to be in the plan?

A: The plan must list current names, addresses, and phone numbers of all persons qualified as emergency coordinators. There must be at least one employee on-site, or close by and on call, who is the designated emergency coordinator. The plan must also include a list of all available emergency equipment located at the facility, including its location, physical description, and a brief outline of the equipment's capabilities (Section 265.52).

Q: Where must the plan be maintained?

A: A copy of the contingency plan must be maintained at the facility and submitted to all local police, fire, hospitals, and emergency-response teams (Section 265.53).

Q: What are the requirements for emergency procedures?

A: Whenever there is an imminent or actual emergency, the emergency coordinator must immediately activate the facility alarm system, notify all facility personnel, and, if needed, notify appropriate state or local agencies [Section 265.56(a)].

In the event of a release, fire, or explosion, the emergency coordinator must identify the source, character, and amount of materials involved. If any hazardous wastes are released into the environment, constituting a Superfund reportable quantity, the owner or operator must contact the National Response Center (1-800-424-8802) immediately [Section 265.56(d)].

Immediately after the incident, the emergency coordinator must provide for the treatment, storage, or disposal of any contaminated material as a result of the emergency [Section 265.56(g)]

Q: If an emergency arises that requires action, are there any reporting requirements?

A: Yes; the generator must report to EPA within 15 days any incident that requires the implementation of the contingency plan [Section 265.56(j)].

Accumulation Units

Q: What are the requirements for units that accumulate hazardous waste?

A: Generators accumulating hazardous waste in containers must comply with Subpart I of Part 265. The requirements of Subpart I are outlined in Chapter 7 of this book.

Generators accumulating hazardous waste in tanks are required to comply with most of the provisions of Subpart J of Part 265 (see Chapter 7), including:

- A one-time assessment of the tank system, including results of an integrity test
- Installation standards for new tank systems
- Design standards, including an assessment of corrosion potential
- Secondary containment phase-in provisions

- Periodic leak testing, if the tank system does not have secondary containment
- Closure
- Response requirements regarding leaks, including reporting to the EPA regional administrator the extent of any release and requirements for repairing or replacing of leaking tanks

However, owners or operators of 90-day accumulation tanks are not required to prepare closure or post-closure plans, contingent closure or post-closure plans, maintain financial responsibility, or conduct waste analysis and trial tests.

Q: Are there special requirements for generators that ship waste off-site in containers?

A: Yes; if hazardous waste is shipped in containers, the generator must use DOT-specification containers that are appropriate for the type and amount of waste being shipped. The DOT-specification-container requirements can be found in 49 CFR Parts 178 through 180. In addition, all containers of 110 gallons or less must have a hazardous waste label as defined in Section 262.32(b).

Q: Are releases of hazardous waste from a generator's 90-day accumulation tanks regulated under RCRA?

A: Such releases are not specifically covered by the RCRA regulations. The generator is not subject to corrective action under Section 3004(u) of RCRA unless he is engaged in other activities that would require a permit. Section 3004(u) applies only to facilities seeking a permit. Section 3008(h) administrative orders apply only to facilities with interim status. Therefore, the existing RCRA corrective action authorities do not apply to releases from 90-day accumulation tanks unless other units at the facility require interim status or a permit. However, a leaking 90-day tank that is not cleaned up could be considered open dumping under RCRA and could be covered by Section 7003, the imminent hazard provision of RCRA, as well as Sections 104, 106, and 107 of Superfund.

Q: Does EPA require generators to conduct a waste analysis (Section 265.13) of the wastes they are shipping to a facility?

A: A generator is only required to conduct a hazardous waste determination per Section 262.11. A generator is not required to do a complete waste analysis as specified in Section 265.13. However, the designated facility is required to *obtain* a detailed chemical and physical analysis of a representative sample of waste that is received. Hence, if the facility requires the generator to complete a waste analysis as a condition of accepting the waste, the generator may need to conduct a partial or complete waste analysis.

Biennial Reports

Q: What is a biennial report?

A: Each generator must prepare and submit a copy of a biennial report (EPA Form 8700-13A) by March 1 of each even-numbered year (Section 262.41). It is important to note that some states require an annual report.

Q: What information is required for a biennial report?

A: Information required for the biennial report includes:

- The facility's EPA identification number
- EPA identification number for each transporter used
- EPA identification number for each designated facility where waste was sent
- A description and quantity of hazardous waste generated
- A report on the reductions achieved by the waste minimization program at that site

Q: What documents must generators use as sources for their biennial reports?

A: For hazardous waste shipped off-site, the information on the manifest used for shipping that waste should be used in preparing the biennial report.

Q: Will generators be required to include in their biennial reports wastes that have hazard class characteristics, as defined by the Department of Transportation but not regulated by RCRA?

A: No; the generator's biennial report need deal only with hazardous wastes as defined in Section 261.3.

Chapter 4

Transportation Provisions

The transportation of hazardous waste is regulated by both the Hazardous Materials Transportation Act (HMTA) and RCRA. Although the regulatory program developed under RCRA is primarily concerned with the safe disposal of hazardous waste, Section 3003 of RCRA required EPA to establish certain standards for transporters and to coordinate regulatory activities with the Department of Transportation (DOT).

The RCRA and HMTA requirements for hazardous waste transportation are outlined in Table 7.

RCRA REQUIREMENTS

General Requirements

Q: What is the definition of a transporter?

A: A *transporter* is any person engaged in the off-site transportation of hazardous waste by air, rail, highway, or water (Section 260.10).

Q: Who is subject to the transporter regulations under Part 263?

A: EPA's transportation regulations under 40 CFR Part 263 apply to any transporter of hazardous waste except for on-site movements (i.e., within a facility's boundaries).

TABLE 7 The RCRA and HMTA requirements for hazardous waste transportation.

Requirements	Agency	CFR citation
Generator/shipper:		
1. Determine if waste is hazardous	EPA	40/262.11
2. Notify EPA and obtain ID number; determine that transporter and designated TSD facility have ID numbers	EPA	40/262.12
3. Identify and classify waste according to DOT's Hazardous Materials Table	DOT	49/172.101
4. Comply with all packaging, marking, and labeling requirements	DOT	49/172-173
5. Determine whether additional shipping requirements are applicable for mode	DOT	49/174-177
6. Complete a hazardous-waste manifest	EPA	40/262.20-23
7. Provide appropriate placards to transporter	DOT	49/172.506
8. Comply with record keeping and reporting	EPA	40/262.40-45
Transporter/carrier:		
1. Notify EPA and obtain ID number	EPA	40/263.11
2. Verify that shipment is properly identified, packaged, marked, labeled, and not leaking	DOT	49/174-177
3. Apply appropriate placards	DOT	49/172.506
4. Comply with the manifest requirements	EPA	40/263.20-21
5. Comply with record keeping and reporting	EPA	40/263.22
6. Take appropriate action (including cleanup) in the event of a release/spill	EPA	40/263.30-31
7. Comply with DOT incident reporting rules	DOT	49/171.15-17

Source: USEPA's Hazardous Waste Transportation *Interface - Guidance Manual, November 1981*

Q: How much waste must one carry to be considered a transporter of a hazardous waste?

A: Any amount of hazardous waste carried causes the transporter to be a transporter of hazardous waste. However, hazardous waste that is excluded from regulations (e.g., small-quantity generator waste) is not subject to RCRA's transportation regulations.

Q: Can a transporter ever be considered a generator?

A: If a transporter imports hazardous waste into the United States or places waste of different DOT shipping names into a common container, the transporter is subject to the Part 262 requirements for generators [Section 263.10(c)].

Q: Who enforces the federal regulations for transporters?

A: EPA has negotiated a memorandum of understanding with DOT to coordinate enforcement of the transportation regulations (OSWER Directive No. 9434.00-6). DOT will conduct compliance monitoring and enforcement for in-transit activities. Although DOT has assumed primary responsibility in this area, EPA retains enforcement authority.

EPA Identification Number

Q: Is a hazardous waste transporter required to have an EPA identification number?

A: Yes; a transporter cannot transport hazardous waste (provided the waste is not excluded from Part 263) without having an EPA identification number.

Q: Is it necessary for a large transportation company with many terminals to notify EPA and obtain an identification number for each terminal?

A: No; such a transporter would have to submit only one notification and obtain only one identification number for its transportation activities consisting of many terminals.

Q: Does a transporter who transports bulk packages of wastes of different DOT shipping descriptions need a unique generator number for the location of this activity?

A: No; Section 263.10(c)(2) says only that a transporter who combines wastes of different shipping descriptions must comply with the Part 262 regulations, not that the transporter is the generator.

Q: A generator transports hazardous waste from the point of generation (or satellite accumulation area) to a 90-day accumulation area. Is the generator required to have an EPA transporter identification number when sending the waste to a different location on-site?

A: No; the Part 263 transporter regulations do not apply to on-site transportation of hazardous wastes [Section 263.10(b)].

Transfer Facilities

Q: Can a transporter (carrier) temporarily hold hazardous waste without a storage permit?

A: Yes; EPA allows transporters to hold wastes in the course of transportation at a transfer facility for up to 10 days, provided the waste is accompanied by a manifest and remains in containers that meet the DOT packaging requirements (Section 263.12).

Q: What is a transfer facility?

A: *Transfer facility* refers to transportation terminals (including vehicle parking areas, loading docks, and other similar areas), break-bulk facilities, or any facility commonly used by transporters (Section 260.10).

Manifest Requirements

Q: What shipments require a manifest?

A: Except for excluded hazardous waste (e.g., SQG waste) all hazardous waste shipments must be manifested. The manifest must accompany the shipment at all times.

Q: What is the manifest system?

A: The initial transporter who accepts the shipment must sign and date the manifest and present a copy of it to the generator before leaving the property. A transporter who delivers the shipment to another transporter or to the designated facility must obtain the date of delivery and signature of the transporter or the designated facility, retain a copy of the signed manifest, and give the remaining copies to either the next transporter or the designated facility. The designated facility will send the generator/shipper the signed copy.

Q: What are the manifest requirements concerning shipments through multiple states?

A: A transporter must comply with the requirements only in the state of origin and the consignment (destination) state. A transporter is not bound to comply with the state manifest requirements of other states through which he travels (49 *FR* 10492, March 20, 1984).

Q: ***A transporter collects unmanifested hazardous waste from several small-quantity generators (<100 kg/mo), which totals over 1,000 kg. Does the transporter have to initiate a manifest?***

A: No; because small-quantity generator waste is not subject to regulation under Parts 262 through 265, the waste does not have to be manifested under RCRA; however, DOT will require a shipping paper.

Q: ***If a transporter consolidates hazardous wastes with the same DOT descriptions from four different generators, must a new manifest be cut to represent the (one) bulk shipment delivered to the designated facility?***

A: No; the transporter could deliver the bulk shipment with the four manifests; the waste volume for the four manifests must add up to the total volume delivered. For such a consolidation, each manifest must be completed in terms of volume of waste, rather than by piece count (e.g., drums).

Q: ***If two small tank trucks combine their wastes into a larger tank truck, should both their manifests be carried by the larger tank truck and/or a new manifest initiated showing the combined volume?***

A: If both manifests are carried, notes could be added to Item 15 of the manifest stating that consolidation has occurred. The quantities on the manifest could be shown in the comment section.

Q: ***What happens if a transporter cannot deliver the hazardous waste shipment to the designated facility or to the alternate designated facility?***

A: The transporter must contact the generator for further instructions and must revise the manifest according to the generator's instructions [Section 263.21(b)].

Q: ***If hazardous waste residue in the heel of a railroad car was rinsed out and added to the last truck carrying the waste to the designated facility, how is the volume increase accounted for?***

A: According to Sections 264/265.72(a)(1), a significant discrepancy in a bulk shipment is 10 percent in weight. If the off-loading increased the entire original rail quantity by more than 10 percent, the facility would have to resolve the manifest discrepancy with the generator/shipper.

Spills

Q: What is required of a transporter if a spill occurs?

A: In the case of a spill by a transporter, the appropriate authorities must be immediately notified, and the transporter must attempt to contain the spill.

If a government official (state, federal, or local) determines that immediate cleanup is required to protect human health or the environment, that official can require appropriate action, such as removal of the waste by transporters who do not have an EPA identification number, manifest, or permit. A person can receive an emergency EPA identification number or permit from a state or EPA regional office [Section 263.30(b)] (see Appendix G).

The transporter must notify the National Response Center (1-800-424-8802) if required by DOT (49 CFR 171.15) or if the spill meets a reportable quantity under Superfund (40 CFR Part 302). The transporter must also submit a written report to DOT within 15 days if required by 49 CFR 171.16.

If the discharge and surrounding contaminated soil are not cleaned up, the transporter could be held liable under Sections 3008(a) and 7003 of RCRA and Sections 104, 106, and 107 of CERCLA.

Q: If a transporter spills a cargo of hazardous waste, what must the transporter do concerning the manifest?

A: Under Section 263.31, a transporter must clean up any hazardous waste discharge that occurs during transport. If the cleanup of the spill generates a waste that has a different DOT shipping name, then a new manifest must be cut. The transporter should also contact the generator and inform him of the spill. If the spill cleanup residue has the same DOT shipping names, but has increased in quantity (i.e., contaminated soil), the original manifest can be adjusted by noting the discrepancy in Item 15 of the manifest.

Q: What if a new hazardous waste is "generated" at a spill site?

A: If a new hazardous waste is generated at a spill site, a new manifest must be initiated by the transporter, and an emergency identification number, obtained from the state or EPA regional office, is necessary (see Appendix G).

HMTA REQUIREMENTS

Applicability

Q: Aren't DOT regulations restricted to hazardous materials, not wastes?

A: No; DOT hazardous materials regulations, contained in 49 CFR Parts 171 through 179, apply to any material regardless of its end use. The fact that a material is considered a waste is irrelevant. Title 49 CFR, Section 171.3, states that "no person may offer for transportation or transport a hazardous waste [defined as any material subject to the manifest requirements of 40 CFR Part 262] in interstate or intrastate except in accordance with the requirements of this subchapter," i.e., the hazardous materials regulations.

Q: If a person has a hazardous waste as defined by RCRA, is it also considered a hazardous material by DOT?

A: Yes; although the table principally addresses hazardous materials, it also incorporates all the hazardous wastes *regulated* by EPA (i.e., those wastes requiring a manifest).

Q: What does the Department of Transportation regulate concerning hazardous waste shipments?

A: DOT has established requirements for shippers (generators) and carriers (transporters). These people must comply with the regulations for proper containers, marking and labeling of containers, placarding of vehicles, and incident reporting. As in RCRA, the most accountability is placed on the generator/shipper.

Q: What are the primary duties for the generator/shipper?

A: A generator/shipper must:

Step 1. Determine the proper shipping name.
Step 2. Determine the hazard class.
Step 3. Select the proper UN/NA identification number.
Step 4. Determine the mode of transport.
Step 5. Select the proper label or labels and apply as required.
Step 6. Determine and select the proper packages.
Step 7. Mark the packages.
Step 8. Determine and provide the proper placards.

Hazardous Materials Table

Q: What is used to determine the various requirements for transportation of hazardous wastes?

A: In determining which DOT regulations apply to a particular waste, it is essential to determine the DOT proper shipping name. The proper shipping names are listed in *The Hazardous Materials Table* (HMT), which is found in 49 CFR, Section 172.101.

Q: What information is contained in the HMTA?

A: The HMTA lists those materials designated by DOT as hazardous for the purposes of transportation. It identifies the proper shipping names, hazard classifications, United Nations/North American identification numbers, and references for the requirements for labeling, packaging, and shipping procedures.

The key feature to the table is that it compiles the regulations into an index that shippers (generators) and carriers (transporters) can use to readily ascertain what procedures they must follow to transport it. (See Table 8 for a sample page from the Hazardous Materials Table.)

To use the table specifically for hazardous wastes, it must first be remembered that DOT considers hazardous wastes as a subset of hazardous materials. Thus, a hazardous waste will always be regarded by DOT as a hazardous material subject to certain additional requirements set forth by both DOT and EPA. One such requirement is the inclusion of the word *waste* as the first word of the DOT proper shipping name of a listed hazardous material when the material is shipped as a waste.

For example; RQ Waste 1,1,1-Trichloroethane, ORMA, UN 2831, (F002) is the DOT shipping name for spent 1,1,1-trichloroethane, which is a RCRA hazardous waste under the F002 listing.

Q: Is the identification of a waste under RCRA sufficient for the HMTA regulations?

A: No; the identification of a hazardous waste under RCRA is a separate procedure from classifying wastes under DOT's regulations. DOT considers hazardous wastes to be a subset of the hazardous materials regulated by HMTA.

TABLE 8 Sample page from the Hazardous Materials Table.

1	2	3	3a	4	Packaging 5		Maximum net quantity 6		Water shipments 7		
+/ A/ W	Hazardous materials descriptions and proper shipping names	Hazard class	Identification number	Label(s) required (if not excepted)	(a) Exceptions	(b) Specific require ments	(a) Passenger carrying aircraft	(b) Cargo only aircraft	(a) Cargo vessel	(b) Pas- senger vessel	(c) Other requirements
	Accumulator, pressurized (pneumatic or hydraulic), containing nonflammable gas	Nonflam- mable gas	NA1956	Nonflam- mable gas	173.306		No limit	No limit	1,2	1,2	
	Acetal	Flammable liquid	UN1088	Flammable liquid	173.118	173.119	1 quart	10 gallons	1,3	4	
	Acetaldehyde (ethyl al- dehyde)	Flammable liquid	UN1089	Flammable liquid	None	173,119	Forbidden	10 gallons	1,3	5	

A	Acetaldehyde ammonia	ORM-A	UN1841	None	173.505	173.510	No limit	No limit	1,2	1,2	
	Acetic acid (aqueous solution)	Corrosive material	UN2790	Corrosive	173.244	173.245	1 quart	10 gallons	1,2	1,2	Stow separate from nitric acid or oxidizing materials
	Acetic acid-glacial	Corrosive material	UN2789	Corrosive	173.244	173.245	1 quart	10 gallons	1,2	1,2	Stow separate from nitric acid or oxidizing materials
	Acetic anhydride	Corrosive material	UN1715	Corrosive	173.244	173.245	1 quart	1 gallon	1,2	1,2	
	Acetone	Flammable liquid	UN1090	Flammable liquid	173.118	173.119	1 quart	10 gallons	1,3	4	
	Acetone cyanohydrin	Poison B	UN1541	Poison	None	173.346 173.3a	Forbidden	55 gallons	1	5	Shade from radiant heat. Stow away from corrosive materials

Source: 49 CFR 172.101

Shipping Name

Q: How is the proper shipping name selected?

A: Column 2 of the HMT lists the proper shipping name of materials designated by DOT as hazardous. When selecting a proper shipping name to describe a waste, you must use the name on the table that most accurately identifies the waste.

Q: What if the listed shipping name is not applicable to a particular waste material?

A: If the correct technical name of a hazardous waste is neither listed nor entirely accurate, selection must then be made based on the characteristics of the waste. It is important to note that the characteristics used by EPA to identify a waste as hazardous are different from DOT's hazard classes. For example, a reactive waste according to RCRA might be an irritating material or explosive according to HMTA, and a waste classified as ignitable by RCRA could either be flammable or combustible by HMTA.

Q: Are there other required modifications to the DOT proper shipping name?

A: Yes; 49 CFR 172.101(c) lists other modifications including:

- If the technical name of a hazardous material that is shipped as a hazardous waste is listed, the proper shipping description must include the word *waste* before the name of the material, e.g., "waste acetone."
- If the proper shipping name for a mixture or solution that is a hazardous waste does not include the name of the hazardous waste or wastes, each hazardous component found in the mixture or solution must be identified in association with the basic description [49 CFR 172.101(c)].
- Except for those proper shipping names preceded by a "+" in column 1, any hazardous waste that meets the definition of a hazard class other than the one specified with its proper shipping name must be described by the shipping name that appropriately corresponds with its actual hazard class. For example, if a compound is listed and identified as "beryllium compound n.o.s., poison B," but actually displays the characteristics of an oxidizer, it must be described as "oxidizer, n.o.s." or "oxidizing material, n.o.s." [49 CFR 172.101(c)].
- If a hazardous waste will be classed under an n.o.s. name because it is a mixture or solution of one or more hazard classes, (e.g., flam-

mable liquid, poison B, n.o.s.), and one of the classes meets the criteria of a poison B, then the generic or technical name of the poison B must be included in the basic description on shipping papers or hazardous waste manifests, e.g., waste flammable liquid, poison B (or poisonous) n.o.s.

- Shipping names may be used in the singular or plural and in either capital or lower case letters.
- The letters "RQ" (reportable quantity) must be used either before or after the description for each waste in a container meeting an RQ.

Hazard Class

Q: What is the hazard class?

A: The hazard class, specified in column 3 of the HMT, is used to reference the requirements for packaging, labeling, and any special requirements that must be met by generators and transporters. Table 9 contains the hazard classes.

Q: What if there are no appropriate hazard classes for a particular waste?

A: In those instances where a hazardous waste is not listed in the HMT or a waste has multiple hazard classes, the generator must evaluate the waste against the criteria for all the hazard classes. The selection of the appropriate hazard class is based on a priority listing provided in 49 CFR Section 173.2.

Q: What is the priority classification system for selection of the appropriate hazard class name?

A: A hazardous waste having more than one hazard must be classified according to the following order of hazards:

1. Radioactive material (except for small quantities)
2. Poison A
3. Flammable gas
4. Nonflammable gas
5. Flammable liquid
6. Oxidizer
7. Flammable solid
8. Corrosive material (liquid)
9. Poison B

10. Corrosive material (solid)
11. Irritating materials
12. Combustible liquid (in containers having capacities exceeding 110 gallons)
13. ORM-B
14. ORM-A
15. Combustible liquid (in containers having capabilitiesof 110 gallons or less)
16. ORM-E

TABLE 9 Hazard classes.

Flammable liquid—Any liquid with a flash point below 100°F as determined by tests listed in 49 CFR 173.115(d). Exceptions are listed in 49 CFR 173.115(a).

Combustible liquid—Any liquid having a flash point at or above 100° and below 200°F as determined by tests listed in 49 CFR 173.115(d). Exceptions are listed in 49 CFR 173.115(b).

Flammable solid—Any solid material, other than an explosive, apt to cause fires through friction or retained heat from manufacturing or processing readily ignitable, a serious transportation hazard because it burns vigorously and persistently (49 CFR 173.150).

Oxidizer—A substance, such as chlorate, permanganate, inorganic peroxide, or a nitrate, that yields oxygen readily to stimulate the combustion of organic matter (49 CFR 173.151).

Organic peroxide—An organic compound containing the bivalent —O—O— structure and which may be considered a derivative of hydrogen peroxide where one or more of the hydrogen atoms have been replaced by organic radicals. Exceptions are listed in 49 CFR 173.151(a).

Corrosive—Liquid or solid that causes visible destruction or irreversible alterations in human skin tissue at the site of contact. Liquids that severely corrode steel are included [49 CFR 173.240(a)].

Flammable gas—A compressed gas, as defined in 49 CFR 173.300(a), that meets certain flammability requirements [49 CFR 173.300(b)].

Nonflammable gas—A compressed gas other than a flammable gas.

Irritating material—A liquid or solid substance that, on contact with fire or when exposed to air gives off dangerous or intensely irritating fumes. Poison A materials excluded (49 CFR 173.381).

Poison A—Extremely dangerous poison gases or liquids belong to this class. Very small amounts of these gases or vapors of these liquids, mixed with air, are dangerous to life (49 CFR 173.326).

Poison B—Substances, liquids, or solids (including pastes and semisolids), other than Poison A or irritating materials, that are known to be toxic to humans. In the absence of adequate data on human toxicity, materials are presumed to be

TABLE 9 *(cont.)*

toxic to humans if they are toxic to laboratory animals exposed under specified conditions (49 CFR 173.343).

Etiologic agents—A viable microorganism, or its toxin, that causes or may cause human disease. These materials are limited to agents listed by the Department of Health and Human Services (49 CFR 173.386, 42 CFR 72.3).

Radioactive material—A material that spontaneously emits ionizing radiation having a specific activity greater than 0.002 microcuries per gram (MCi/g). Further classifications are made within this category according to levels of radioactivity (49 CFR 173, Subpart I).

Explosive—Any chemical compound, mixture, or device the primary or common purpose of which is to function by explosion, unless such compound, mixture, or device is otherwise classified (49 CFR 173.50).

Explosives are divided into three subclasses:

Class A explosives are detonating explosives (49 CFR 173.53);

Class B explosives generally function by rapid combustion, rather than detonation (49 CFR 173.88); and

Class C explosives are manufactured articles, such as small arms ammunition, that contain restricted quantities of Class A and/or Class B explosives and certain types of fireworks (49 CFR 173.100).

Blasting agent—A material designed for blasting, but so insensitive that there is very little probability of ignition during transport (49 CFR 173.114(a).

ORM (Other Regulated Materials)—Any material that does not meet the definition of the other hazard classes. ORMs are divided into five substances:

ORM-A is a material with an anesthetic, irritating, noxious, toxic, or other similar property that can cause extreme annoyance or discomfort to passengers and crew in the event of leakage during transportation [49 CFR 173.500(a)(1)].

ORM-B is a material capable of causing significant damage to a transport vehicle or vessel if leaked. This class includes materials that may be corrosive to aluminum [49 CFR 173.500(a)(2)].

ORM-C is a material that has other inherent characteristics not described as an ORM-A or ORM-B, that make it unsuitable for shipment unless properly identified and prepared for transportation. Each ORM-C material is specifically named in the Hazardous Materials Table in 49 CFR 172.101 [49 CFR 173.500(a)(3)].

ORM-D is a material, such as a consumer commodity, that, although otherwise subject to regulation, presents a limited hazard during transportation due to its form, quantity, and packaging [49 CFR 173.500(a)(4)].

ORM-E is a material that is not included in any other hazard class, but is subject to the requirements of this subchapter. Materials in this class include hazardous wastes and hazardous substances [49 CFR 173.500(a)(5)].

Source: 49 CFR 172.101 and 173

UN/NA Identification Number

Q: What is the UN/NA identification number?

A: United Nations/North America (UN/NA) numbers consist of the prefix "UN" or "NA" followed by a four-digit number. UN/NA numbers were adopted by the U.S. Department of Transportation in 1980 to facilitate international transportation of hazardous materials. The UN numbers are based on an international system developed by the United Nations Committee of Experts on the Transport of Dangerous Goods. The NA numbers identify materials not recognized for international shipment by the U.N. committee, except for transport between the United States and Canada. The UN/NA numbers for each assigned hazardous waste are located in column 3(a) of the Hazardous Materials Table.

Q: Are the UN/NA numbers the same as EPA's waste code numbers?

A: No; the four-digit UN/NA numbers should not be confused with the hazardous waste code numbers assigned by EPA. The two systems are unrelated.

Labels

Q: What are labels?

A: Labels are symbolic representations of the hazards associated with a particular material.

Q: What are the requirements for using labels?

A: Column 4 of the HMT indicates which materials require labels. Shipments of limited quantities of certain hazardous materials may not require labeling; these exceptions are referenced in column 5(a) under packaging exceptions. The labels, when required, must be printed or affixed near the marked shipping name (49 CFR, Section 172.406).

Packaging

Q: What is required for proper packaging?

A: The RCRA regulations require that all generators of hazardous wastes transport their wastes in packaging that complies with the DOT shipping and packaging regulations in Parts 173, 178, and 179 of Title 49. Part 173 of the regulations addresses the general requirements for shipments and packaging of hazardous wastes, while Parts 178 and 179 outline the specifications for both shipping containers and tank cars, respectively.

Q: Where are the specific packaging requirements identified?

A: Once the proper shipping name and hazard class of a particular hazardous waste has been determined, column 5 of the HMT identifies the specific packaging regulations that apply to that waste. Column 5(a) specifies whether any exceptions may be taken for limited-quantity shipments, and column 5 outlines the general packaging requirements for non-excepted shipments of the waste. In addition, columns 6 and 7 highlight any further restrictions that may be placed on the transportation of the waste material by aircraft, passenger rail, or vessel.

Q: What criteria are used to select proper packaging?

A: In addition to the packaging requirements specified in column 5 of the HMT, 49 CFR Section 173.2 lists the hazard classes in a priority order based on the extent of danger a waste may present in transportation.

Q: What is the priority order of hazards?

A: The priority order of hazards is:

(1) Radioactive Material
(2) Poison A
(3) Flammable Gas
(4) Nonflammable Gas
(5) Flammable liquid
(6) Oxidizer
(7) Flammable Solid
(8) Corrosive material (liquid)
(9) Poison B
(10) Corrosive material (solid)
(11) Irritating materials
(12) Combustible liquid (in containers more than 110 gallons)
(13) ORM-B
(14) ORM-A
(15) Combustible liquid (in containers less than 110 gallons)
(16) ORM-E

Q: How is a hazardous waste packaged if it has more than one hazard?

A: The waste must be classified and packaged according to the above priority order. For example, a waste that is both a flammable gas and an irritating material must be classed and packaged according to the requirements of a flammable gas, because it appears 3rd on the list, and an irritating material appears 11th.

Q: What is required of the generator in preparing a waste for transportation?

A: Before offering a hazardous waste for transportation, a shipper (generator) must be certain that packages and containers meet all the packaging and shipping requirements, in accordance with Section 173.22. A shipper (generator) must determine that all containers used are assembled with all parts or fittings in their proper place and properly secured and marked in accordance with the applicable specifications prescribed in Parts 178 and 179. In determining whether a specification container is manufactured in accordance with the applicable specifications, the shipper (generator) may accept a manufacturer's certification or specification marking. The shipper (generator) must also be certain that the packagings used are not of the type prohibited in 49 CFR 173.21.

In addition to the basic responsibilities outlined above, the shipper (generator) must recognize further the general standards and requirements applicable to all packagings and containers used to ship hazardous wastes.

Q: What are these general requirements?

A: These requirements are as follows: (1) no "significant" release of a hazardous waste material can occur into the environment, (2) the effectiveness of a packaging or container cannot deteriorate, and (3) no mixture of gases can occur that, through a credible spontaneous increase of heat or pressure or through an explosion, might reduce the effectiveness of the packaging (49 CFR 173.24(a), (b), (c). In addition, all packagings used to transport hazardous wastes must conform to the design and construction criteria specified in 49 CFR 173.24.

Q: Can a container be reused to ship hazardous wastes?

A: Yes; if the regulations do not limit a packaging or container as nonreusable (NRC) or single-trip (STC), its reuse is authorized, but subject to the special requirements of 49 CFR Section 173.28. Although this section allows for the reuse of containers, the general responsibility for assuring the safety of the container is placed on the shipper (generator) by Sections 173.22, 173.24, and 173.28. Therefore, any shipper (generator) who plans to reuse a container or packaging for the transportation of a hazardous waste must make certain that the packaging "complies in all respects with the prescribed requirements" for the original container (49 CFR 173.28). Thorough visual inspections and nondestructive testing may be necessary to comply with these requirements. Part 173 specifies when inspections and retests must be done.

Q: What are the specific packaging and shipping requirements for hazardous wastes?

A: Subparts C through O of Part 173 provide the specific packaging and shipping requirements for each of the various DOT classes of hazardous wastes and for the individual materials within each class. Prior to each subpart's instructions on shipment and packaging is a definition of the characteristics of each hazard class. In determining which packaging is appropriate for a particular waste, the shipper should consult the Hazardous Materials Table for the specific regulation section listed in column 5. Table 10 delineates the classes of hazardous wastes for shipping and packaging and lists the sections in which they are regulated.

Marking

Q: What is meant by marking?

A: Marking means placing on the outside of a shipping container, one or more of the following: the descriptive name, instructions, caution, or weight. Marking also includes any required specification marks on the inside or outside shipping container.

TABLE 10 Classes of hazardous material regulations for shipping and packaging.

Class	Subpart	Section
Explosives	C	173.50–173.114a
Flammable, combustible, and pyrophoric liquids	D	173.115–173.149a
Flammable solids, oxidizers, and organic peroxides	E	173.150–173.239a
Corrosive Materials	F	173.240–173.299a
Gases	G	173.300–173.320
Poisonous materials and etiologic agents	H	173.325–173.388
Radioactive materials	I	173.401–173.478
ORM A, B, C, D, and E	J,K,L,M,N,O	173.500–173.1300

Q: What are the marking requirements?

A: DOT has established marking requirements for packages, freight containers, and transport vehicles. Shippers (generators) are required to mark all packages with a capacity of 110 gallons or less with the proper shipping name, the UN/NA identification number, and the name and address of either the shipper or the designated facility (49 CFR Section 172.300-304).

Q: Are there other additional requirements?

A: Yes; additional requirements are specified for liquid wastes. This requirement includes that the packages containing liquid hazardous waste must be marked "THIS SIDE UP" or "THIS END UP" and have an arrow symbol on the outside packaging to show the correct position for the package (49 CFR Sections 172.312, .316, and .324).

In addition, EPA also requires special markings for packages of hazardous wastes identifying the shipper and indicating that federal law prohibits improper disposal of wastes [40 CFR, Section 262.32(b)].

Placards

Q: What are placards?

A: Placards are color-coded signs that are placed on the ends and sides of transport vehicles indicating the hazards of the cargo.

Q: When is placarding required?

A: All motor vehicles, rail cars, and freight containers carrying any hazardous waste meeting the hazard class identified in Table 11 or a hazardous waste over 1,000 pounds of those identified in Table 12 must be placarded (49 CFR Section 172.504).

Q: What are the placarding requirements when a transport vehicle has more than one hazard class?

A: If a transport vehicle contains a waste that falls under a hazard class identified under Table 11, it must have the appropriate placard regardless of quantity. A "DANGEROUS" placard may be used in lieu of the specified placards for combined shipments of wastes identified in Table 12,

TABLE 11 DOT Placarding Table 1.

If the transport vehicle or freight container contains a material classed (described) as:	The transport vehicle or freight container must be placarded on each side and each end:
Class A explosives	EXPLOSIVES A
Class B explosives	EXPLOSIVES B
Poison A	POISON GAS
Flammable solid (DANGEROUS WHEN WET label only)	FLAMMABLE SOLID W
Radioactive material	RADIOACTIVE
Radioactive material:	Uranium hexafluoride, fissile (containing >1% U^{235})
Uranium hexafluoride, low specific activity (1.0% or less U^{235})	RADIOACTIVE & CORROSIVE

Source: 49 CFR Section 172.504

provided that the total weight of any *one* waste does not exceed 5,000 pounds. In the event that one of the hazardous wastes in a mixed load does exceed 5,000 pounds, then a separate placard would be required for that waste. However, the "DANGEROUS" placard could still be displayed for the remaining wastes in the load.

TABLE 12 DOT Placarding table 2.

If the transport vehicle or freight container contains a material classed (described) as:	The transport vehicle or freight container must be placarded on each side and each end:
Class C explosives	DANGEROUS
Blasting agents	BLASTING AGENTS
Nonflammable gas	NONFLAMMABLE GAS
Nonflammable gas (chlorine)	CHLORINE
Nonflammable gas (fluorine)	POISON
Nonflammable gas (Oxy, cryogenic liquid)	OXYGEN
Flammable gas	FLAMMABLE GAS
Combustible liquid	COMBUSTIBLE
Flammable liquid	FLAMMABLE
Flammable solid	FLAMMABLE SOLID
Oxidizer	OXIDIZER
Organic peroxide	ORGANIC PEROXIDE
Poison B	POISON
Corrosive material	CORROSIVE
Irritating material	DANGEROUS

Source: 49 CFR Section 172.504

Q: What are the requirements for using placards?

A: All placards required on motor vehicles, rail cars, and freight containers must be readily visible and placed on the front, back, and sides. They must be located away from such apparatus as ladders, pipes, and doors and be clear of any marking, such as advertising, that could substantially reduce their effectiveness. In addition, placards should be securely affixed and maintained in such a manner that the format, legibility, color, and visibility are not damaged, deteriorated, or obscured by dirt or other matter. The appropriate UN or NA identification numbers must also be displayed on portable tanks, cargo tanks, and tank cars, either on the designated placard or on a separate orange panel described in 49 CFR Section 172.332.

Q: Who is responsible for placarding?

A: Placarding hazardous wastes is the joint responsibility of the shipper (generator) and the carrier (transporter). Shippers who offer a hazardous waste for transport by highway must provide the carrier (transporter) with the placards required for the waste prior to or at the time it is offered for transport, unless the carrier's (transporter's) vehicle is already properly placarded for the waste material.

Incidents (Spills)

Q: What is required if an incident occurs?

A: In accordance with 49 Section 171.15, at the earliest practicable moment, a carrier must give notice to the National Response Center after each incident that occurs during the course of transportation (including loading, unloading, and temporary storage) in which, as a direct result:

- A person is killed
- A person is injured such that he requires hospitalization
- Estimated carrier or property damage exceeds $50,000
- Fire, breakage, spillage, or suspected radioactive or etiologic contamination occurs
- It is determined by the carrier that a continuing danger to life exists at the scene of the incident

Q: Are there reporting requirements beyond the verbal notification?

A: Yes; 49 CFR Section 171.16 requires each carrier to file a written report (DOT Form F 5800.1) to DOT in Washington, DC, within 15 days concerning any incident requiring verbal notification addressed in the previous question.

In addition, the carrier must include a copy of the manifest and, in Part H of Form F 5800.1, include an estimate of the quantity of waste removed from the scene, the name and address of the facility to which it was taken, and the manner of disposition of any unremoved waste.

Emergency Response Information

Q: Are there additional requirements for emergency response information?

A: Yes; under 49 CFR 172-600, a copy of an Emergency Response Information form must be attached to each manifest outlining the following items:

1. DOT basic description of the waste
2. Immediate health hazards
3. Risks
4. Immediate precautions to be taken
5. Immediate methods for handling fires
6. Methods for handling leaks
7. Preliminary first aid

The form must be prepared by the shipper. However, a copy of the actual DOT Emergency Response Guide may be attached instead. In addition, a telephone number must be provided with each individual shipment, designated as the EMERGENCY CONTACT.

Chapter 5

General Standards for TSD Facilities

A person who treats, stores, or disposes of hazardous waste (except for those persons excluded, see page 106) is classified an owner or operator of a treatment, storage, or disposal (TSD) facility (also referred to as a *hazardous waste management facility*) and is subject to the requirements of Part 264 or 265. The Part 264 requirements are applicable to new facilities and facilities that have a permit (operator or post-closure). The Part 265 requirements are for facilities with interim status.

The General Standards are divided into subparts as follows:

Subpart A - General Requirements
Subpart B - General Facility Standards
Subpart C - Preparedness and Prevention
Subpart D - Contingency Plan and Emergency Procedures
Subpart E - Manifest System, Record Keeping, and Reporting
Subpart F - Groundwater Monitoring (addressed in Chapter 6)
Subpart G - Closure and Post-closure
Subpart H - Financial Responsibility

GENERAL REQUIREMENTS

Applicability

Q: Who is subject to the general standards for hazardous waste management facilities?

A: The general standards of Parts 264 and 265 are a set of requirements that all treatment, storage, and disposal facilities must follow. In addition, selected portions of the general standards are applicable to generators accumulating hazardous waste, as specified in Section 262.34.

Q: What is the definition of treatment?

A: *Treatment* is any method, technique, or process, including neutralization, designed to change the physical, chemical, or biological character or composition of any hazardous waste so as to neutralize such waste, or so as to recover energy or material resources from the waste, or so as to render such waste nonhazardous or less hazardous; safer to transport, store, or dispose of; or amenable for recovery, amenable for storage, or reduced in volume (Section 260.10).

Q: What is the definition of storage?

A: *Storage* is the holding of hazardous waste for a temporary period, at the end of which the hazardous waste is treated, disposed of, or stored elsewhere (Section 260.10).

Q: What is the definition of disposal?

A: *Disposal* is the discharge, deposit, injection, dumping, spilling, leaking, or placing of any solid waste or hazardous waste into or on any land or water, so that such solid waste or hazardous waste or any constituent thereof may enter the environment or be emitted into the air or discharged into any waters, including groundwaters (Section 260.10).

Q: What is the difference between Parts 264 and 265?

A: The basic difference between Part 265 (interim status standards) and Part 264 (final operating standards) is that the interim status standards were written for facilities treating, storing, or disposing of RCRA hazardous waste when the Phase I regulations first went into effect on November 19, 1980. Congress wanted EPA to establish interim standards that would allow facilities to continue to operate as though they had a permit, while EPA developed the more stringent Part 264 standards for new and existing facilities. Additionally, the Part 265 standards are self-implementing, whereas the Part 264 standards are established as part of the conditions of a permit.

Q: What is an approved hazardous-waste facility?

A: The term *approved* is not used in RCRA or the regulations. When the term is used, EPA takes it to mean a facility that has received a RCRA permit from EPA or an authorized state or has interim status.

Q: Is a series of impoundments on different portions of a site considered one facility?

A: Yes; the definition of facility in Section 260.10 makes it clear that a single "facility" may consist of several similar or different types of hazardous waste units, such as several impoundments, an incinerator, a landfill, and storage areas.

Q: How did a facility receive interim status?

A: A facility received interim status by filing a RCRA Section 3010 notification (EPA Form 8700-12, *Notification of Hazardous Waste Activity*) by August 26, 1980, and a Part A permit application by November 19, 1980. (Refer to Chapter 9 for a more detailed discussion of interim status.) An owner or operator that qualified for and obtained interim status remains subject to the Part 265 standards until the final administrative disposition of the facility's permit application is made.

Q: Is the owner or the operator of a hazardous waste facility responsible for complying with the RCRA regulations?

A: Both the owner and the operator are responsible. If EPA detects a violation of the regulations, it can proceed with legal action against the owner, the operator, or both.

Q: What activities or persons are excluded from Parts 264 or 265?

A: No; pursuant to Sections 264/265.1, the following persons or activities are excluded from Parts 264 and 265:

- Facilities that are state-approved to handle small-quantity generator waste (<100 kg/mo) exclusively
- A totally enclosed facility, as defined in Section 260.10
- A generator accumulating hazardous waste in compliance with Section 262.34
- A farmer disposing of waste pesticides in compliance with Section 262.51
- The management of wastes in an elementary neutralization unit or a wastewater treatment unit, as defined in Section 260.10

- A person engaged in the immediate treatment or containment of a discharge of hazardous waste
- A transporter storing manifested waste at a transfer facility in compliance with Section 262.30
- The addition of absorbent material to waste or the addition of waste to absorbent material, provided it occurs at the time the waste is first generated

Although these persons or activities are for the most part excluded from Parts 264 and 265, there are specific regulatory references to these parts for some of these persons or activities. For example, generators accumulating hazardous waste on-site in accordance with Section 262.34 must comply with specified sections of Part 265 (e.g., personnel training, emergency procedures).

Q: Is the neutralization of an acidic or basic waste that is a hazardous waste considered treatment of hazardous waste?

A: Yes; it is considered treatment of a hazardous waste, because the definition of "treatment" (Section 260.10) includes neutralization. However, if a unit meets the definition of an elementary neutralization unit, it is excluded from regulation under Sections 264.1(g)(6) and 265.1(c)(10).

Q: Would dewatering hazardous waste solutions to reduce volume constitute treatment and thus require a RCRA permit?

A: The definition of treatment in Section 260.10 includes volume reduction and changes in the physical character and composition of a waste. Dewatering typically reduces the volume of the waste and changes the physical character of the waste; thus, dewatering is "treatment." However, a permit may not be required if it is done by a generator accumulating waste in compliance with Section 262.34.

Q: Do portable treatment units connected to a process unit meet the totally enclosed treatment exclusion?

A: Yes; if the unit (when connected to a process) is in compliance with the Regulatory Interpretive Letter (RIL 84), which specifies the parameter of a totally enclosed treatment facility. Thus, portable treatment units could be used by multiple facilities and be excluded from regulations by Sections 264.1(g)(5) and 265.1(c)(9).

Q: If hazardous waste is transferred from one container to another container and at the same time an absorbent material is added to the waste, is the act

of placing the waste in the second container and adding an absorbent covered by the exemptions of Sections 264.1(g)(10) and 265.1(c)(10)?

A: Yes; this regulation was intended to reduce the amount of free liquids in containerized wastes by allowing anyone to add absorbent material to the waste at the time the waste is first placed in the container. Thus, a hazardous waste can be transferred from one container to another container and absorbent material added at the time of transfer, and this activity would not be subject to the permit requirements for treatment under Sections 264.1(g)(10) and 270.1(c)(2)(vii) (47 *FR* 8304, February 25, 1982).

Q: Leachate from a sanitary (nonhazardous) landfill is collected and pumped back into the landfill. This leachate is hazardous because of the EP toxic characteristic. Is the landfill a RCRA TSD facility?

A: Yes; once the leachate is collected, its subsequent management is regulated by RCRA. Thus, because the leachate is hazardous, the landfill is classified as a TSD facility [OSWER Directive No. 9441.08(83)].

Q: An activated carbon filtration unit is attached to a hazardous waste storage tank's vent pipe to capture the waste vapors. Is this filtration unit considered a treatment unit subject to Parts 264 or 265?

A: The activated-carbon filtration unit is viewed as an appurtenance to the storage tank and is not looked at individually during permitting. The carbon filtration unit is treating a hazardous waste, and RCRA has jurisdiction over its activity. The carbon would be a solid waste when discarded and a hazardous waste if it exhibited a characteristic or if it contained a listed waste. If required (e.g., accumulating waste in violation of Section 262.34), a permit would be issued for the tank simply for storage.

Q: Would a publicly owned treatment work (POTW) that accepts hazardous waste by truck be classified as a hazardous waste management facility?

A: Yes; these facilities are considered hazardous waste management facilities if they accept *off-site* shipments of hazardous waste. Such a facility is eligible for a RCRA permit-by-rule under Section 270.60 (see Chapter 9).

Q: A container leaked 10 pounds of a P-code commercial chemical product that was subsequently cleaned up with an absorber and held for 120 days before it was shipped off-site for disposal. Would the company be considered a storage facility under RCRA?

A: Yes; if a material listed in Section 261.33(e) is discarded or intended to be discarded, it is a hazardous waste. In this case the absorbed material is intended to be discarded. If more than 1 kilogram (2.2 pounds) per month of that hazardous waste is generated, it is subject to the full set of regulations. In this case, the 10-pound leak is assumed to be a single event or, at a minimum, a leak that occurred at a rate greater than 1 kilogram per month; thus, Parts 262, 264, 265, and 270 apply.

If the generator of a hazardous waste accumulates (or holds) the waste for more than 90 days before shipping it off-site, the generator becomes an operator of a storage facility [see Section 262.34(b)]. In this case, the generator held the waste 120 days and, therefore, is the operator of a storage facility.

GENERAL FACILITY STANDARDS

EPA Identification Numbers

Q: Are TSD facilities required to obtain an EPA identification number?

A: Yes; an owner or operator of a hazardous waste management facility must have an EPA identification number to accept hazardous waste from off-site sources.

Q: How is an identification number obtained?

A: The EPA identification number is obtained by submitting EPA Form 8700-12 to the EPA regional office or authorized state.

Q: Because most TSD facilities are also generators because of the derived-from rule, how is the identification number addressed?

A: Facilities that are also generators must have their site-specific EPA identification number amended to include generator status. Because each facility has one identification number only, the original number will remain the same.

Required Notices

Q: What are the required notices for TSD facilities?

A: Section 264.12(b) requires the owner or operator of any permitted hazardous waste management facility receiving waste from an off-site source to inform the generator in writing that the facility has all appropriate permits and will accept the waste that the generator is shipping. The facility must keep a copy of this letter as part of its operating record. Interim status facilities are not required to comply with this section.

Q: Is a generator required to receive this written notice prior to transport?

A: No; Section 262.20(b) states that wastes transported off-site must be sent to a facility permitted to handle that waste; a generator is not required to ask for or receive a written notice from the facility owner or operator. However, a written notice would assure that the generator is in compliance with Section 262.20(b). A written notice would also avoid the potential problem of a generator sending waste to a facility that has the proper permits, but has not agreed to accept the waste.

Q: If an owner transfers ownership of a facility, are there any notification requirements?

A: Yes; before transferring ownership or operational control of a hazardous waste management facility during its operating life, or of a land disposal facility during its post-closure care period, the owner or operator must notify the new owner or operator in writing of the requirements of Subtitle C of RCRA [Sections 264.12(c) and 265.12(b)].

Q: What happens if the former owner or operator fails to properly notify the new owner or operator of the requirements of Subtitle C?

A: If the former owner or operator fails to comply with this requirement, the new owner or operator is still required to comply with all applicable RCRA regulations (45 *FR* 33179, May 19, 1980).

Waste Analysis

Q: Who must provide the detailed chemical analysis of a waste before it is treated, stored, or disposed of?

A: The owner or operator of a hazardous waste treatment, storage, or disposal facility is required to obtain a waste analysis in accordance with a required plan, which usually will involve a detailed chemical analysis before treating, storing, or disposing of the waste (Sections 264/265.13). The owner or operator, however, may require the generator to perform part or all of the required waste analysis as a condition of doing business with the generator. This would be a business arrangement, not a requirement of the regulations.

Q: What is the purpose of a waste analysis plan?

A: The purpose of a waste analysis plan is to describe the procedures that will be undertaken to obtain sufficient waste information to operate a hazardous waste management facility in accordance with its permit or interim status requirements (i.e., to ensure that wastes accepted by the facility fall within the scope of the facility's permit or Part A permit application, and that the process performance standards are met). The waste analysis plan establishes hazardous waste sampling and analysis procedures that will be routinely conducted as a requirement of the permit (Sections 264/265.13).

Q: What is the required format for a plan?

A: The regulations do not require a specific format for the waste analysis plan. The permit applicant should consult EPA's *Waste Analysis Plans, A Guidance Manual,* EPA/530-SW-84-012. This document contains various model waste analysis plans.

Q: What information is required in the waste analysis plan?

A: There are four issues that the owner or operator should address:

1. The specific wastes or types of wastes that will be managed within each process
2. The waste-associated properties that are of concern in ensuring safe and effective management (e.g., kilocalories per gram, percentage of water)
3. The specific waste parameters required to satisfy data needs
4. The procedures for obtaining the necessary data, including sampling and analysis procedures and quality control/quality assurance procedures

For owners or operators seeking a permit, there are other portions of the RCRA permit application that require an in-depth description of the facility and the processes to be permitted. Those sections will establish the types and the characteristics of wastes to be managed and any process constraints. The waste analysis plan should reference these other sections of the application, and it is suggested that the applicant summarize those points that are particularly germane to the plan to assist the permit reviewer.

Q: Are there different requirements for off-site and on-site facilities?

A: Yes; the regulations distinguish between on-site and off-site facilities. On-site facilities manage only those hazardous wastes that are generated on their own geographic sites, as defined in Section 260.10. An off-site facility receives and manages hazardous wastes that are generated outside the facility.

The objectives are the same for both on-site and off-site facilities. However, EPA believes that a generator owned and operated facility will tend to know more about the waste generation process than would a facility not owned and operated by the waste's generator. Thus, off-site facilities are required by the regulations to conduct more frequent checks on wastes than on-site facilities [Sections 264/265.13(c)].

Q: If an owner or operator of a facility accepts hazardous waste from small-quantity generators, must the waste be addressed in the facility's waste analysis plan?

A: No; the owner or operator would not have to address the wastes from small-quantity generators in the waste analysis plan. Sections 264/265.1(b) state that all the Part 265 and 264 standards do not apply if otherwise excluded in Sections 265.(1)(c) or 264.(1)(f) and (g) or in Part 261. Section 261.5(b) states that a small-quantity generator's hazardous wastes are not subject to regulation under Parts 262-265 and Parts 270 and 124, if the small-quantity generator complies with the Section 261.5 standards. Hence, hazardous wastes from small-quantity generators in compliance with 261.5 are not subject to Part 265 or 264 standards, including Sections 264/265.13 for waste analysis.

Q: Are there other waste analysis requirements for specific units?

A: Yes; in addition to the above general requirements, there are specific waste analysis requirements for each type of unit (see Chapter 7).

Prevention of Ignition or Reaction of Wastes

Q: What is the intent of this section?

A: The intent of this section is to reduce the potential for accidental ignition or reaction of ignitable and/or reactive wastes, or the mixing of incompatible wastes by ensuring that facility personnel are thoroughly familiar with the prevailing dangers, and, more significantly, that the design of the facility and the devices utilized for treating, storing, and transporting wastes account for these dangers (Sections 264/265.17).

Q: What are the requirements?

A: Sections 264/265.17(b) require any owner or operator of a facility that handles ignitable or reactive wastes or that mixes incompatible wastes or wastes and materials that are incompatible to take precautions against reactions that:

- Generate extreme heat or pressure, fire or explosions, or violent reactions
- Produce uncontrolled toxic mists, fumes, dusts, gases in sufficient quantities to threaten human health or the environment
- Produce uncontrolled flammable fumes or gases that, in sufficient quantities, pose a risk of fire or explosion
- Damage the structural integrity of the device or facility
- Through other means threaten human health or the environment

Q: What are adequate precautions?

A: The adequate precautions taken to prevent these reactions would generally encompass:

- Identification of ignitable, reactive, and incompatible wastes
- Identification of the ways that incompatible wastes are combined
- Analysis of the storage devices used, and their placement in the facility, should the wastes be stored as they are
- Analysis of methods of treatment that would render wastes unreactive or nonignitable, and subsequent testing to ensure that the wastes are no longer ignitable or reactive

Q: Must ignitable or reactive wastes be separated from potential ignition sources?

A: Yes; the regulations [Sections 264/235.17(a)] list several potential sources of ignition or reaction from which the ignitable or reactive wastes must be separated and protected, including but not limited to:

- Open flames
- Smoking
- Cutting and welding
- Hot surfaces
- Frictional heat
- Radiant heat

- Sparks (static, electrical, or mechanical)
- Spontaneous ignition (e.g., heat-producing chemical reactions)

Location Standards

Q: What is the purpose of the location standards?

A: Because the physical location of a waste management facility directly influences the potential for impacting human health and the environment, EPA has established some minimally accepted location standards (Section 264.18). Physical location refers to the geologic, hydrologic, and pedologic characteristics of a site, as well as adjoining lands, surface water, and groundwater that may be impacted in the event hazardous wastes or constituents are released from the facility.

Q: What are the location standards?

A: Currently, the location standards are applicable only to facilities under Part 264. Additionally, the fault restrictions are applicable only to new facilities.

The location restrictions are:

- The facility must be at least 200 feet from an active (during the last 10,000 years) Holocene fault.
- Facilities are not allowed in a 100-year floodplain unless one of the three conditions are met [Section 264.18(b):
  1. The facility is protected, using dikes or equivalent measures, from washout during a 100-year flood.
  2. All hazardous wastes can be removed to safe ground prior to flooding.
  3. It can be demonstrated that no adverse effects to human health or the environment will occur should flood waters reach the hazardous wastes.

Q: How does a permit applicant determine if there is an active fault zone?

A: Only those facilities located in certain political jurisdictions listed in Appendix VI to Part 264 are required to demonstrate compliance with these standards. These jurisdictions include areas in Alaska, Arizona, and Colorado, and the entire state of California [Section 264.18(a)].

Q: Are there other location standards?

A: Yes; in addition to a location that satisfies the floodplain and seismic standards, Executive Order 11990 (Protection of Wetlands) must be considered for facilities located only on federally owned lands that may potentially have an impact on wetlands.

Q: Are there location standards for sensitive hydrogeologic conditions?

A: No; site-specific location standards based solely on hydrogeologic considerations are nonexistent, although the groundwater protection standards, as well as the general design and operating requirements, contain performance standards that implicitly involve hydrologic and geologic factors. Current regulations do not provide the legal basis to deny a RCRA permit based on sensitive locations, such as vulnerable groundwater formations, although a RCRA Section 7003 order can be used if there may be an imminent threat to health or the environment.

However, HSWA contains specific amendments that require EPA to promulgate location regulations and to provide location guidance concerning the potential vulnerability of groundwater to contamination based on hydrogeologic conditions at a site. EPA has issued location guidance concerning vulnerable hydrogeology.

Q: Where can this guidance be found?

A: For further information concerning current location standards and guidance, consult EPA's *Permit Writer's Guidance Manual for Hazardous Waste Land Storage and Disposal Facilities,* February 1985 (OSWER Directive No. 9472-004).

Q: What will be included in the upcoming new location standards?

A: More stringent location standards, as mandated by HSWA, will be proposed soon. The purpose of these standards will be to create minimum national requirements for the location of hazardous waste management facilities. These requirements will be contained under the new Subpart Z to Part 264 and will include location restrictions for new and existing facilities seeking a permit based on proximity to populations, vulnerable hydrogeology, seismic zones, 100-year floodplains, poor foundation areas, subsidence-prone areas, landslide-prone areas, wetlands, karst terranes (limestone areas with fissures, sinkholes, underground streams, and caverns), salt dome formations, salt bed formations, and underground mines and caves.

Security

Q: What is the intent of the security requirements?

A: The intent of the security requirements contained in Sections 264/265.14, are to prevent the unknowing and/or unauthorized entry of persons or livestock to active portions of the facility.

Q: What are adequate security measures?

A: Adequate security measures are a 24-hour surveillance system and signs around the active portion of the facility, or artificial or natural barriers with controlled access points and signs around the active portion of the facility [Sections 264/265.14(b)].

Q: If the active portion is contained within a large complex that already has a security system, are separate security measures required?

A: No; the requirements are satisfied if the active portion is located within a facility or plant that itself has a surveillance system, or a barrier and a means to control entry, that complies with the requirements. Thus, a separate security system around only the active portion may not be required. However, the requirements for signs around the active portion would still be applicable [Sections 264/265.14(b)(ii)].

Q: What is the intent of a controlled-access system?

A: The intent is that there should be a finite number of entrances into the facility and that attempted or inadvertent entry to the facility at points other than the identified, controlled entrances will be physically hindered or, at least, immediately identified (OSWER Directive No. 9523.00-10).

Q: For a facility to meet the regulatory requirement for a 24-hour surveillance system, must the facility have guards whose sole function is to control access to the active portion of the site?

A: No; if a facility is operated continuously so that the active portion is always within view and control of employees other than guards, then these employees can be considered to perform the same function as guards. Enough employees have to be at the site, however, to minimize the possibility that an unauthorized person will slip in unnoticed (OSWER Directive No. 9523.00-10).

Q: What about the requirement for posting signs?

A: Regardless of the methods employed to limit entry to the active portion of the facility, the owner or operator must post signs in accordance with the requirements of Sections 264/265.14(c).

Q: What are the requirements for these signs?

A: There are four requirements for these signs. They are:

1. A legend that makes it plain that unauthorized persons are in danger and are not allowed into the active portion of the facility, i.e., "DANGER-UNAUTHORIZED PERSONNEL KEEP OUT"
2. Legibility from a distance of at least 25 feet
3. Visibility from any approach to the active portion of the facility
4. Legend in English and in any other language predominant in the area surrounding the facility (e.g., French and Spanish)

Q: Is there an exemption or waiver to the security measure available?

A: Yes; the provisions of Sections 264/265.14(a) provide for an exemption/waiver from the security provisions.

Q: What are the requirements for such a waiver?

A: An owner or operator must be able to demonstrate that unknowing or unauthorized persons or livestock would not injure themselves or cause a RCRA violation upon entering the active portion of a facility.

Q: How are these requirements demonstrated?

A: These requirements may be demonstrated by showing that the nature and duration of the hazard potential from the hazardous waste on-site do not warrant the required security procedure or equipment. In addition, if an owner or operator can show that the facility provides certain features, such as cover materials or containers, that would prevent contact with equipment or structures, certain security procedures and equipment might not be needed. Finally, a waiver justification could show that safety or operating practices related to equipment and structures would eliminate the potential for an intruder to cause a spill, mix incompatible wastes, ignite ignitable wastes, damage containment or monitoring systems, etc. The circumstances under which a waiver will be granted are limited (OSWER Directive No. 9523.00-10).

Personnel Training

Q: What is the purpose of the personnel training requirements?

A: The intent of the personnel training requirements is to reduce the potential for mistakes that might threaten human health or the environment by ensuring that facility personnel working in jobs where they handle hazardous waste will be thoroughly familiar with their duties and responsibilities, especially in the areas of safety and emergency response.

Q: What are the requirements of the personnel training program?

A: Specifically, the requirements of Sections 264/265.16 are:

- Facility personnel must successfully complete a training program that ensures the facility's compliance with the requirements of Parts 264 and 265.
- The training program must be directed by a person trained in hazardous waste management procedures.
- The training program must be designed to ensure at a minimum that facility personnel are able to respond with familiarity during an emergency situation.
- Facility personnel must successfully complete the program within 6 months of their assignment to a facility.
- Facility personnel must take part in an annual review of the training program.
- The owner or operator must maintain documentation at the facility.
- Training records on current personnel must be kept until closure of the facility.

Q: What is meant by facility personnel?

A: *Facility personnel* are defined in Section 260.10 as: "All persons who work at, or oversee the operations of a hazardous waste facility, and whose actions or failure to act may result in noncompliance with the requirements of Parts 264 or 265 of this Chapter." In other words, all personnel (supervisors and nonsupervisory personnel) who are actively engaged in the operation of the facility require the type of training specified.

Q: What type of training program is required?

A: Under the regulations, all personnel associated with the handling of hazardous wastes are required to "successfully" complete a program of classroom instruction or on-the-job training that teaches them to perform their duties in a way that ensures the facility's compliance with the applicable requirements of Parts 264 and 265 [Sections 264/265.16(a)(1)].

It is required that the training programs be specific to the various positions performed at the facility. Training should be structured so that it parallels as realistically as possible the actual job, in order that the real world activities are approximated as much as possible. Any training programs must also take into account the educational level of the class.

Q: If an industrial plant has on-site facilities to treat hazardous wastes, do the training requirements extend to personnel in the production area?

A: No; the training requirements apply only to personnel involved in those aspects of the facility's operation that relate to the management of hazardous waste.

Q: When must facility personnel be trained?

A: Employees must not work in unsupervised positions requiring them to handle hazardous wastes until they have completed their training programs. New employees may handle hazardous wastes, but only under the supervision of trained employees. In any case, personnel must successfully complete the program within 6 months after their employment or assignment to a new position [Sections 264/265.16(b)].

Q: Is supplemental training required?

A: Yes; the emergency procedures taught in the original training program must be reviewed annually to keep personnel up to date with any changes, such as the characteristics of new wastes managed at the facility. With new and more sophisticated technologies being developed for hazardous waste, facilities may have to periodically change certain procedures to remain current with these new technologies. Also, due to changes in facility processes or emergency equipment, or with the types of wastes being accepted at the facility, the facility's contingency plan may need to be modified. Therefore, the contingency plan should also be included in the annual review [Section 264/265.16(c)].

Emergency Response Training

Q: What facility personnel must be trained in emergency procedures?

A: All facility personnel, regardless of their position, must be familiarized with the facility's contingency plan so they will be able to respond effectively in an emergency. The majority of employees will probably be responsible for vacating the premises in a predetermined manner, while other facility personnel (those who have been properly trained) will have higher levels of responsibility. Some may be responsible for containing the spill, informing local officials (i.e., police and fire), or employing fire-fighting equipment (OSWER Directive No. 9523.00-10).

Q: What specifically should facility personnel be familiarized with concerning emergency procedures?

A: At a minimum, the training program must familiarize facility personnel with emergency procedures, emergency equipment, and emergency systems applicable to their positions. Emergency response procedures that should be taught to selected facility personnel, as required by Sections 264/265.16(a)(3), include:

- Procedures for using, inspecting, repairing, and replacing facility emergency and monitoring equipment
- Key parameters for automatic waste feed cutoff systems
- Communications or alarm systems
- Response to fires or explosions
- Response to groundwater contamination incidents
- Shutdown of operations

Record Keeping

Q: Are records required to be kept by the facility concerning personnel training?

A: Yes; records must be kept at the facility for examination by EPA upon request. Maintenance of facility personnel training records acts as a certification program [Sections 264/265.16(e)].

Q: What information must be contained in these records?

A: According to Sections 264/265.16(d), the following information must be included in the records:

- A job title for each position at the facility that is related to hazardous waste management (i.e., excluding clerical or janitorial positions) and the names of the employees filling those positions
- A job description for each of those positions, including skill, education, or other qualifications needed by employees to fill each po-

sition at the facility and duties of employees assigned to each position

- A description of the type and amount of introductory and continued training that will be given to each employee

Q: Specifically, what information is required concerning the documentation of training?

A: The records must be documented to demonstrate that the proper training has been given to and completed by facility personnel. Therefore, records must be kept of the dates on which employees received their initial training and are scheduled for their annual review [Sections 264/265.16(d)(4)].

In addition, for each job description, the type of training to be given must be included and the length of the program. For example, if employees are sent to a formal training program, written documentation must be kept stating the types of hazardous waste management practices being taught and the length of time involved.

Similarly, if a facility has designed its own training programs to be conducted in-house or on-the-job, a detailed written account of the material to be presented for each position must be kept. It must also include the techniques to be used and a schedule to be followed by the instructors. The training records must also contain the type and amount of training that will be given to fulfill the annual review requirements (OSWER Directive No. 9523.00-10).

Q: If a facility is seeking a permit, must all the personnel training records accompany the Part B?

A: No; although training records for each employee must be maintained at the facility, these records and other required paperwork are not required to be submitted with the Part B application [Section 270.14(b)(12)]. The only items in regard to personnel training that must accompany the application are an outline of introductory and continuing training programs and a brief description of how job tasks relate to the training. However, these items should be specific enough to demonstrate that the training program complies with the requirements of Section 264.16.

Q: What are the requirements for record retention?

A: The training records for current personnel must be kept on file at the facility until the facility certifies closure. The training records of former employees must be kept for at least 3 years from their last date of employment at the facility. If a person is transferred within the same company, his training records remain the same [Sections 264/265.16(e)].

Inspections

Q: What are the inspection requirements for a facility?

A: The inspection requirements for a facility are divided into two sections, general inspection requirements and unit specific requirements.

General Requirements

Q: What are the general inspection requirements?

A: The general inspection regulations contained in Sections 264/265.15 require a facility owner or operator to:

- Identify the specific equipment to be inspected
- Identify how each item will be inspected
- Determine a justifiable frequency for inspecting each item
- Develop a schedule for additional inspections required by the regulations, depending on the type of facility you are operating

Q: What is required in developing an inspection schedule?

A: The first step in developing a schedule is identification of all equipment, devices, and structures that will be inspected. Identification can be organized either by equipment type (e.g., safety equipment, security devices, monitoring equipment) or by facility operating area (e.g., receiving area, treatment area, office area) or both. The level of detail required in the inspection varies depending on the function of the equipment or area inspected. Emergency equipment should be thoroughly inspected to ensure that it will perform its contingency action and, thus, prevent dangerous conditions. The detail required in inspection of noncontingency equipment will in many cases be as stringent, but generally not as detailed (OSWER Directive No. 9523.00-10).

Q: What inspection procedures are required?

A: As soon as all the equipment, structures, and areas to be checked during the inspections have been identified, the next step is to document how each item on the list is checked or inspected. Simply using words like *check* or *observe* is inadequate. Rather, use phrases, such as "check ----- for ----- by -----" or "observe ----- for signs of -----."

Many structures or areas will not lend themselves to a quantitative measurement, which is otherwise possible for a piece of mechanical equipment. In these cases, the inspection procedures should indicate a visual inspection as a minimum (OSWER Directive No. 9523.00-10).

Q: What are the frequency requirements?

A: The regulations outlined in Sections 264/265.15(b)(4) state that the frequency should be based on the rate of possible deterioration and the probability of an incident if the deterioration or operator error goes undetected between inspections. In general, equipment, structures, or areas that are subjected to daily or continual use or stress should be inspected more frequently than backup or emergency components. Items inspected less frequently should be subjected to more rigorous inspection. However, if the failure of equipment used continuously would present a hazardous situation of any kind, the inspection should be rigorous, as well as frequent.

Q: If an area or device is continuously used, does it still have to be inspected?

A: Yes; continual or frequent successful use of an area or equipment item does not substitute for inspection (OSWER Directive 9523.00-10).

Unit-Specific Requirements

Q: Are there additional inspection requirements for specific units?

A: Yes; there are additional inspection requirements for containers, tank systems, surface impoundments, waste piles, landfills, and incinerators.

Q: What are the inspection requirements for containers?

A: The owner or operator of any facility that stores containers of hazardous waste must, at least weekly, inspect the areas where containers are stored, looking for leaking containers and for deterioration of containers and the containment system caused by corrosion or other factors.

Q: What are the inspection requirements for tank systems?

A: At least once each operating day the owner or operator must inspect:

- Above-ground portions of the tank to detect corrosion or releases
- Data gathered from monitoring and leak-detection equipment
- Construction materials and surrounding areas of the tank system for visible erosion or releases
- Overfill and spill control equipment (Part 265 and 90-day accumulation tanks only); permitted tanks must develop a schedule that will be specified in the permit for inspection of the spill control equipment

Q: What are the inspection requirements for surface impoundments?

A: The regulations require inspections at two distinct times; during and immediately after construction, and weekly during operation.

Q: What must be inspected during construction and installation?

A: During construction and installation, liners and cover systems (e.g., membranes, sheets, or coatings) must be inspected for uniformity, damage, and imperfections (e.g., holes, cracks, thin spots, or foreign materials).

Immediately after construction or installation:

- Synthetic liners and covers must be inspected to ensure tight seams and joints and the absence of tears, punctures, or blisters.
- Soil-based and admixed liners and covers must be inspected for imperfections, including lenses, cracks, channels, root holes, or other structural nonuniformities that may cause an increase in the permeability of the liner or cover.

Q: What are the inspection requirements for a surface impoundment in operation?

A: While a surface impoundment is in operation, it must be inspected weekly and after storms to detect evidence of any:

- Deterioration, malfunctions, or improper operation of overtopping control systems
- Sudden drops in the level of the impoundment's contents
- The presence of liquids in leak detection systems
- Severe erosion or other signs of deterioration in dikes or other containment devices

Q: What are the inspection requirements for waste piles and landfills?

A: The inspection requirements for waste piles and landfills, including the construction phase, are nearly identical to those of surface impoundments.

Q: What are the inspection requirements for incinerators?

A: The combustion and emission parameters and the incinerator and associated equipment (e.g., pumps,valves, conveyors) for leaks, spills, and/or fugitive emissions.

PREPAREDNESS AND PREVENTION

Applicability

Q: What is the intent of the preparedness and prevention requirements?

A: The intent of these requirements is to ensure that each facility has a complete description of the preparedness and prevention measures that will be implemented at the facility to minimize the possibility of a fire, explosion, or any unplanned sudden or nonsudden release of hazardous waste or hazardous waste constituents to air, soil, or surface water, which could threaten human health or the environment (Sections 264/265.31).

Q: What are the requirements for preparedness and prevention?

A: The general requirements for preparedness and prevention are:

- Facility design and operation
- Equipment
- Equipment testing and maintenance
- Communications or alarm systems
- Aisle space
- Arrangements with local authorities

Q: What are the requirements for facility operation and maintenance for the purposes of preparedness and prevention?

A: All hazardous waste facilities must be operated and maintained in a manner that minimizes the possibility of a fire, explosion, or any unplanned sudden or nonsudden release of hazardous waste or constituents into the environment (Sections 264/265.31).

Emergency Equipment

Q: What emergency equipment is required at a facility?

A: The regulations (Sections 264/265.32) specifically identify three categories of required emergency equipment: internal communication system, external communication system, and fire-fighting/response equipment.

Internal Communication System

Q: What is the purpose of the internal communication system?

A: The primary purpose of the internal communication system or alarm system, is to provide emergency instruction to enable rapid evacuation of the affected areas. A secondary but equally important purpose is to initiate the emergency response plan as part of the contingency plan [Sections 264/265.32(a)].

Q: What items are considered adequate communication devices?

A: Intercoms, internal telephones, or two-way radios can serve as communication devices. The alarm system should be a siren, buzzer, or similar device audible from all areas of the facility. Alarms can be either manually or automatically activated, but automatic systems should also provide for manual activation. Switches should be readily accessible to all personnel involved in waste handling operations (OSWER Directive No. 9523.00-10).

External Communication System

Q: What is required of an external communication system?

A: Should an emergency situation arise, the facility will need some means of summoning assistance from local police departments, fire departments, and state and local emergency response teams. Both a telephone and a hand-held two-way radio are acceptable communication devices. Such devices must be immediately available at the scene of operations [Sections 264/265.32(b)].

Q: What are the requirements for immediate access to outside assistance?

A: All personnel involved in hazardous waste handling operations must have immediate access to an emergency communication device or internal alarm system [Sections 264/265.34(a)].

Q: What constitutes access?

A: Access can either be direct (e.g., telephone, hand-held two-way radio) or through visual or voice contact with another employee [264/265.34(a)].

Fire-Fighting/Response Equipment

Q: What type of equipment is required?

A: Equipment must include portable fire extinguishers, fire control equipment (including any special extinguishing equipment, such as foams, inert gas, or dry chemicals), spill control equipment (e.g., pumps, absorbents, vacuum cleaners, etc.) and decontamination equipment. A description of the number and type of each piece of emergency equipment and its location and function relative to process operations must be provided [Sections 264/265.32(f)].

Q: What type of water supply is required?

A: An adequate fire-fighting supply for hoses, foam-producing equipment, or automatic sprinkler system is required [Sections 264/265.32(d)].

Q: What is the minimum aisle space required?

A: There is no specified minimum aisle space required. The facility owner or operator must have sufficient aisle space to allow unobstructed movement of emergency equipment or personnel to any location at the facility. The dimensions should be determined by consulting with the local emergency response organizations.

Arrangements with Local Authorities

Q: What is required when making arrangements with local authorities and emergency services?

A: The facility owner or operator is required to attempt to make arrangements with appropriate local agencies in an effort to coordinate responses to emergency situations [Sections 264/265.37(A)].

Q: What information is to be provided to the local authorities?

A: The owner or operator must provide the police, fire, and emergency response teams with a layout of the facility, as well as identifying types of hazardous wastes handled, places where workers normally work, facility entrances, roads inside the facility, and possible evacuation routes. Detailed site maps and floor plans showing water supplies, process operations, and access routes should be provided. Indicate any special hazardous situations that might encountered (e.g., explosions, toxic fumes, reactive chemicals), and discuss any special equipment needed to protect personnel (e.g., protective clothing, respirators). It would also be a good idea to have fire department officials visit the plant to inspect fire protection capabilities on-site [Sections 264/265.37(a)].

Q: What if the local authorities or emergency services decline to enter into an arrangement?

A: Declination of any response organization or local authority to enter into such arrangement must be documented and noted in the facility's operating record [Sections 264/265.37(b)].

CONTINGENCY PLAN AND EMERGENCY PROCEDURES

Contingency Plan

Q: What is the purpose of the contingency plan?

A: The contingency plan is designed to minimize the hazards to human health or the environment from fires, explosions, or any unplanned sudden or nonsudden release of hazardous waste or constituents into the environment [Sections 264/265.51(a)].

Q: When must the contingency plan be implemented?

A: The contingency plan must be implemented immediately whenever there is a fire, explosion, or release of hazardous waste or constituents that could threaten human health or the environment [Sections 264/265.51(b)].

Q: What is to be included in the contingency plan?

A: The contingency plan must describe the actions facility personnel must take in response to fires, explosions, or releases [Sections 264/265.51(a)].

Q: Is there a specified format for the contingency plan?

A: No; but EPA provides a sample table of contents for a contingency plan in OSWER Directive No. 9523.00-10:

1. Purpose
2. Scope
3. Responsibilities
4. Organization and Duties
5. Coordinated Emergency Services
6. Training
7. Routine Surveillance to Detect Potential Hazards
8. Emergency Procedures
   a. Fire
   b. Explosion
   c. Hazardous Waste Release
9. Evacuation Plans
10. Record Keeping and Incident Reports

Appendix A: List of Emergency Coordinators; names, addresses and phone numbers
Appendix B: List of Emergency Equipment
Appendix C: Incident Reports

Q: If an owner or operator has already prepared a Spill Prevention, Control, and Countermeasures (SPCC) plan, must a separate contingency plan be developed?

A: No; if an owner or operator has already prepared an SPCC plan or other emergency or contingency plan, that plan need only be amended to incorporate hazardous waste management provisions that are sufficient to comply with the requirements of this Subpart [Sections 264/265.52(b)].

Q: Sections 264/265.52(b) refer to the Spill Prevention, Control, and Countermeasures (SPCC) Plan and also Part 1510 of Chapter V. What is Part 1510 of Chapter V?

A: Part 1510 of Chapter V, Title 40, is a reference to the National Contingency Plan (NCP) as originally prepared by the Council on Environmental

Quality. NCP responsibility was subsequently delegated to the EPA by Executive Order 12316 (March 12, 1982). As a result of this change, the NCP was redesignated under Chapter I of Title 40 CFR, Part 300.

Q: Where is the contingency plan required to be maintained?

A: Copies of the contingency plan and all revisions to the plan must be maintained at the facility and submitted to all local police departments, fire departments, hospitals, and state and local emergency response teams that may provide emergency services to the facility (Sections 264/265.53).

Q: Does the contingency plan ever have to be amended?

A: Yes; according to Sections 264/265.54, the contingency plan must be reviewed and immediately amended, if necessary, whenever:

- The facility permit is revised.
- The plan fails in an emergency.
- The list of emergency coordinators changes.
- The list of emergency equipment changes.
- The facility design, construction, operation, maintenance, or other circumstances change to increase the potential for fires, explosions, or releases of hazardous waste or hazardous waste constituents, or change the response necessary in an emergency.

Contents of the Contingency Plan

Q: What else must be included in the contingency plan?

A: In addition, a coordinated emergency services plan, a list of emergency coordinators, a list of emergency equipment, and an evacuation plan must also be included in the contingency plan.

Q: What is the emergency services plan?

A: A coordinated emergency services plan must be arranged and agreed to by local police departments, fire departments, hospitals, contractors, state and local emergency response teams, and the facility owner or operator. The arrangements with local authorities and emergency response organizations required under the preparedness and prevention subpart should serve as the basis for this requirement [Sections 264/265.52(c)].

Q: What is required in a list of emergency coordinators?

A: A list of names, addresses, and phone numbers (office and home) of all persons qualified to act as emergency coordinators in the order in which they will assume responsibility [Sections 264/265.52(d)].

Q: What constitutes an emergency coordinator?

A: The facility owner or operator should select at least one employee who is either on the facility premises during peak operational periods (preferred) or available to respond to an emergency by reaching the facility within a short time. This employee should be designated the primary emergency coordinator. The emergency coordinator is responsible for coordinating all emergency response measures and being thoroughly familiar with:

- The facility's contingency plan
- All operations and activities at the facility
- The location and characteristics of waste handled
- The location of all records within the facility
- The physical layout of the facility

The selected emergency coordinator should have the authority to expend funds and recruit employees to implement the contingency plan. The owner/operator should also select alternative employees to act as emergency coordinators if for some reason the designated emergency coordinator is unavailable [Sections 264/265.52(d)].

Q: What information is required in the list of emergency equipment?

A: This information should be the same as developed for the preparedness and prevention requirements.

Q: What is the required content of the evacuation plan?

A: Information to be contained in the evacuation plan includes recognizable signals to commence an evacuation, evacuation routes, alternative evacuation routes (in case primary exit routes are blocked by releases of hazardous waste or fires), and safe assembly areas to account for all evacuated personnel. Maps clearly delineating evacuation routes, fire-fighting equipment, alarms, and assembly areas should be prepared. In addition, the applicant should note the presence of employee-training sessions, fire drills, and the placement of evacuation maps posted at the facility [Sections 264/265.52(f)].

Emergency Procedures

Q: When are the emergency procedures required to be implemented?

A: The emergency procedures are required to be implemented whenever there is a fire, explosion, or release of hazardous waste. There are separate requirements for the emergency coordinator and the facility owner or operator [Sections 264/265.56(a)].

Q: What emergency procedures are to be implemented by the emergency coordinator?

A: According to Sections 264/265.56(h), the emergency coordinator must:

- Activate facility alarms and notify appropriate state or local agencies.
- Identify the character, exact source, amount, and areal extent of any released material.
- Assess possible direct and indirect hazards to human health or the environment that may result from the release, fire, or explosion.
- Determine if evacuation of local areas is required, and immediately notify either the government official designated as on-scene coordinator or the National Response Center.
- Ensure that fires, explosions, and releases do not occur, recur, or spread to other hazardous waste at the facility.
- Monitor for leaks, pressure buildup, gas generation, or ruptures in valves, pipes, or other equipment if facility operations cease.
- Provide treatment, storage, and disposal of any material that results from a release, fire, or explosion immediately after an emergency.
- Ensure that no waste incompatible with the released material is processed until cleanup procedures are completed and all emergency equipment listed in the contingency plan is cleaned and fit for its intended use.

Q: What emergency procedures are to be implemented by the facility owner or operator?

A: According to Sections 264/265.56(j), the emergency procedures to be implemented by the owner or operator of the facility are:

- Notify the EPA regional administrator and state and local authorities that the facility is in compliance with Sections 264/265.56(h), e.g., the emergency coordinator's duties, before operations commence.

- Record the time, date, and details of any incident that requires implementing the contingency plan and submit a written report on the incident to the EPA regional administrator within 15 days of the incident.

MANIFEST SYSTEM, RECORD KEEPING, AND REPORTING

Manifest System

Q: Are all TSD facilities subject to the manifest requirements?

A: No; the regulations make a distinction between facilities that treat, store, or dispose of wastes generated on the same property (on-site facilities) and facilities that treat, store, or dispose of wastes that were not generated on the facility property (off-site facilities). Specifically, owners and operators of on-site facilities are not required to utilize a manifest system if they do not receive any hazardous waste from off-site sources. However, they are still required to comply with the manifest system standards contained in Subpart B of Part 262 if they ship hazardous waste off their property (Sections 264/265.70).

Q: What are the requirements for handling a manifest?

A: The procedures for handling a manifest include:

- Sign to certify receipt of the waste.
- Note, on the document, discrepancies between the document and the wastes received.
- Immediately give a signed copy to the person delivering the waste.
- Send a signed copy to the generator within 30 days (or 60 days for medium-quantity generators).
- Keep a signed copy on file for at least 3 years from date of receipt. However, because a person's liability under RCRA does not end after 3 years, it is recommended that these records be maintained indefinitely.

Manifest Discrepancies

Q: What constitutes a manifest discrepancy?

A: A *manifest discrepancy* is a significant difference between the type or quantity of waste received and the type or quantity of waste described on the manifest. Significant discrepancies include variations of 10 percent or more by weight of bulk waste, any variation in piece count (e.g., 1 drum), or difference in waste description [Sections 264/265.72(a)].

Q: Does a waste analysis have to be conducted immediately upon receipt of the manifest to check for a discrepancy?

A: No; waste analyses do not have to be performed immediately upon receipt to identify discrepancies.The discrepancies that should be noted on the manifest or shipping papers are obvious ones that can be immediately determined by counting or measuring the waste received and comparing the manifest or shipping papers with labels on the waste [Sections 264/265.71(a)(2)].

Q: What happens if there is a difference in the weight of bulk waste but it is less than 10 percent?

A: Any difference between the amount of the bulk shipment stated on the manifest or shipping papers and the amount actually received should be noted on the documents at the time of receipt. The receiver should then decide whether or not to take the actions identified in Sections 264/265.72(b) based on the quantity of difference and not the percentage of difference for bulk shipments.

Q: What must be done if a significant discrepancy occurs?

A: Upon discovery of a significant discrepancy involving information contained in the manifest, the owner or operator must reconcile with the generator or transporter verbally. If the discrepancy is not resolved within 15 days after receipt of the waste, the owner or operator must notify EPA in writing describing the discrepancy [Sections 264/265.72(b)].

Unmanifested Waste

Q: What are the requirements concerning the receipt of an unmanifested shipment of hazardous waste?

A: If a facility receives a shipment of unmanifested waste that is not excluded (e.g., SQG), the owner or operator must prepare and submit to EPA an unmanifested waste report (EPA Form 8700-13B) (Sections 264/265.76).

Record Keeping

Q: What are the record keeping requirements?

A: A facility must maintain an operating record, which is a written log of facility activities that contain specified information (Sections 264/265.73).

Q: What are the information requirements?

A: According to Sections 264/265.73(b), the following information must be recorded in the operating record:

- Descriptions of wastes received, quantity, date, and method of treatment, storage, or disposal
- Records of locations of wastes within the facility
- Waste analyses results
- Records of contingency plan implementation
- Waste minimization reports
- Inspection records
- Groundwater monitoring, testing, or analytical data
- Notices of permits and waste acceptance to generators
- Closure and post-closure estimates

Q: How long does the operating record have to be maintained at the facility?

A: The operating record must be kept until certification of closure [Sections 264/265.73(b)].

Q: Is a facility required to keep copies of manifests and biennial reports on-site?

A: Yes; Sections 264/265.71(a)(5) require facilities to retain copies of manifests on site for at least 3 years from the date of delivery. Sections 264/265.74(a) state that all required records must be furnished upon request and made available for inspection by EPA personnel. Biennial reports are required records.

Q: What are the requirements for handling records?

A: The requirements of Sections 264/265.74 place stipulations on the handling (i.e., availability, retention, and disposition) of all records relative to Parts 264 and 265. Specifically, they require that EPA personnel be allowed to inspect the records, that records be retained for longer than required by the regulations if the facility is involved in enforcement actions or is requested to retain them by EPA, and that copies of the records be submitted as required by Sections 264/265.73(b)(2) to the local land authority and EPA when the facility is closed.

Reporting

Q: What are the reporting requirements?

A: The reporting requirements include notification by disposal facilities and biennial reports.

Q: What are the notification requirements concerning disposal facilities?

A: Disposal facilities must submit copies of the records of waste disposal locations and the quantities at each location to EPA and the local land authorities upon closure [Sections 264/265.74(c)].

Q: What are the requirements concerning the biennial report?

A: The owner or operator must submit a biennial report before March 1 of each even-numbered year. It is important to note that many states require annual reporting. The required contents of the report are:

- The facility's EPA identification number
- The EPA identification number of each generator that sent waste to the facility
- The quantity and description of each hazardous waste received
- The method of treatment, storage, or disposal for each waste
- The most recent closure and post-closure cost estimate
- A signed statement certifying the accuracy of the information

Q: Are there any additional reporting requirements?

A: Yes; Sections 264/265.77 identify additional reports that must be made to EPA. They include reports of hazardous waste releases, fires, explosions, closures, or any other reports that may be required under Parts 264 and 265.

CLOSURE AND POST-CLOSURE

Applicability

Q: What is the purpose of closure and post-closure?

A: The primary purpose of the closure and post-closure standards is to ensure that all hazardous waste management facilities are closed in a manner that to the extent necessary (1) protects human health and the environment; and (2) controls, minimizes, or eliminates post-closure escape of hazardous waste, hazardous constituents, leachate, or contaminated prescription run-off, or waste decomposition to the ground or atmosphere (Sections 264/265.11).

The closure and post-closure requirements are divided into general standards applicable to all hazardous waste management facilities and technical standards specific to the type of waste management unit.

Q: Are all facilities required to conduct post-closure care?

A: No; all facilities are subject to the closure requirements; only land disposal facilities and any facility that cannot *clean close,* are subject to the post-closure care requirements (Sections 264/265.110).

Closure Requirements

Q: What is closure?

A: Closure is the period after which wastes are no longer accepted and during which the owner or operator completes all treatment, storage, or disposal operations.

Q: Must a facility owner or operator close the entire facility, or can discrete units be closed systematically?

A: Yes; an owner or operator can conduct partial closure of the facility. *Partial closure* is closure of a hazardous waste management unit at a facility that contains other operating hazardous waste management units. Closure of the last unit constitutes final closure of the facility.

Closure Plan

Q: What is required to document the procedures to be used for closure?

A: All hazardous waste management facilities must prepare and maintain a written closure plan and, if required, a contingent closure plan. The plan must identify steps necessary to perform partial or final closure at any time during the facility's active life (Sections 264/265.112).

Q: What facilities are required to have contingent closure plans?

A: Permitted storage surface impoundments and waste piles that do not have liners that meet the requirements of Sections 264.221(a) and 264.251(a) and permitted and interim status tanks without secondary closure (assuming that they are not exempt) must prepare closure plans that describe activities necessary to conduct closure under two sets of conditions. The first plan must describe how the unit will be closed by removing all hazardous waste and hazardous constituents. The second is a *contingent closure plan* that outlines the closure activities to be undertaken if the unit cannot remove all hazardous waste and constituents and must be closed as a landfill.

Q: What are the required contents of a closure plan?

A: Pursuant to Sections 264/265.112(b), the closure plan must, at a minimum, contain the following:

- A description of how each hazardous waste management unit will be closed in accordance with applicable closure performance standards (i.e., cleanup levels)
- A description of how final closure of the entire facility will be conducted, including a description of the maximum extent of the operations that will remain unclosed during the facility's active life
- An estimate of the maximum inventory of hazardous wastes onsite at any time during the active life of the facility, including a detailed description of the methods to be used during partial and final closure, including removing, treating, storing, and disposing of all hazardous waste
- A description of the steps needed to remove or decontaminate all hazardous waste residues from any of the facility components, equipment, or structures
- A sampling and analysis plan for testing surrounding soils to determine the extent of decontamination required to meet the closure performance standards (i.e., cleanup levels)
- A description of all other activities during closure, including groundwater monitoring, leachate collection, and precipitation control (i.e., run-on/run-off)
- A schedule for closure of each hazardous waste management unit and for final closure of the facility
- An estimate of the expected year of closure

Q: Sections 264/265.112(b)(3) require a facility's closure plan to include an estimate of the maximum inventory of waste in storage and in treatment at any time during the life of the facility. What is used to estimate the maximum inventory?

A: According to OSWER Directive No. 9476.00-5, the estimate of the maximum inventory of wastes should include:

- The maximum amount of hazardous wastes, including residues, in all treatment, storage, and disposal units
- The maximum amount of contaminated soils and residues from drips and spills from routine operations
- If applicable, the maximum amount of hazardous wastes from manufacturing/process areas and raw material/product storage and handling areas

Q: A closure plan must include a description of the steps needed to decontaminate facility equipment during closure [264/265.112(a)(4)]. Does EPA provide any guidance regarding specific levels of decontamination?

A: No; EPA does not specify levels of decontamination; however, EPA has recommended decontamination techniques. If a sludge is present, it must be removed. If hard scale is present, it must be scraped, chiseled, sandblasted, or brushed with a wire brush to remove any visible material. All surfaces must be rinsed with an appropriate solvent. All solid wastes from decontamination must be treated as hazardous wastes unless the owner/operator can demonstrate otherwise in accordance with Section 261.3(d) (HMR).

Q: Are there additional information requirements to be included in the plan for an owner or operator who intends to clean-close a unit?

A: Yes; an owner or operator who intends to clean-close a unit must include specific details of how the owner or operator expects to make the necessary demonstration. Specific details that are required include sampling protocols, schedules, and the cleanup levels that are intended to be used as a standard for assessing whether removal or decontamination is achieved (52 *FR* 8706, March 19, 1987).

Q: Where is the closure plan required to be kept?

A: For interim status facilities, a copy of the closure plan, including all revisions to the plan, must be kept at the facility until closure is certified to be complete. The plan must be furnished to EPA upon request, including a request by mail. In addition, for facilities without approved plans (i.e., they have been reviewed and approved by EPA), the plan must be provided on the day of a site inspection to any duly designated representative of EPA [Section 265.112(a)].

For permitted facilities, EPA should already have a copy of the plan as part of the Part B permit application.

Q: When must the closure plan be submitted to EPA?

A: The owner or operator of an interim status facility must submit the closure plan to EPA at least 180 days prior to the expected date on which closure will begin for the first land disposal unit (i.e., surface impoundments, waste piles, land treatment, or landfills (includes cells) or final closure if it involves such a unit. An owner or operator of a facility that has only incinerators, tanks, or container storage units must submit the closure plan at least 45 days before the expected date of the commencement of final closure of the facility [Section 265.112(d)].

If a facility's interim status is terminated for reasons other than the issuance of a permit, the owner or operator must submit a closure plan within 15 days, unless the facility is issued either a judicial decree or a compliance order to close.

Q: When is the expected date of closure?

A: The *expected date of closure* must either be within 30 days after the date on which any hazardous waste management unit receives the known final volume of hazardous wastes or, if there is "reasonable possibility" that the hazardous waste management unit will be receiving additional hazardous wastes, no later than 1 year after the date the unit received the most recent volume of hazardous waste. However, this date can be extended upon approval by EPA [Sections 264/265.112(d)(2)].

Q: When must a permitted facility submit a closure plan?

A: A permitted facility was required to submit the closure plan as part of its permit application [Section 270.14(13)].

Closure Plan Approval

Q: What is the requirement for closure plan approval?

A: Owners and operators of interim status facilities must submit their plans (closure and, if required, post-closure) for review and approval to EPA prior to final closure of the facility or closure of the first disposal unit.

Q: How are closure plans approved?

A: Within 30 days of receipt of a closure plan, EPA must provide the owner or operator and the public the opportunity to submit written comments or requests for modifications to the plan by publishing a newspaper notice. A public hearing may be held if one is requested. EPA must then approve, modify, or disapprove a closure plan of an interim status facility within 90 days of its receipt [Section 265.112(d)(4)].

A permitted facility was to have included the closure plan as part of the permit application [Section 270.14(13)]. When the permit is issued, the closure plan, which is now a condition of the permit, is the *approved* closure plan.

Q: What happens if EPA does not approve the closure plan of an interim status facility?

A: If the plan is not approved, EPA must provide a detailed written statement to the owner or operator with the reasons for disapproval. The owner or operator then has 30 days after receiving the written statement to modify the plan or submit a new one. EPA must approve or modify the new or resubmitted plan within 60 days. If EPA modifies this plan, it becomes the approved closure plan [Section 265.112(d)(4)].

Q: What avenue of appeal is available to an owner or operator who wants to contest modifications made by EPA to a final closure plan?

A: Currently, there are no provisions under RCRA that allow an owner or operator to appeal the final closure plan issued by EPA. The owner or operator would have to pursue other legal recourse outside the RCRA regulations to appeal provisions in a final closure plan.

Q: If an owner or operator has an* approved *closure plan, does EPA have to be notified concerning closure?

A: Yes; all facilities that have approved closure plans must notify EPA at least 60 days prior to closure for any land disposal unit or 45 days for final closure of facilities with only container storage, tank, or incinerator units [Sections 264/265.112(d)].

Q: Can a facility amend or modify the closure plan?

A: Yes; facilities operating under interim status may amend the closure plan at any time prior to notification of partial or final closure. A written request for approval of any amendments to an *approved* closure plan must be submitted to EPA for authorization [Section 265.112(c)].

Permitted facilities must seek an appropriate permit modification to amend or modify its closure plan [Section 264.112(c)] (see Chapter 9).

Q: Are there instances when a facility must amend the closure plan?

A: Yes; the owner or operator of a facility with either an approved or nonapproved plan must amend the closure plan whenever:

- Changes in the facility design or operation affect the closure plan
- A change in the expected year of closure occurs
- Unexpected events occur during closure

The plan must be amended at least 60 days prior to a proposed change and no later than 60 days after an unexpected change occurs. If an unexpected change occurs during closure, the plan must be amended within 30 days. For example, if a surface impoundment or waste pile originally intended to clean close but cannot, and must close as a landfill, the closure plan must be amended within 30 days [Sections 264.112(c)(3) and 265.112(c)(2)].

Closure Period

Q: How long does an owner or operator have to conduct closure?

A: Within 90 days after receiving the final volume of hazardous waste or 90 days after approval of the closure plan, whichever is later, all hazardous waste must be treated, disposed of, or removed. The owner or operator has an additional 90 days to complete decontaminating the facility, dismantling equipment, etc. Thus, closure must be completed within 180 days after final receipt of hazardous waste or plan approval. However, a time extension can be granted by EPA under specified conditions (outlined below) no later than 30 days prior to the expiration of either that 90-day or 180-day period (Sections 264/265.113).

Q: On what conditions can an extension be granted?

A: According to Sections 264/265.113(b), the conditions for a time extension during closure include:

- The required closure activities will, if necessary, take longer than the applicable 90 or 180 days.
- The facility has the capacity to receive additional wastes.
- There is reasonable likelihood that a person other than the owner or operator will recommence operation of the site.

- Closure of the facility would be incompatible with continued operation of the site.

An approval of a time extension for closure is contingent on whether the owner or operator has taken and will continue to take all steps necessary to prevent threats to human health and the environment.

Cleanup Requirements

Q: ***What are the cleanup requirements for closure?***

A: During partial or final closure, all contaminated liners, equipment, structures, and subsoils must be removed and properly disposed of or decontaminated; or close the unit as a landfill (Sections 264/265.114).

Q: ***Must contaminated groundwater also be removed or decontaminated before clean-closure is possible?***

A: Yes; EPA interprets the term *contaminated subsoils* to include contaminated groundwater (53 *FR* 9944, March 28, 1988).

Q: ***What happens if during closure, new TSD units are required?***

A: To satisfy the closure requirements, it may be necessary to create new treatment, storage, or disposal units (e.g., a waste pile). There is no exemption from the permitting requirements for a facility that is subject to the closure requirements. The Part 264 standards are applicable to new units added during closure, as well as to new operating units. However, the addition of new units may constitute an allowable change if the facility is operating under interim status. According to Section 270.72(c), changes in processes or addition of processes may be allowed if a revised Part A permit application and justification are submitted and EPA approves the change (see Chapter 9).

Clean Closure

Q: ***What is clean closure?***

A: At closure, owners and operators of hazardous waste management units can choose between removing/decontaminating all hazardous wastes and waste residues *(clean closure)* and terminating further regulatory responsibility under RCRA for the unit; or closing the unit with hazardous waste or waste residue remaining in place *(dirty closure)* and instituting post-closure care similar to landfills.

Q: What are the cleanup requirements for clean closure?

A: Previously, owners and operators of interim status facilities attempting a clean closure were required to remove wastes from a unit to the point that wastes were no longer *hazardous*. The criteria for this determination depended on whether the wastes in the unit were hazardous because they were listed or because they exhibited a characteristic. Thus, if a surface impoundment contained ignitable hazardous waste only, the owner or operator could cease the removal of materials if that material no longer exhibited the ignitability characteristic. In addition, the owner or operator's responsibility under Subtitle C of RCRA ceased at the time of certification of clean closure (45 *FR* 33203-4, May 19, 1980). Consequently, the clean closure standard allowed facilities to be relieved of their RCRA responsibility even though there may have been contamination remaining (assuming the contamination was not defined as a hazardous waste).

To address this contamination, EPA promulgated final regulations on March 19, 1987 (52 *FR* 8704), and December 1, 1987 (52 *FR* 45788), that significantly strengthened the clean-closure requirements for interim status units. The interim status requirements are now nearly identical to the Part 264 requirements. An important aspect of these regulations is that they are applicable to any land disposal unit that received waste after July 26, 1982, or certified closure after January 26, 1983. Thus, even if a unit had previously successfully clean closed, the facility owner or operator will have to demonstrate that the unit has met the Part 264 clean closure requirements (52 *FR* 45795, December 1, 1987).

For example, on June 13, 1983, an owner of a surface impoundment (which was used to treat ignitable waste only) clean closed the surface impoundment in accordance with the approved closure plan. The owner of the impoundment ceased removing contaminated subsoils at the point that the soil did not exhibit the ignitability characteristic, in accordance with Section 265.228(b). However, the recent changes to the closure rules (52 *FR* 8704 and 52 *FR* 45788) now require that owner to either remove *all* wastes and waste residues (i.e., hazardous constituents) from the previously closed impoundment to attain a clean closure or close the impoundment as a landfill and commence the post-closure care period.

Q: How does EPA interpret remove and decontaminate?

A: EPA interprets *remove* and *decontaminate* to describe the amount of removal or decontamination that obviates the need for post-closure care

(52 *FR* 8706, March 19, 1987). This means that an owner or operator must remove all hazardous waste or waste residue (i.e., all hazardous constituents) that pose a "substantial present or potential threat to human health or the environment." EPA intends to review site-specific demonstrations submitted by the owner or operator to determine if the removal or decontamination is sufficient. The closure demonstrations to be submitted must document that the contaminants left in the soil and/or groundwater, surface water, or atmosphere, in excess of EPA-recommended limits or factors, and direct contact through dermal exposure, inhalation, or ingestion will not result in a threat to human health or the environment.

Q: How does EPA interpret EPA-recommended limits or factors?

A: *EPA-recommended limits or factors* are those that have undergone peer review by EPA. At the present time, these include Federal Water Quality Criteria, Verified Reference Doses, Carcinogenic Potency Factors, and Health Assessment Documents. If no EPA-recommended exposure limit exists for a particular constituent, then the owner or operator must either remove the constituent to background levels, submit data of sufficient quality for EPA to determine the environmental and health effects of the constituent in accordance with TSCA (40 CFR Parts 797 and 798), or follow landfill closure and post-closure requirements (52 *FR* 8706, March 19, 1987).

Q: How does an owner or operator demonstrate that a clean closure is successful?

A: The demonstration can be accomplished by either submitting a Part B permit application for a post-closure permit or by petitioning EPA for an *equivalency determination,* which is a petition that attempts to demonstrate that a post-closure permit is not required because the owner or operator has met the applicable Part 264 closure standards for that unit (52 *FR* 45795, December 1, 1987). If an owner or operator seeks an equivalency demonstration under Section 270.1(c)(5), EPA must provide for a 30-day public comment period.

Q: How does EPA determine if the demonstration is adequate?

A: EPA must make a determination as to whether the unit has met the removal or decontamination requirements within 90 days of receiving a demonstration. If EPA finds that the closure did not meet the applicable clean-up requirements (i.e., hazardous constituents above EPA recommended limits or factors remain), it must provide a written statement of the reasons why the closure was not successful. The owner or operator can submit additional information in support of the demonstration within 30 days after receiving such a written statement. EPA must review this additional information and make a final determination within 60 days. If EPA determines that the facility did not close in accordance with the clean-closure requirements, the facility is subject to the post-closure care and permitting requirements.

Q: Aren't there implications if an owner or operator uses a Part B for the demonstration?

A: Yes; if a Part B permit application is used to demonstrate a successful clean closure, the 3004(u) corrective action for solid waste management units (SWMUs) provision must be addressed. Thus, the facility owner or operator will have to address releases from SWMUs at the entire facility.

Q: How are wastes generated as a result of closure activities handled?

A: During the closure process, all hazardous wastes, waste residues, contaminated subsoils (including groundwater), and equipment must be managed as a hazardous waste unless the material is delisted (if it was a listed waste) or if the waste does not exhibit a characteristic of hazardous waste. This means that even though contaminated material at a surface impoundment is not (by definition) a hazardous waste, it must be removed, but it can be managed as a Subtitle D waste subject to state rules and regulations (Sections 264/265.114).

Dirty Closure

Q: What are the closure requirements if a unit cannot clean close?

A: If an owner or operator either chooses not to conduct a clean closure, or fails to do so, that owner or operator must provide post-closure care similar to a landfill (Sections 264/264.310), including:

- Eliminate all free liquids by either removing the liquid wastes/residues from the impoundment or solidifying them.
- Stabilize the remaining waste and waste residues to a bearing capacity sufficient to support a final cover.
- Install a final cover that provides long-term minimization of infiltration into the closed unit, functions with minimum maintenance, promotes drainage, and minimizes erosion.
- Perform post-closure care and groundwater monitoring.

Q: Are there additional closure requirements for "dirty closure"?

A: Yes; there are unit-specific standards that provide specific instructions. In addition, the general closure performance standards (Sections 264/265.111) apply to activities that are not otherwise addressed by the process-specific standards but are necessary to ensure that the facility is closed in a manner that will ensure protection of human health and the environment. For example, under the closure performance standard, an owner or operator can be required to install source control (e.g., leachate collection and run-on/run-off control) not otherwise specifically required (OSWER Directive No. 9476.00-13)

Certification Of Closure

Q: What is certification of closure?

A: Certification of closure is a required documentation, prepared by the owner or operator and a qualified, independent, registered professional engineer, that certifies that closure has been conducted in accordance with the approved closure plan.

Q: The owner or operator of a surface impoundment hires a contractor to conduct the closure. Can the engineer who is employed by the contractor performing the closure, certify the closure of the facility?

A: Yes; OSWER Directive No. 9476.00-5 clarifies that an "independent" engineer cannot be directly employed by the owner or operator of the unit. Also, the May 2, 1986, *Federal Register* (51 *FR* 16433) states that ". . . the certification should be made by a person who is least subject to conscious or subconscious pressures to certify to the adequacy of a closure that in fact is not in accordance with the approved closure plan."

Q: When must the certification be prepared?

A: Within 60 days of closure completion for each individual land-disposal unit at a facility or 60 days of closure completion for an incinerator, tank, or container storage facility, the certification of closure must be prepared and submitted to EPA (Sections 264/265.115).

Q: What occurs after the closure certification?

A: Within 60 days after receiving the closure certification, EPA will notify the owner or operator in writing that under RCRA, the owner or operator is no longer required to maintain financial assurance for closure and liability for that particular facility unless EPA has reason to believe that closure has not been done in accordance with the closure plan [Sections 264/265.115 and 264/265.147(e)].

Closure Notices

Q: What closure notices are required?

A: For all land disposal units, an owner or operator is responsible for submitting a survey plat and a deed notation/certification (Sections 264/265.116).

Q: What are the requirements for a survey plat?

A: Within 60 days of the closure certification, the owner or operator of a land disposal unit must submit a survey plat to the local zoning authority. It must include the location and dimensions of landfill cells or other land disposal units with respect to permanently surveyed benchmarks. The plat must be prepared and certified by a professional land surveyor. The survey plat, when deposited with the local zoning authority, must contain a prominently displayed note stating the obligation to restrict disturbances of the disposal units (Sections 264/265.116).

Q: What are the requirements for deed notation and certification?

A: Within 60 days of closure certification, a permanent notation must be made on the deed stating that:

- Hazardous waste management occurred on the property.
- Its use is restricted under RCRA.
- The survey plat and other applicable information is available at the local zoning authority.

A certification by the owner or operator that the notification was placed on the deed and a copy of the deed must be submitted to EPA.

If the owner or operator or any subsequent owner desires to remove any waste from the closed facility, a modification to the approved post-closure plan must be obtained. A notation indicating that waste was removed may be added to the deed at a later time (Sections 264/265.116).

Post-Closure

Q: What is post-closure?

A: Facilities at which hazardous wastes or waste residues will remain above EPA-recommended limits or factors after partial or final closure are subject to the post-closure requirements, which require various monitoring and maintenance activities during the post-closure care period [Sections 264/265.117(a)(1)].

Q: How long is the post-closure period?

A: The post-closure period begins with the certification of closure and continues for 30 years; however, the time can be extended or reduced by EPA when appropriate. In addition, an owner or operator can petition EPA to modify the time as well as any required post-closure activities [Sections 264/265.117(a)(1)].

Q: What units are required to conduct post-closure care?

A: All land disposal units that received waste after July 26, 1982, or certified closure after January 26, 1983, are subject to the post-closure requirements. Thus, an interim status unit that is a regulated unit and entering its post-closure period can be required to obtain a post-closure permit at any time during its post-closure care period. An owner or operator of a facility that clean-closes must also have a post-closure permit unless the owner or operator successfully demonstrates closure by removal or decontamination under Sections 270.1(c)(5) and (6) and Sections 264/265.117(a)(1).

Q: Does the issuance of a post-closure permit affect corrective action?

A: Yes; if a facility receives a post-closure permit, it is subject to both the Section 264.100 corrective action requirements for groundwater and the 3004(u) provisions [the 3004(u) provisions are initiated by a permit application], which may require corrective action for any solid waste management unit (SWMU) at the entire facility, regardless of when the SWMU closed.

Post-Closure Plan

Q: What facilities are required to have a post-closure plan?

A: Facilities with land disposal units must prepare and maintain a written post-closure plan at the facility. Facilities with land disposal units that intend to clean close must maintain a contingent post-closure plan in case the clean closure is not successful. This plan must be kept until final closure at facilities without an approved post-closure plan; a copy of the most current plan must be furnished to EPA upon request [Sections 264/265.118(b)].

Q: What is the purpose of the post-closure plan?

A: The purpose of the post-closure plan is to describe the frequency of monitoring and maintenance activities to be conducted after closure of each disposal unit [Sections 264/265.118(b)].

Q: What are the required contents of the post-closure plan?

A: In accordance with Sections 264/265.118(b), the post-closure plan must include, at a minimum:

- A description of the planned groundwater monitoring requirements, including the frequency of sampling
- The integrity of the cap and final cover or other containment systems
- A description of the planned maintenance activities for the cap, containment, and monitoring equipment
- The name and address of the facility contact person overseeing post-closure care

The plan should also include provisions for the kinds of monitoring and maintenance activities that reasonably can be expected during the post-closure care period. Because of the difficulty in predicting what may be required, the plan should include a range of possible alternatives, thus avoiding a potential permit modification.

Q: When is the post-closure plan required to be submitted?

A: For interim status facilities, the plan must be submitted to EPA at least 180 days before the owner or operator expects to begin partial or final closure of the first hazardous waste disposal unit (the same requirement for closure plans).

Facilities that intend to clean close are not required to have a post-closure plan, but are required to have a contingent post-closure plan. If an owner or operator intended to clean close but cannot and must close as a landfill, that owner or operator must submit a post-closure plan within 90 days of making that determination [Section 265.118(e)].

Facilities seeking an operating permit must submit the post-closure plan as part of the permit application in accordance with Section 270.14(b)(18).

Q: What is the process for approving a post-closure plan?

A: EPA must approve, modify, or disapprove a post-closure plan within 90 days of its receipt. Upon receipt of a post-closure plan, EPA will provide the owner or operator and the public the opportunity to submit written comments or requests for modifications to the plan. A public hearing may be held if one is requested and at the discretion of EPA [Section 265.118(f)].

Q: What happens if a post-closure plan is not approved?

A: If the plan is not approved, EPA must provide a detailed written statement to the owner or operator with the reasons for disapproval. The owner or operator then has 30 days after receiving the written statement to modify the plan or submit a new one. EPA must approve or modify the new or resubmitted plan within 60 days. If EPA modifies this plan, the modified plan becomes the *approved* post-closure plan. All post-closure activities must conform to the approved post-closure plan [Section 265.118(f)].

Q: Can the post-closure plan be amended?

A: Yes; the plan may be amended by the owner or operator at any time during the active life or post-closure care period. An owner or operator with an approved post-closure plan must submit a written request to EPA for authorization to modify the approved plan [Sections 264/265.118(d)].

Q: Are there instances where the post-closure plan must be amended?

A: Yes; the plan must be amended whenever operating plans or facility designs affect the post-closure plan, whenever any event occurs that affects the plan, and at least 60 days before a change in operations or 60 days after an unexpected event. If the plan needs to be amended and the facility has a permit, an appropriate modification to the permit must be sought [Sections 264/265.118(d)].

Post-closure Certification

Q: What are the requirements for post-closure certification?

A: Within 60 days after the completion of the post-closure care period for each hazardous waste disposal unit, the owner or operator must submit, by registered mail, a certification stating that post-closure care was performed in accordance with the post-closure plan. The certification must be signed by both the owner or operator, and an independent, qualified, registered professional engineer (Sections 264/265.120).

FINANCIAL RESPONSIBILITY REQUIREMENTS

Applicability

Q: What are the financial responsibility requirements?

A: The financial responsibility requirements were established to ensure that owners and operators have sufficient funds to pay for properly closing a facility, for maintaining post-closure care at disposal facilities, and for compensation to third parties for bodily injury and property damage caused by sudden and nonsudden accidents related to a facility's operation (liability).

Q: What facilities are subject to the financial responsibility requirements?

A: All waste management facilities must have adequate financial assurance for closure and liability. Land disposal units and tank systems that cannot clean-close [Sections 264/265.140(b)] are required to maintain financial assurance for post-closure care.

Q: Are government owned facilities subject to the financial responsibility requirements?

A: No; federal- and state-owned facilities are exempt from these requirements [Sections 264/265.140(c)]. However, this exclusion does not cover county- or municipality-owned facilities (45 *FR* 33199, May 19, 1980).

Financial Assurance for Closure and Post-Closure

Q: What is the intent of the financial requirements for closure and post-closure?

A: The financial requirements for closure and post-closure are to ensure that there is sufficient money available for proper closure and post-closure care of hazardous waste management units.

Q: How are the closure costs determined?

A: The owner or operator must have a detailed written estimate for facility closure in current dollars. The closure cost estimate must equal the cost of final closure at the point during the active life when closure would be the most expensive and shall be based on third-party costs (i.e., contractor costs) and not on the use of facility personnel. The costs for closure are to be based on the description of closure contained in the closure plan [Sections 264/265.142(a)].

Q: Can the owner or operator reduce the closure cost by including the recycling of waste at the facility or the sale of equipment?

A: No; Sections 264/265.142(a)(3) prohibit the owner or operator from taking into account any value from salvageable items including waste.

Q: How are the post-closure costs determined?

A: The post-closure cost estimate must be in current dollars and based on the annual cost required for proper post-closure maintenance according to the post-closure plan. The annual cost is then multiplied by the established post-closure care period (e.g., 30 years). Facilities that intend to clean close are not required to maintain a post-closure cost estimate (Sections 264/265.144).

Q: Does the closure and post-closure cost estimate have to be updated?

A: Yes; the closure and post-closure cost estimate must be revised yearly to account for inflation using the annual inflation factor. This value may be obtained from either the RCRA/Superfund hotline (see Appendix G) or by dividing the latest annual implicit price deflator for the gross national product by the previous annual deflator. The deflators are published by the U.S. Department of Commerce in its *Survey of Current Business* [Sections 264/265.142(b) and 264/265.144(b)].

Q: When must these cost updates occur?

A: The closure and post-closure cost estimate must be updated:

- Within 30 days after the close of the firm's fiscal year for facilities using the financial test or corporate guarantee
- Within 60 days of the anniversary of when the first cost estimate was made for all other financial mechanisms

Q: What types of financial assurance mechanisms can be used?

A: Owners or operators may use any of several mechanisms to satisfy the requirements for financial assurance of closure and post-closure care (Sections 264/265.143 and 264/265.145). These include:

- Trust funds
- Surety bonds
- Letters of credit
- Closure and post-closure insurance
- Financial test and corporate guarantee
- State-required mechanisms and state guarantees

Q: Can more than one mechanism be used to satisfy the requirements?

A: Yes; more than one mechanism may be used to provide financial assurance for a facility. All the mechanisms may be used to cover a portion of a closure or post-closure cost estimate except the following: one type of surety bond (the bond guaranteeing performance of closure or post-closure care) and the financial test and corporate guarantee; these mechanisms, if used, must always be for the entire closure or post-closure cost estimate [Sections 264/265.143(g) and 264/265.145(g)].

Q: What amount must the mechanism cover?

A: Assurance must be established for the whole amount of the current closure or post-closure cost estimate except when a trust fund is used. A trust fund may be built up to the amount of the cost estimate over a "pay-in" period. Following an increase in the cost estimate, the owner or operator must increase the amount of funds assured.

Trust Fund

Q: What is a trust fund?

A: A trust is an arrangement in which one party, the grantor, transfers money to another party, the trustee, who manages the money for the benefit of one or more beneficiaries. For the purposes of this section, the facility owner or operator is the grantor; the financial institution is the trustee; and EPA is the beneficiary.

These trusts are irrevocable; they cannot be altered or terminated by the owner or operator without the consent of EPA and the financial institution. The trust is established when the trust agreement is signed by the grantor and the financial institution [Sections 264/265.143(a) and 264/265.145(a)].

Q: What are the trustee's requirements concerning a trust fund?

A: The trustee must carry out responsibilities that are described in the trust agreement. The agreement and the accompanying "certificate of acknowledgment" (a notary's statement attesting to the identity of the owner's or operator's representative signing the trust agreement) must be submitted to EPA.

Among the trustee's responsibilities are: investing the funds, providing an annual valuation to the owner or operator and to EPA, notifying EPA if a payment into the fund is not made during the pay-in period, and releasing funds as directed by EPA. In investing the funds, the trustee will follow the general guidance of the owner or operator as long as it is in accordance with the trust provisions.

Q: How long is the pay-in period?

A: For permitted facilities, payments must be made annually over the term of the initial RCRA permit (10 years). For interm status facilities, payments are made annually over 20 years or the remaining operating life of the facility, whichever is shorter. The operating life of the facility is determined by using the expected year of closure, which should be identified in the closure plan [Sections 264/265.143(g) and 264/265.145(g)(3)].

Q: If an interim status facility receives a permit during the pay-in period, does the length of the pay-in period change?

A: Yes; if the facility subsequently receives a permit, the pay-in period is adjusted accordingly, e.g., 10 years.

Q: If an owner or operator completes the pay-in period, are other payments required?

A: After the pay-in period is complete, more payments must be made if the closure cost estimate increases or inflation factoring shows a greater amount is needed. However, if the amount in the trust fund is greater than needed for closure, the owner or operator may submit to EPA a written request to release excess money.

After beginning partial or final closure, an owner or operator may submit a request for reimbursement for closure expenditures to EPA, using itemized bills [Sections 264/265.143(g)(6) and 264/265.145(g)(6)].

Surety Bond

Q: What is a surety bond?

A: A surety bond is a guarantee by a surety company that obligations in the bond, such as closure and/or post-closure, will be fulfilled.

Q: What are the requirements for using a surety bond?

A: If an owner or operator uses a surety bond or a letter of credit, a standby trust fund (essentially the same as the trust fund) must be established. In most cases, a standby trust fund is established with an initial nominal fee agreed on by the owner or operator and the trustee. Further payments into this fund are not required until the standby trust is funded by a surety company as required. The surety company must be listed as an acceptable surety in *Circular 570* of the U.S. Department of Treasury [Sections 264/265.143(b) and 264/265.145(b)].

If an administrative order is issued to compel closure, or closure is required by an appropriate court order, the amount equal to the penalty sum must be placed into the standby trust fund within 15 days of issuance.

Letter of Credit

Q: What is a letter of credit?

A: A letter of credit is a letter certifying that a named person is entitled to draw on the person's credit up to a specified amount.

Q: What are the requirements for using a letter of credit?

A: In accordance with Sections 264/265.143(c) and 264/265.145(c), the letter must be from a bank or other institution with authority to issue letters of credit and whose letter-of-credit operations are regulated and examined by a state or federal agency. The letter will entitle the EPA regional administrator to direct the issuing institution to deposit funds into the owner's or operator's standby trust fund if the owner or operator fails to fulfill closure or post-closure requirements.

The letter of credit must be irrevocable and issued for 1 year. The letter must have an automatic extension unless the issuing institution notifies both EPA and the facility owner or operator at least 120 days prior to the expiration date that the letter will not be extended. The letter of credit must also establish a standby trust fund.

An owner or operator using a letter of credit for financial assurance must also submit to EPA a standby trust agreement letter identifying the facilities and estimated costs covered by the letter of credit.

If the facility fails to perform final closure in accordance with the approved closure plan, EPA may draw upon the letter of credit to ensure proper closure.

Q: The operator of a hazardous waste management facility established a letter of credit and a standby trust fund (containing $1 to keep it active) in accordance with the financial responsibility requirements. The trustee (i.e., the bank) then levied a $1,500 per annum service charge on the standby trust fund. Does RCRA prescribe service charge rates for standby trust funds or control the service charge in any way?

A: No; RCRA only prescribes the mechanisms that can be used to meet the financial requirements.

Insurance

Q: Can insurance be used for closure and post-closure?

A: Yes; insurance may be purchased that assures funds for closure and post-closure care. It must be issued by a company that is licensed to transact the business of insurance or eligible as an excess or surplus lines insurer in one or more states. As evidence of this insurance, the owner or operator must submit to EPA a certificate of insurance issued by the insurer [Sections 264/265.143(e) and 264/265.145(e)].

Q: What are the requirements for using insurance?

A: The policy must be issued with a face amount equal to at least the current cost estimate for closure or post-closure care, unless the policy covers only part of the estimated cost and the rest is covered by another financial instrument. When the cost estimate increases, the face amount of the policy must be increased by the owner or operator, unless the increase is covered by another instrument. When the estimate decreases, the face amount may be decreased following written approval by EPA.

The owner or operator must continue to make premium payments that are due unless alternate financial assurance, as specified in the regulations, is substituted. Failure to pay the premium without alternate financial assurance will constitute violation of the regulations, a violation that begins upon receipt by EPA of a notice of cancellation, termination, or nonrenewal [Sections 264/265.143(e) and 264/265.145(e)].

Q: Can the insurer cancel or terminate a policy for a facility?

A: Yes; the insurer may cancel, terminate, or fail to renew the policy only if the premium is not paid. The automatic renewal of the policy must, at a minimum, provide the insured with the option of renewal at the face amount of the expiring policy. If the cost estimates to which the policy applies have increased, the insurer and insured may agree to cover that increase in the renewal policy [Sections 264/265.143(e)(6) and 264/265.145(e)(6)].

Q: What is required if the insurer wants to cancel or terminate a policy?

A: To cancel, terminate, or not renew the policy upon nonpayment of premium, the insurer must provide 120 days' notice to the owner or operator and EPA, by certified mail; however, cancellation, termination, or nonrenewal may not occur if by the expiration date: EPA deems the facility to be abandoned; EPA terminates interim status or the permit, whichever is in effect; closure is ordered by EPA or a U.S. district court or other court of competent jurisdiction; the owner or operator is named as a debtor in bankruptcy proceedings; or the premium is paid [Sections 264/265.143(e)(8) and Sections 264/265.145(e)(8)].

Q: Can a facility owner or operator cancel an insurance policy?

A: Yes; the owner or operator may cancel the policy if EPA gives written consent based on his receipt of alternate financial assurance that meets the requirements of the regulations or on completion of the closure or post-closure obligations [Sections 264/265.143(e)(11) and 264/265.145(e)(11)].

*Q: **What is the difference between** cancellation **and** termination **of an insurance policy?***

A: Cancellation occurs during the active life of the policy (i.e., cancellation for nonpayment of the premium). Termination occurs when a policy runs its course and is not renewed.

Financial Test and Guarantee

Q: What is the financial test and guarantee?

A: The owner or operator of a facility can assure, through means of a financial test, that sufficient funds exist within the company to pay for closure and post-closure activities.

Q: What is required for the financial test?

A: Evidence of passing a test of financial strength uses two alternative sets of test criteria, as specified in Sections 264/265.143(f) and 264/265.145(f).

The first set of criteria is:

A. Two of the following three ratios:
   1. Total liabilities/net worth = less than 2
   2. Net income plus depreciation, depletion, and amortization/total liabilities = greater than 0.1
   3. Current assets/current liabilities = greater than 1.5
B. Net working capital and tangible net worth each at least six times the sum of closure and post-closure cost estimates covered by the test
C. Tangible net worth of at least $10 million
D. U.S. assets amounting to at least 90 percent of total assets or at least six times the sum of closure and post-closure cost estimates covered by the test

The alternative set of criteria is:

A. Bond rating of most recent bond issuance within the highest four categories of ratings by Standard and Poor's or Moody's
B. Tangible net worth at least six times the sum of the closure and post-closure cost estimates covered by the test
C. Tangible net worth of at least $10 million
D. U.S. assets amounting to at least 90 percent of total assets or at least six times the sum of closure and post-closure cost estimates covered by the test

Q: How is the evidence documented?

A: To demonstrate that the owner or operator meets this test, data must be submitted each year from independently audited financial statements, a special report from an independent auditor confirming that the data is consistent with those in the audited statements, and a copy of the standard auditor's report accompanying the annual financial statements. Qualifications in the auditor's opinion may be grounds for disallowance. Upon failing the test or being disallowed from using the test, the owner or operator must provide substitute assurance within the periods specified in the regulations [Sections 264/265.143(e)(3) and 264/265.145(e)(3).

Q: What is the corporate guarantee?

A: If an owner's or operator's parent corporation passes the financial test, the parent's guarantee of closure or post-closure costs may be used as financial assurance [Section 264/265.143(e)(10) and 264/265.145(e)(3)].

Q: If a facility becomes a separate company, completely autonomous from the parent company, may the exparent company provide financial assurance for the owner or operator of the newly independent company?

A: No; the exparent company may not provide financial assurance for the newly independent company. Sections 264.143(f)(10) and 265. 143(e)(10) state: "The guarantor must be the parent corporation of the owner or operator." Therefore, the newly independent company must establish its own financial assurance, because its exparent company can no longer function as its guarantor. This financial assurance must be in place upon independence.

State-Required Mechanisms

Q: Can a facility use state-required mechanisms if they satisfy the federal dollar requirements?

A: Yes; states in which EPA is administering the RCRA financial requirements may have their own requirements for financial assurance of closure or post-closure care. The federal regulations allow owners or operators to use a state-required mechanism to satisfy the federal requirements if it provides assurance equivalent to that of mechanisms specified in the federal regulations. The owner or operator must submit to EPA evidence of having established the state-required mechanism and a letter requesting that it be considered acceptable for meeting the federal requirements.

Similarly, if a state assumes responsibility for closure or post-closure care or for the costs of these obligations, such guarantees may be used by owners or operators to satisfy the federal requirements, if they provide assurance equivalent to that of mechanisms specified in the regulations. The owner or operator must submit a letter from the state describing the assumption of responsibility and a letter requesting that the state's guarantee be considered acceptable for satisfying the federal requirements (Sections 264/265.149).

Q: If a state does not have a required financial mechanism, but has an approved financial mechanism, can a facility owner or operator use the approved mechanism?

A: Yes; Sections 264/265.149 allow the substitution of a financial mechanism that the state requires for one of the EPA-approved mechanisms. This substitution requires the approval of EPA. A state-approved (but not required) mechanism can also be used in lieu of the federal mechanism if the facility owner or operator receives approval from EPA.

Liability Coverage

Q: What is required for liability coverage?

A: Owners and operators of hazardous waste management facilities must demonstrate liability coverage during the operating life of a facility for bodily injury and property damage to third parties resulting from facility operations. Under the liability coverage regulations (Sections 264/265.147), owners and operators of all hazardous waste management facilities are required to demonstrate, on an owner/operator (per firm) basis, adequate liability coverage.

Q: If an owner or operator has multiple hazardous waste management facilities, does the coverage stay the same?

A: Yes; the requirements are for a per firm basis, thus, the coverage stays the same regardless of the number of waste management facilities (47 *FR* 16546, April 16, 1982). For example, an owner of four container storage facilities would still need only sudden accidental coverage of $1 million per occurrence with an annual aggregate of $2 million.

Q: What are the dollar requirements for liability coverage?

A: Hazardous waste management facilities must have liability coverage for sudden accidental occurrences in the amount of $1 million per occurrence, with an annual aggregate of at least $2 million, exclusive of legal defense costs. In addition, any facility with a surface impoundment, landfill, or land treatment must have liability coverage for nonsudden accidental occurrences in the amount of $3 million per occurrence and $6 million annual aggregate, exclusive of legal defense costs [Sections 264/265.147(b)].

Q: Can an owner or operator of a facility with a surface impoundment, landfill, or land treatment facility combine coverage for sudden and nonsudden accidental occurrences?

A: Yes; according to Sections 264/265.147(b), owners or operators may combine coverage levels for sudden and nonsudden accidental occurrences by maintaining liability coverage in the amount of at least $4 million per occurrence and $8 million aggregate.

Q: What is an accidental occurrence? Nonsudden accidental occurrence? Sudden accidental occurrence?

A: An *accidental occurrence* is an accident that is neither expected nor intended; a *nonsudden accidental occurrence* is an accidental occurrence that takes place over a span of time and involves continuous or repeated exposure, such as a hazardous waste leaching into groundwater; and a *sudden accidental occurrence* is an accidental occurrence that is not continuous or repeated, such as a fire or an explosion.

Q: What is the period of coverage for liability?

A: Liability coverage must be continuously provided for a facility until the certification of closure is received by EPA [Sections 264/265.147(e)].

Q: What are the coverage options for liability?

A: An owner or operator may meet the requirements of liability coverage by:

- Financial test
- Corporate guarantee
- Insurance
- Letter of credit
- Surety bond
- Trust fund
- Guarantee by firms that are not the direct parents of facility owners or operators

Q: Can an owner or operator use a combination of these financial mechanisms?

A: Yes; owners or operators may use any combination of these financial mechanisms. However, owners and operators must specify which of the combined instruments should be drawn upon first in the case of a claim by designating instruments as "primary" or "excess" coverage.

Q: Are the financial mechanisms for liability the same as for closure and post-closure care?

A: No; in contrast to the mechanisms authorized under Subtitle C for closure and post-closure care, the liability coverage mechanisms do not name EPA as their beneficiary. The issuer of the mechanism assumes the obligation to satisfy third-party liability claims for personal injury or property damage arising from operation of the facilities covered by the mechanism if the owner or operator does not do so. Third parties, and not EPA, are designated beneficiaries to ensure that the third parties are paid directly for liability claims without involvement by EPA. The issuer of the mechanism must honor all valid certified claims or judgments upon the mechanism up to the limit of the amount covered. In addition, unlike the requirements for closure and post-closure care and corrective action, EPA is not requiring the establishment of a standby trust for mechanisms issued for liability coverage. A standby trust is necessary when funds are payable to EPA, because by law, monies paid to the federal government must be deposited in the United States Treasury. Because the mechanisms will pay third parties directly, a standby trust is not necessary for liability coverage.

Chapter 6

Groundwater Monitoring

The goal of the groundwater monitoring program is to detect, identify, and clean up any releases of hazardous waste or constituents from land-disposal units that have entered an underlying aquifer in sufficient quantities to cause a "significant" change in groundwater quality.

The interim status regulations (Part 265) establish a two-stage groundwater program designed to detect and characterize the release of any hazardous waste or constituents. The Part 264 regulations, however, establish a three-stage program designed to detect, evaluate, and correct groundwater contamination from land-disposal units.

GROUNDWATER MONITORING AT INTERIM STATUS UNITS

Applicability

Q: Which owners or operators are required to implement a Part 265 groundwater monitoring program?

A: Owners or operators of interim status hazardous waste management facilities using land treatment, landfills, waste piles, and/or surface impoundments.

Q: What is the goal of the interim status groundwater monitoring program?

A: The goal of the interim status (Part 265) groundwater program is to ensure that owners and operators evaluate the impact of their facility on the groundwater quality underlying the site [Section 265.90(a)].

Q: Can a facility be required to monitor "perched" water underlying the unit?

A: Yes; a "perched" hydrologic unit can be considered the uppermost aquifer it it is *capable of yielding significant amounts of water*. This is limited by the rate (gallons per day), as well as the total storage capacity of that perched zone. Whenever both of these are locally *significant,* the monitoring requirements must be enforced in the perched aquifer, as well as all hydraulically interconnected lower aquifers.

Q: What is "significant amount" or "significant yield" of groundwater?

A: Significant yield has not been assigned a discrete number, because significance can vary from location to location. Significant yield is dependent, in part, on geologic and hydrologic conditions. For instance, one location may have abundant surface and groundwater resources, with uppermost geologic strata yielding only very small amounts. Another location may have similar upper strata, but without such rich resources. Decisions on the "significance" of the yield from these similar strata must be made in light of such regional considerations. Because of this variability, EPA has not established a minimum significant yield figure. Some EPA regions have found 20 gallons per day (gpd) to be appropriate. Other regions have used local definitions or ranges (e.g., 5–50 gpd). A discussion of significant yield is in the July 26, 1982, *Federal Register* (47 *FR* 32289).

Q: What if a facility is underlain by a hydrologic unit that does not meet the regulatory definition of an aquifer?

A: For a facility that is situated above saturated zones that do not meet the regulatory definition of *aquifer* (Section 260.10), the nonaquifer saturated zones are to be monitored. This is because the intent of the regulations is to monitor for the first sign of groundwater contamination (OSWER Directive No. 9481.02).

Q: How long is the program required to be conducted?

A: The groundwater monitoring program is to be carried out during the active life of the hazardous waste management unit. The term *active life* includes the closure and post-closure period. The post-closure period uses 30 years as a baseline, but this period can be reduced or extended by EPA when appropriate [Section 265.117(a)(1)].

Waivers

Q: Are there any waivers from the groundwater monitoring requirements?

A: Yes; some facilities may qualify for a partial or complete waiver [Section 265.90(c)] of the monitoring requirements if the owner or operator can demonstrate that there is a low potential for hazardous waste (or waste constituents) to migrate from the facility via the uppermost aquifer to surface water or water-supply wells. Only a small percentage of interim status facilities would be expected to qualify for a waiver of all monitoring requirements based on the underlying hydrogeology.

Q: Must the waiver be submitted to EPA?

A: No; the owner or operator claiming such a waiver must keep a detailed written demonstration, certified by a qualified geologist or geotechnical engineer at the facility [Section 265.90(c)].

Q: What must be contained in the demonstration?

A: Pursuant to Section 265.90(k), the written demonstration must establish the following:

- The potential for migration of hazardous waste or hazardous-waste constituents from the facility to the uppermost aquifer by an evaluation of:
  1. A water balance of precipitation, evapotranspiration, runoff, and infiltration
  2. Unsaturated zone characteristics (i.e., geologic materials, physical properties, and depth to ground water)
- The potential for hazardous waste or hazardous waste constituents that enter the uppermost aquifer to migrate to a water-supply well or surface water, by an evaluation of:
  1. Saturated zone characteristics (i.e., geologic materials, physical properties, and rate of groundwater flow)
  2. The proximity of the facility to water-supply wells or surface water

Q: What is the other waiver?

A: An additional waiver from the groundwater-monitoring program is for surface impoundments used solely to neutralize corrosive wastes. Such a waiver is not based on hydrogeological factors. This waiver is based on the premise and documentation that only corrosive wastes will be added to the impoundment and that the neutralization occurs so rapidly that the waste is no longer corrosive if it migrates out of the impoundment. This waiver must also be certified by "a qualified professional" and maintained at the facility [Section 265.90(e)].

Sampling and Analysis Plan

Q: Is an owner or operator required to establish a sampling and analysis plan?

A: Yes; the owner or operator must develop and follow a groundwater sampling and analysis plan [Section 265.92(a)]. This plan must be maintained at the facility. Owners or operators must also simultaneously prepare and maintain an outline of a *groundwater quality assessment program* [Section 265.93(a)].

Q: What are the required contents of the plan?

A: The contents of the plan must include procedures and techniques for:

- Sample collection
- Sample preservation
- Sample shipment
- Analytical procedures
- Chain-of-custody control

Q: What are the essential elements of a chain-of-custody program?

A: A chain-of-custody program should include sample labels, sample seals, a field log book, a chain-of-custody record, sample analysis request sheets, and a laboratory log book. A detailed description of each, including sample forms, is available in EPA's *Test Methods for Evaluating Solid Waste: SW-846*.

Q: What is required for the groundwater assessment outline?

A: In accordance with Section 265.93(a), the outline must describe a more comprehensive program than the detection monitoring program that is capable of determining:

- Whether hazardous waste or constituents have entered the groundwater
- The rate and extent of migration of hazardous waste or constituents in the groundwater
- The concentration of hazardous waste or constituents in the groundwater

Monitoring System

***Q:* What are the requirements for a groundwater monitoring system?**

A: A groundwater monitoring system must be installed in a manner that will yield good quality groundwater samples for analysis.

The monitoring wells must be cased in a manner that maintains the integrity of the monitoring well bore hole. This casing must be screened or perforated and, if necessary, packed with sand or gravel to enable sample collection at depths where appropriate aquifer flow zones exist. The annular space (the space between the bore hole and the well casing) above the selected sampling depth must be sealed with a suitable material, such as bentonite slurry or grout [Section 265.91(a)].

***Q:* Are there regulations for constrution methods and materials?**

A: No; but EPA provides guidance on location, construction methods and materials in its *RCRA Ground Water Monitoring Technical Enforcement Guidance Document,* OSWER Directive No. 9950.1A.

***Q:* If a facility has several waste management units adjacent to each other, must each unit have its own groundwater monitoring system?**

A: No; if a facility has more than one hazardous waste management unit, a separate groundwater monitoring system is not required provided the system is adequate to detect any discharge from any of the units. The *waste management area* in this situation could be described by an imaginary line that surrounds all the units [Section 265.91(b)].

Upgradient Wells

***Q:* What is required for upgradient wells?**

A: A minimum of one monitoring well must be installed hydraulically upgradient from the waste management area. The number, location, and depth of the well or wells must be sufficient to yield samples that are representative of the background groundwater quality in the uppermost aquifer and that are not affected by the facility [Section 265.91(a)(1)].

Q: A facility's upgradient well is drilled into bedrock and is seasonally dry. What course of action should the facility take to comply with quarterly groundwater monitoring requirements if the well is sometimes dry?

A: Because it is the owner or operator's responsibility to establish operable wells, the first consideration is whether to drill a deeper well, which is operable throughout the year (realizing that the upgradient well should be at the same relative depth as the downgradient wells, so that the same groundwater flow is monitored). As stated in John Skinner's November 30, 1983, Guidance Memorandum from EPA's Office of Solid Waste in Washington, DC, if one quarter's values are missing (at least three quarters are required), a statistical comparison should still be made by assigning values for the four missing replicates. These values may be obtained by averaging values of the other three quarters. If more than one quarter is missing, no comparison should be made.

Downgradient Wells

Q: What is required for downgradient wells?

A: A minimum of three monitoring wells must be installed hydraulically downgradient of the waste management area. Their number, location, and depth must ensure the immediate detection of any statistically significant amounts of hazardous waste or hazardous waste constituents that migrate from the facility to the uppermost aquifer [Section 265.91(a)(2)].

Q: Section 265.91 requires at least one upgradient and least three downgradient wells for a groundwater monitoring system. Is this minimum number of wells sufficient to detect contamination at most facilities?

A: In general, the required minimum number of wells (four) is not sufficient to detect contamination at typical facilities. There are many conditions that can complicate the detection of contaminants in the groundwater. Large, multicomponent, waste management areas consisting of several landfills, surface impoundments, or land treatment zones may require a greater number of monitoring wells. The minimum number of wells would be adequate only for a small facility in which the contaminants and hydrogeology were simple and well documented.

Q: A facility has two shallow downgradient groundwater monitoring wells and one deep downgradient well, which is adjacent to one of the shallow wells. Does this satisfy the requirement in Section 265.91(a)(2) to have at least three downgradient monitoring wells?

A: Because the requirement is to have at least three downgradient wells in the uppermost aquifer that will immediately detect any significant release, the "deep" well would be acceptable as the third downgradient well if it were screened within the uppermost aquifer (monitoring the heavy fraction of the plume). The "deep" well would not be acceptable if it were screened in another aquifer below the uppermost aquifer.

Detection Monitoring

Q: What is detection monitoring?

A: *Detection monitoring,* the first stage in interim status groundwater monitoring, is to determine whether a land disposal facility has leaked hazardous waste or constituents into an underlying aquifer in quantities sufficient to cause a significant change in groundwater quality.

Q: What are the requirements of detection monitoring?

A: Facilities not qualifying for a waiver from the groundwater monitoring program are required to install a basic detection monitoring system. For 1 year, the facility owner or operator must conduct quarterly sampling of wells upgradient and downgradient of the facility to account for seasonal variation. The owner or operator must analyze samples for three sets of parameters: drinking water suitability, groundwater quality, indicators of groundwater contamination, and general groundwater elevation [Section 265.92(b)]. The first year of monitoring is intended to establish baseline information (background data) on the underlying aquifer for future statistical comparison, as outlined in Figure 1.

Q: If an owner or operator assumes or knows that a unit is leaking, must the 1 year background monitoring be conducted?

A: No; if an owner or operator assumes or knows that a statistically significant increase (or pH decrease) in one or more of the specified indicator parameters could occur, the owner or operator may install, operate, and maintain an *alternate groundwater monitoring program* [Section 265.90(d)].

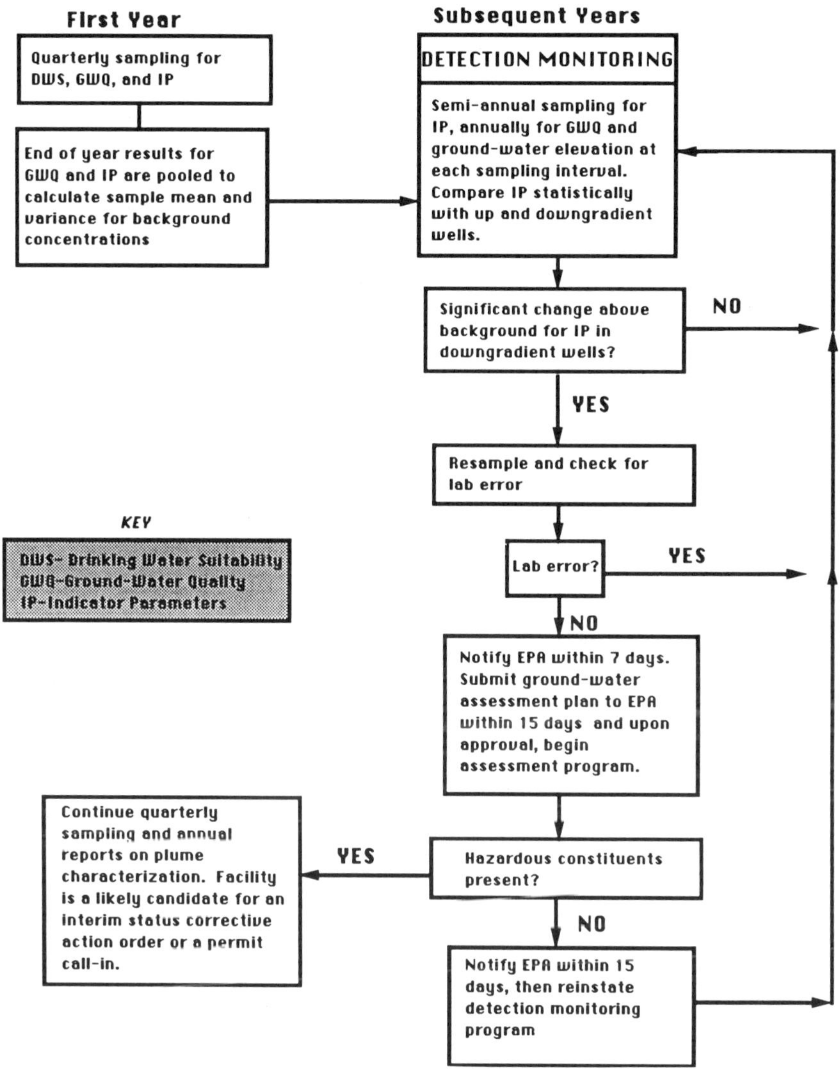

Figure 1. Interim Status Ground-Water Monitoring Program
Source: 40 CFR Part 265, Subpart F

This program allows facilities suspected or known to be discharging hazardous waste or waste constituents to groundwater to enter immediately into the assessment phase, rather than delay the assessment program a year by doing detection monitoring and background comparisons. An owner or operator who chooses the alternate groundwater monitoring program must submit to EPA a groundwater assessment plan as outlined in Section 265.93(d)(3).

Monitoring Parameters

***Q:** What parameters are used to determine if a release has occurred?*

A: To determine if a release has occurred, the owner or operator must compare monitoring data from downgradient wells to background concentration levels established at the end of the first year for drinking water suitability, groundwater quality, and indicator parameters based on quarterly measurements.

***Q:** What are the parameters for drinking water suitability?*

A: The parameters of drinking water suitability, which assess the suitability of the aquifer as a drinking water supply, are arsenic; barium; cadmium; chromium; fluoride; lead; mercury; nitrate; selenium; silver; endrin; lindane; methoxychlor; toxaphene; 2,4-D; 2,4,5-TP (Silvex); radium; gross alpha; gross beta; and coliform bacteria [Section 265.92(b)(1)].

***Q:** What are the requirements for analyzing for drinking water suitability during the first year?*

A: Within 15 days of completing each quarterly analysis, the owner or operator must report to EPA the concentrations of drinking water suitability parameters [Section 265.94(a)(2)(i)]. It must be noted in the report if any of these constituents exceed the levels established in Appendix III of Part 265.

***Q:** What are the monitoring requirements for subsequent years?*

A: After the first year, the owner or operator is not required to analyze for drinking water suitability. However, the first year's data will have to be submitted if an assessment plan is required [Section 265.93(d)(3)(iii)].

***Q:** Fluoride in Appendix III of Part 265 has a range from 1.4 to 2.4 mg/l. Are these minimum and maximum concentration levels?*

A: No; the range for fluoride was given because the measurement is temperature-dependent. The fluoride value will depend on the temperature of the sample.

Q: What happens if the first year background (upgradient) values exceed the drinking water suitability parameters?

A: The facility would be required to activate its groundwater quality assessment program and determine the source of groundwater contamination. If the source of contamination is on-site, as opposed to regionally high background values or an upgradient contamination source, enforcement action against the facility may be taken.

Q: What are the parameters of groundwater quality?

A: The parameters of groundwater quality, which are used to assess the suitability of the groundwater for other nondrinking purposes, are chloride, iron, manganese, phenols, sodium, and sulfate [Section 265.92 (b)(2)].

Q: What are the requirements for monitoring these parameters during the first year?

A: During the first year, the owner or operator must establish the initial background concentrations based on quarterly measurements.

Q: What are the requirements for subsequent years?

A: After the first year, the owner or operator must analyze samples of both upgradient and downgradient wells annually for groundwater quality. Although there is no requirement for the statistical evaluation of these parameters, the data is to be used as a basis for comparison if a groundwater quality assessment program is required by Section 265.93(d).

Q: What are the indicator parameters?

A: The indicator parameters, which are used as gross indicators of whether contamination has occurred, are pH, specific conductance, total organic carbon (TOC), and total organic halogen (TOX) [Section 265.92(b)(3)].

Q: Why are TOX and TOC both used as indicator parameters when TOX is a subset of TOC?

A: TOX was selected to measure only those organic compounds containing halogens. Despite the more limited use of TOX, it is an important indicator because of the prevalence of hazardous wastes that contain halogenated hydrocarbons. Measurement of total organic carbon (TOC) cannot discern fluctuations in TOX. Conversely, TOX analysis does not account for the full range of organic compounds likely to be encountered in the field.

Q: ***What are the requirements for the first year of monitoring for the indicator parameters?***

A: For each quarterly sampling during the first year, at least four replicate measurements must be obtained for each sample from the upgradient well or wells; then the initial background arithmetic mean and variance must be determined by pooling the replicate measurements for the respective indicator parameter concentrations. For the downgradient wells, initial background concentrations must be obtained based on quarterly sampling [Section 265.92(c)(2)].

Q: ***When four replicate measurements are required for each groundwater monitoring well sample, does this require four separate samples from the well or one sample split into four aliquots?***

A: The intent of the replicate samples is to check analytical accuracy, so one water sample should be split into four aliquots.

Q: ***Section 265.93(b) requires an initial background arithmetic mean and variance for the indicator parameters. How should the calculations be done if three of the four values are less than the detectable limits of the test?***

A: The desired detection limit for total organic carbon is 1 mg/l and for total organic halogen, 5 mg/l. These values are acceptable substitutes (HMR).

Q: ***Section 265.93 states that an arithmetic mean and variance must be calculated for each indicator parameter at each well sampled and compared with its background mean and variance. Does this mean that the comparison is made between initial values and subsequent semiannual results for a given well?***

A: No; statistical comparison is made between the initial background mean and variance of each indicator parameter derived from the first year's quarterly sampling of the upgradient well and subsequent means and variances of each indicator parameter derived from all wells, including the upgradient well (HMR).

Q: What are the monitoring requirements for subsequent years?

A: In subsequent years of monitoring, all monitoring wells must be sampled for indicator parameters semiannually. A mean and variance, based on four replicate measurements, must be determined for each of the indicator parameters.

By March 1 of each calendar year, the concentrations of each of the indicator parameters (as well as the groundwater elevations) must be reported to EPA for each monitoring well. Each monitoring well must be compared statistically with its initial background arithmetic mean established in the first year's monitoring [Section 265.93(b)]. The comparison must consider each of the wells individually in the monitoring system.

Q: Are the values obtained at each well compared with that same well's initial background value?

A: No; after background values have been established from the first year, subsequent comparisons of both upgradient and downgradient indicator parameter values are made with the initial background values of the upgradient well only.

Statistical Comparisons

Q: What is used to determine statistically significant changes?

A: Pursuant to Section 265.93(b), the owner or operator must use *Student's* t *test* to determine statistically significant changes in the concentrations of an indicator parameter in groundwater samples compared with the initial background concentration of that indicator parameter.

The comparison must consider individually each of the wells in the monitoring system. For three of the indicators' parameters (specific conductance, total organic carbon, and total organic halogen) a single-tailed Student's t test must be used at the 0.01 level of significance for increases over background. The test for pH uses a two-tailed test because pH increases or decreases can be significant.

Q: Is there a specified Student's t ***test that must be used?***

A: No; there are different types of Student's t tests available. The interim status regulations do not require the use of a specific method.

Q: What if a significant increase occurs?

A: If the comparison for the downgradient well shows a significant increase (or any change in pH), the affected wells must be resampled and samples split into two for a qualitative check for possible laboratory error. The sample may be split into four replicates and another *t* test run, but this is not required [Section 265.93(c)(2)].

If the comparison for the upgradient wells shows a significant increase (or pH decrease), the owner or operator must submit this information to EPA without a qualitative laboratory check pursuant to Section 265.94 (a)(2)(ii).

Q: What if resampling determines that the change is due to laboratory error?

A: If this resampling indicates that the change was due to laboratory error, the owner or operator should continue with the detection monitoring program [Section 265.93(d)(1)].

Q: What if resampling shows that the significant change is not a result of laboratory error?

A: If resampling shows that the significant change was not a result of laboratory error, the owner or operator must notify EPA within 7 days that contamination may have occurred. Within 15 days of that notification, the owner or operator must submit to EPA a groundwater quality assessment plan, based on the required assessment outline, certified by a qualified geologist or geotechnical engineer [Section 265.93(d)(1)].

Q: Is groundwater quality assessment triggered if there is a statistically significant change in the indicator parameters in the upgradient well?

A: No; but the owner or operator must submit this information to EPA [Section 265.93(c)(1)].

Q: What happens if a facility owner or operator never detects a significant increase?

A: If a facility never detects a statistically significant increase in indicator parameters, and an assessment plan is never done, detection monitoring for groundwater quality and indicator parameters continues on an annual and semiannual basis, respectively, through closure and throughout the 30-year post-closure period.

Assessment Monitoring

Q: What is assessment monitoring and what are the basic requirements?

A: If a significant change in water quality is discovered during the detection monitoring phase, the owner or operator must undertake a more aggressive groundwater program called *assessment monitoring*.

Because detection monitoring parameters are nonspecific, a statistically significant change in a parameter may not necessarily represent leakage into the aquifer. Thus, the first step is to determine whether hazardous waste constituents have indeed migrated into the groundwater. Then, the owner or operator must determine the vertical and horizontal concentration profiles of all hazardous waste constituents in the plume or plumes leaking from the waste management area. In addition, the owner or operator must establish the rate and extent of contaminant migration [Section 265.93(d)(3)].

Q: What is the information developed by assessment monitoring used for?

A: The information developed from assessment monitoring is used by EPA to evaluate the need for corrective action at the facility or may form the basis for issuing an enforcement order compelling corrective action prior to issuance of a permit.

Q: How is the assessment monitoring program implemented?

A: Owners or operators required to conduct plume characterization activities for the assessment program are required to have a written assessment monitoring plan.

Q: What is required in the plan?

A: According to Section 265.93(d)(3), the required elements of the assessment monitoring plan include:

- Number, location, and depth of wells
- Sampling and analytical methods to be used
- Evaluation procedures of groundwater data
- A schedule of assessment monitoring implementation

Q: What if the assessment shows that hazardous constituents have entered the groundwater?

A: If the assessment shows that hazardous constituents have entered the groundwater, the owner or operator must continue the assessment quarterly until closure or until a permit is issued [Section 265.93(d)(7)].

By March 1 of each calendar year, the owner or operator must submit to EPA a report containing the results of the facility's assessment program, which must include the calculated (or measured) rate of migration of the hazardous waste or hazardous waste constituents for the reporting period.

If the initial assessment confirming contamination was performed during post-closure, no subsequent quarterly monitoring is required [Section 265.93(d)(7)(ii)]. However, this type of facility would be a likely candidate for a post-closure permit or an interim status corrective action order (see Chapter 11). In applying for a post-closure permit, the owner or operator is required to conduct a comprehensive groundwater characterization program to generate the information that is required for the permit application found in Section 270.14(c)(8).

Q: What if the assessment shows that no hazardous constituents have entered the groundwater?

A: If this assessment shows that no hazardous constituents have entered the groundwater, EPA must be notified within 15 days of the determination with a written report. The owner or operator may then reinstate the detection monitoring program or enter into a consent agreement with EPA to follow a revised protocol designed to avoid false triggers [Section 265.93(d)(6)].

Q: If a groundwater quality assessment program is initiated at a facility during its active life, how long does the program continue?

A: According to Section 265.93(d)(7)(i), the assessment must continue quarterly until final closure of the facility. There are no post-closure assessment requirements.

Q: Is there a corrective action or remediation program requirement for interim status facilities with groundwater contamination?

A: No; under interim status there is only detection monitoring and groundwater quality assessment. However, corrective action could be implemented if the facility were issued a permit and becomes subject to Part 264, Subpart F. Another possibility is corrective action through an enforcement action through RCRA Sections 3008(h) and 7003, as well as CERCLA Section 106.

GROUNDWATER MONITORING AT PERMITTED UNITS

Applicability

Q: What is the objective of the groundwater monitoring program for regulated units under Part 264?

A: Whereas the principal objective of the Part 265 groundwater monitoring program is to identify releases from land disposal units, the objective of the Part 264 groundwater monitoring program (for existing facilities) is to establish individual monitoring programs to characterize any leachate and to ensure that corrective action is taken to prevent leachate migration beyond the facility boundary. To achieve this goal, the regulations establish a three-step program designed to detect, evaluate, and correct groundwater contamination arising from leaks or discharges from *regulated units,* as outlined in Figure 2.

Q: What is a regulated unit?

A: A *regulated unit* is any land disposal unit that accepted waste after July 26, 1982. All regulated units are subject to the Subpart F, Part 264, groundwater monitoring requirements [Section 265.90(a)].

The previous action date for classification as a regulated unit was January 26, 1983. However, HSWA changed this date to July 26, 1982, to force facilities that had previously closed under Part 265 to be subject to the Part 264 standards, because they are more stringent than the Part 265 groundwater monitoring program.

Q: When do the Part 264 requirements apply?

A: An interim status facility must comply with the Part 265 requirements until the final administrative disposition is made on the facility's permit. When a permit is issued for a unit, it is then subject to its permit conditions, which are based on the Part 264 standards. Thus, an interim status unit that is a regulated unit follows the Part 265 groundwater monitoring program requirements only until a permit is issued (Section 264.3).

Q: How is the Part 264 program implemented?

A: The groundwater protection requirements of Part 264 define a general set of responsibilities that the owner or operator must meet; however,

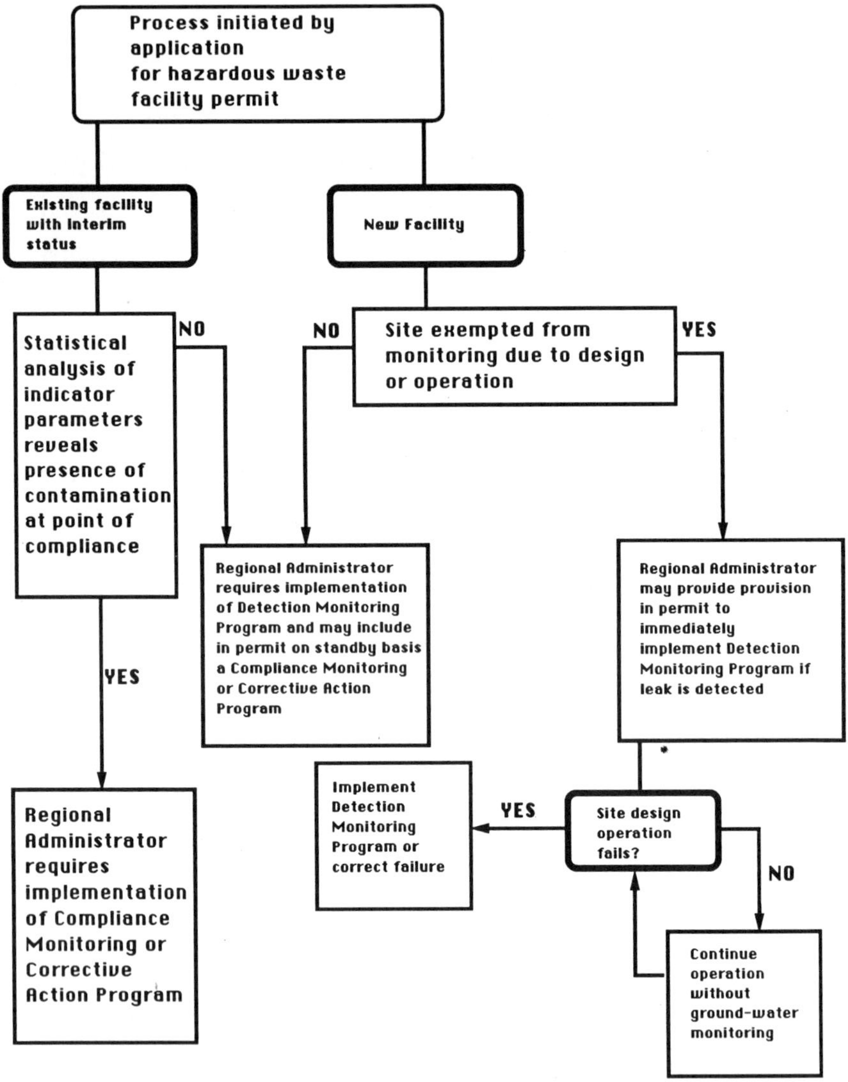

Figure 2. Part 264 Ground-Water Monitoring Program

Source: 40 CFR Part 264, Subpart F

the specific requirements are tailored to the individual facility through its permit. The permit provisions concerning groundwater are based on Sections 264.90 through 264.100 and Sections 270.14(6), (7), and (8).

The nature of the monitoring program in the permit will depend on the information available at the time of permitting. The key question is whether a regulated unit has begun to leak. For new units this is not an issue. For existing units, there should be a reliable base of information that can be used to determine whether hazardous constituents have entered the groundwater.

Variances

Q: Are there variances from the groundwater monitoring requirements of Part 264?

A: Yes; Section 264.90(b) contains four variances from the Part 264 groundwater monitoring requirements. The variances are applied for through the submittal of the Part B permit application.

Q: What are the four variances?

A: The variances are:

1. EPA finds that a regulated unit is an engineered structure, does not receive or contain free liquids or wastes containing free liquids, is designed and operated to exclude liquid, precipitation, and other run-on and run-off, has both inner and outer layers of containment enclosing the waste, has a leak detection system built into each containment layer, the owner or operator will provide continuing operation and maintenance of the leak detection systems throughout the facility's "active life," and to a reasonable degree of certainty, will not allow hazardous constituents to migrate beyond the outer containment layer prior to the end of the post-closure period [Section 264.90(b)(2)].
2. EPA finds, pursuant to Section 264.280(d), that the treatment zone of a land treatment unit does not contain hazardous constituents above background levels.
3. EPA finds that there is no potential for migration of liquid from regulated units to the uppermost aquifer during the facility's active life [Section 264.90(b)(4)].
4. The unit is a waste pile and is operated in compliance with Section 264.250(c) (a waste pile that is enclosed).

General Requirements

Q: What are the general requirements of the Part 264 groundwater program?

A: The facility must have a sufficient number of wells that represent the background water quality not affected by leakage from a unit and represent the water quality passing the point of compliance [Section 265.97(a)].

Q: ***If a facility has more than one regulated unit, are separate groundwater monitoring systems required?***

A: No; if the facility has more than one regulated unit, separate groundwater monitoring systems are not necessarily required, provided the system is adequate for all units. The *waste management area* could be described by an imaginary line circumscribing all the regulated units [Section 264.97(b)].

Q: ***What are the requirements for the sampling program?***

A: The sampling program must ensure a reliable indication of groundwater quality. The program must also include a chain-of-custody control, sample collection procedures, sample preservation and shipment, and analytical procedures [Section 264.97(d)].

Q: ***What statistical analysis is used to determine if a unit is leaking?***

A: EPA has specified five different statistical methods that an owner or operator can use [Section 264.97(h)]. They are:

1. A parametric analysis of variance (ANOVA) followed by multiple comparisons procedures to identify statistically significant evidence of contamination. The method must include estimation and testing of the contrasts between each compliance well's mean and the background mean levels for each constituent.
2. An analysis of variance (ANOVA) based on ranks followed by multiple comparisons procedures to identify statistically significant evidence of contamination. The method must include estimation and testing of the contrasts between each compliance well's median and the background median levels for each constituent.
3. A tolerance or prediction interval procedure in which an interval for each constituent is established from the distribution of the background data, and the level of each constituent in each compliance well is compared to the upper tolerance or prediction limit.
4. A control chart approach that gives control limits for each constituent.
5. Another statistical test method submitted by the owner or operator and approved by the EPA regional administrator.

Q: Isn't the **Cochran's Approximation to the Behrens-Fisher Student's t Test** *the required statistical test?*

A: No; on October 11, 1988 (53 *FR* 39720), EPA stated that there were too many problems with the test and established the five statistical tests listed above.

Detection Monitoring

Q: What is detection monitoring?

A: Detection monitoring, the first stage of the program, is composed of sampling for *indicator* parameters to determine if the unit is leaking.

Q: What indicator parameters are used?

A: Indicator parameters used may include specific conductance, total organic carbon, total organic halogen, waste constituents, or reaction products. The selected constituents, established in the permit, are determined by considering:

- The types, quantities, and concentrations of constituents in waste at the regulated unit
- The mobility, stability, and persistence of waste constituents and reaction products in the underlying unsaturated zone
- The detectability of the parameters
- The concentration or values and coefficients of variation of the parameters in the groundwater background

Q: Do all facilities start with the detection monitoring phase?

A: No; detection monitoring is implemented at units where no hazardous constituents are known to have migrated from the unit to the groundwater. Applicants who are seeking permits for new units and for interim status units that have not triggered into the assessment phase will generally qualify for the detection monitoring phase. Units that are in the assessment program of Part 265 would start with compliance monitoring [Section 264.91(a)(1)].

Q: What are the requirements for detection monitoring?

A: The detection monitoring program, like its interim status counterpart, is based on 1 year of background groundwater monitoring. Background, or upgradient, wells must be sampled quarterly for 1 year. The permittee then routinely monitors for a select set of indicator parameters specified in the permit, rather than the four indicator parameters used in the Part 265 program.

Q: Are there specific sampling requirements for detection monitoring?

A: Yes; the number and types of samples (to be specified in the permit) collected to establish background levels must be appropriate for the form of statistical test employed, following generally accepted statistical principles [Section 264.97(g)].

Q: Must units in the detection monitoring period continue to monitor their upgradient wells after the 1st year?

A: No; values from the upgradient wells are used to establish background levels during the 1st year of monitoring only. However, the downgradient wells (at the point of compliance) must continue to be monitored after the 1st year. The regulations do not explicitly require continued monitoring of the upgradient wells.

Q: What if a significant change in the indicator parameters occurs?

A: In accordance with Section 264.98(h), if it is determined that there is significant change above background levels for any of the indicator parameters, the owner or operator must:

- Notify EPA in writing within 7 days of the indicators that have changed
- Run a complete Appendix IX (see Appendix E of this book) scan for all the monitoring wells to determine the chemical composition of the leachate
- Establish a background level for each Appendix IX constituent found at each well
- Submit a permit modification application to EPA within 90 days to establish a compliance monitoring program
- Submit an engineering feasibility plan for a corrective action program

Q: If a significant increase occurs, can the owner or operator resample?

A: Yes; the owner or operator can, within 90 days, attempt to demonstrate that a source other than the regulated unit is the cause of the increase or that there is an error in sampling or analysis. EPA must be notified while the owner or operator is attempting to demonstrate these options [Section 264.98(i)].

Compliance Monitoring

Q: What is the goal of compliance monitoring?

A: The goal of the compliance monitoring program is to ensure that leakage of hazardous constituents (Appendix IX of Part 264) into the groundwater does not exceed acceptable levels.

Q: What is used to determine if constituents have exceeded acceptable levels?

A: The permit establishes the framework for a compliance monitoring program by incorporating a groundwater protection standard into the permit.

Groundwater Protection Standard

Q: What is the groundwater protection standard?

A: The *groundwater protection standard* (GWPS), established in the permit, is a standard that places a limit on the leachate from a regulated unit at the point of compliance. In other words, it is used to determine if corrective action is required.

Q: When is the groundwater protection standard established at a facility?

A: EPA will establish the groundwater protection standard in the facility's permit when hazardous constituents have entered the groundwater from a regulated unit. A groundwater protection standard is not established at a regulated unit where groundwater contamination has not been detected (Section 264.92).

Q: What are the elements of the groundwater protection standard?

A: The GWPS consists of four elements, each of which is specified in the permit. The elements are:

1. A listing of selected Appendix VIII (Part 261) hazardous constituents that could reasonably have been derived from the waste at

the facility and that are present in the groundwater. The burden of demonstrating that a particular constituent could not reasonably be derived from the facility or is incapable of posing a substantial present or potential hazard to human health or the environment lies with the owner or operator, in accordance with Sections 264.93(a) and (b).

2. The establishment of concentration limits for each hazardous constituent listed in accordance with the above. Where possible, concentration limits must be based on well-established numerical concentration limits, so as to prevent degradation of water quality unless the owner or operator can demonstrate that a higher limit will not adversely affect public health or the environment [Section 264.94(a)].

   The concentration limits are set at either:
   - Maximum contaminant levels (MCLs)
   - Alternate concentration limits (ACLs)
   - Background levels

   (Background levels are rarely used, because the GWPS is implemented only when contamination is detected. Thus, ACLs and MCLs would generally be less stringent than background levels.)
3. Establishment of the point of compliance. The *point of compliance* is the location at which the GWPS is measured. It is a vertical plane located at the hydraulically downgradient limit of the waste management area that extends down into the uppermost aquifer underlying the regulated unit (Section 264.95).
4. Establishment of the compliance period during which the GWPS applies. The compliance period is to be the number of years equal to the active life of the waste management area, including the closure period (Section 264.99).

Q: Is it possible for the compliance period to extend beyond the post-closure period?

A: Yes; the compliance period may extend beyond the post-closure period if corrective action has been initiated but not completed. If the owner or operator is engaged in corrective action, the compliance period is extended until the owner or operator can demonstrate that the groundwater protection standard has not been exceeded for 3 consecutive years [Section 264.96(c)].

Q: What are alternate concentration limits?

A: Alternate concentration limits (ACLs) generally may be established when the levels of hazardous constituents in the groundwater are found above background.

An ACL can be established provided that it will not pose a substantial or potential hazard to human health or the environment as long as the ACLs are not exceeded at the point of compliance. The ACL demonstration must justify all claims regarding the potential effects of groundwater contaminants on human health and the environment. In terms of human considerations, the regulations require assessments of toxicity, exposure pathways, and exposed populations [Section 264.94(b)].

Q: What is the general policy regarding the use of ACLs at a facility?

A: There are three basic policy guidelines established (OSWER No. 9481.00-6c) for ACLs at facilities with "usable" groundwater:

1. Groundwater containment plumes should not increase in size or concentration above allowable health or environmental exposure levels.
2. Increased facility property holdings should not be used to allow a greater ACL.
3. ACLs should not be established so as to contaminate off-site groundwater above allowable health or environmental exposure levels.

Q: In establishing an ACL, what information is required?

A: The information required to be considered in establishing ACLs [Section 264.94(b)] is:

1. Potential adverse effects on groundwater quality, considering:
   - The physical and chemical characteristics of the waste in the regulated unit, including its potential for migration
   - The hydrogeological characteristics of the facility and surrounding land
   - The quantity of groundwater and the direction of groundwater flow
   - The proximity and withdrawal rates of groundwater users
   - The current and future uses of groundwater in the area
   - The existing quality of groundwater, including other sources of contamination and their cumulative impact on the groundwater quality
   - The potential for health risks caused by human exposure to waste constituents
   - The potential damage to wildlife, crops, vegetation, and physical structures caused by exposure to waste constituents
   - The persistence and permanence of potential adverse effects
2. Potential adverse effects on hydraulically connected surface water quality, considering:

- The volume and physical and chemical characteristics of the waste in the regulated unit
- The hydrogeological characteristics of the facility and surrounding land
- The quantity and quality of groundwater, and the direction of groundwater flow
- The patterns of rainfall in the region
- The proximity of the regulated unit to surface waters
- The current and future uses of surface waters in the area and any water quality standards established for those surface waters
- The existing quality of surface water, including other sources of contamination and the cumulative impact on surface water quality
- The potential for health risks caused by human exposure to waste constituents
- The potential damage to wildlife, crops, vegetation, and physical structures caused by exposure to waste constituents
- The persistence and permanence of the potential adverse effects

Q: Section 264.94(b)(i)(iv) states that for establishing an ACL, the proximity and withdrawal rates of groundwater users must be considered. How far from the regulated unit must the owner or operator look to determine the proximity of groundwater users?

A: There is not a set distance for potential user consideration. The owner or operator must demonstrate that an ACL will be attenuated or diluted over a certain distance, so there will be no adverse impact on any potential users or on the environment (HMR).

Requirements of Compliance Monitoring

Q: What are the general requirements of compliance monitoring?

A: In accordance with Section 264.99(a), during the compliance monitoring program, the owner or operator must:

- Determine whether the regulated units are in compliance with the groundwater protection standard
- Determine the concentration of constituents at the point of compliance at least quarterly
- Determine the groundwater flow rate and direction in the uppermost aquifer at least annually

- Conduct a complete Appendix IX scan at least annually to determine if there are any new constituents at the point of compliance

Q: In the compliance monitoring program, with what frequency must the owner or operator sample downgradient wells for hazardous constituents?

A: At least quarterly during the compliance period [Section 264.99(d)].

Q: What happens if it is determined that the groundwater protection standard is being exceeded?

A: According to Section 264.99(i), if during compliance monitoring it is determined that the groundwater protection standard is being exceeded at any monitoring well, the owner or operator must:

- Notify EPA in writing within 7 days
- Submit a required permit modification application to EPA within 180 days to establish a corrective action program
- Prepare a plan for a groundwater monitoring program that will demonstrate the effectiveness of the corrective action program

Q: If a facility is in compliance monitoring, can it ever revert to detection monitoring?

A: Once a facility is triggered into compliance monitoring, it must continue for the active life of the facility, including closure (see Sections 264.99(a)(4) and 264.96). During post-closure, the facility may switch back to detection monitoring, but likely only if the levels in compliance monitoring have consistently reached background. Otherwise, the facility must remain in compliance monitoring.

Corrective Action

Q: What is the goal of the corrective action program?

A: The goal of corrective action is to bring the facility back into compliance with its groundwater protection standard (Section 264.100).

Q: What are the elements of the corrective action program?

A: The elements of the corrective action program include:

- Implementation of corrective measures to remove or treat the constituents as specified by the permit

- The time span for implementing the corrective action program as specified in the permit
- Termination of the corrective action program only upon the demonstration of meeting the groundwater protection standard
- Submittal of a written report semiannually to EPA on the effectiveness of the corrective action program

Q: When a facility is in the corrective action program, is the owner or operator still required to monitor groundwater quality?

A: Yes; in conjunction with a corrective action program, the owner or operator must implement a groundwater monitoring program at least as effective as the compliance monitoring program (in determining compliance with the groundwater protection standard) to demonstrate the effectiveness of the corrective action program [Section 264.100(d)].

Q: If the owner or operator is engaged in corrective action at the end of the specified compliance period, can the corrective action be terminated when the compliance period ends?

A: No; the compliance period is extended, and corrective action will continue until the owner or operator can demonstrate that the groundwater protection standard has not been exceeded for a period of 3 consecutive years [Sections 264.96(c) and 264.100(f)].

Q: If, during the compliance period, compliance with the groundwater protection standard is achieved, can the facility cease corrective action and revert to compliance monitoring?

A: Yes; once contamination has been reduced below the concentration limits set in the GWPS, the facility may discontinue corrective action measures and monitoring and return to the compliance monitoring program [Section 264.100(f)].

Chapter 7

Technical Standards For Waste Management Units

In addition to the general standards applicable to all hazardous waste management facilities, RCRA establishes operating requirements that are unique to each type of unit. These units include container storage; tank systems; surface impoundments; waste piles; land treatment areas; landfills; incinerators; thermal treatment processes; chemical, physical, and biological treatment processes; and miscellaneous units.

Generally, it is not the process that is regulated per se, but the type of unit through which the process occurs. For example, a facility may employ many processes that occur in tanks to treat various wastes; however, the facility would comply with the same standards for all the tanks regardless of which process is being employed.

CONTAINER STORAGE UNITS

Applicability

Q: What is the definition of a container?

A: A *container* is any portable device in which a material is stored, transported, treated, disposed of, or otherwise handled (Section 260.10).

Q. What are the requirements for managing hazardous waste in containers?

A: A container storing waste must always be closed, unless hazardous waste is being removed or added. A container must not be opened, handled, or stored in a way that might cause the contents to leak. If a container leaks or is in poor condition, the contents of that container must be placed into a sound container (Sections 264/265.171).

Q: Can drums of hazardous waste be stored in the open?

A: Yes; they can be stored in this way, provided that the drums of hazardous waste are not stored in such a manner that they may rupture or begin to leak (Sections 264/265.173).

Interim Status Requirements

Q: What are the management requirements for interim status container units?

A: The owner or operator must inspect areas where containers are stored, at least weekly, for leaks and corrosion or other indications of potential container failure. Results of these inspections are to be recorded in the facility's operating record (Section 265.174).

Waste must be compatible with a container, and waste cannot be placed in an unwashed container if it previously held an incompatible waste. Incompatible wastes cannot be placed in the same container. A container storing waste that is incompatible with other materials must be separated from these materials by a physical structure (e.g., dike, wall) or removed from the area (Section 265.172).

In addition, containers holding ignitable or reactive wastes must be stored at least 50 feet from the facility's property boundary (Section 265.176).

Q: What are the closure requirements?

A: At closure, all hazardous wastes and residues must be removed from the system. Remaining containers, liners, bases, or contaminated soil must be either removed or decontaminated (Section 265.111).

Permitted Units

Q: Are there more stringent requirements for permitted container storage units?

A: Yes; according to Section 264.175, a permitted container storage unit must have a containment system that includes:

- An impervious flooring that will allow the collection of any leakage
- Flooring that is sloped to collect any leaks or spills
- A containment area that has sufficient capacity to contain 10 percent of the volume of containers or the volume of the largest container, whichever is greater
- A system to prevent precipitation run-on into the containment system, unless the containment system has excess containment capacity
- A collection and removal system for any leaked or spilled waste

Q: Is there an exclusion from the containment system?

A: Yes; container storage areas that store only wastes that have no free liquids (i.e., that pass the paint filter test) are not required [Section 264.175(c)] to have a containment system described in the above paragraph, provided:

- The storage area is sloped to drain and remove any liquid resulting from precipitation
- The containers are elevated or protected from contact with any accumulation

However, dioxin-containing wastes must have a containment system that complies with the storage requirements for free-liquid wastes regardless of whether free liquid is present [Section 264.175(d)].

TANK SYSTEMS

Applicability

Q: What is the definition of a tank?

A: A *tank* is a stationary device, designed to contain an accumulation of hazardous waste, that is constructed primarily of nonearthen materials (e.g., wood, steel, plastic) that provide structural support (Section 260.10). The Subpart J regulations are applicable to *tank systems* (Sections 264/265.190).

Q: Weren't the regulations previously applicable only to hazardous waste tanks?

A: Yes; however, on July 14, 1986 (51 *FR* 25422), EPA promulgated sweeping regulatory changes to the existing hazardous waste tank regulations under Subpart J. These regulations established new and revised standards for *accumulation tank systems* (i.e., for on-site generators regulated under Section 262.34), interim status tank systems, and permitted tank systems.

Q: What is the definition of a tank system?

A: A *tank system* includes a hazardous waste storage or treatment tank and its ancillary equipment and secondary containment system. *Ancillary equipment* means any device including but not limited to such devices as piping, fittings, flanges, valves, and pumps that are used to distribute, meter, or control the flow of hazardous waste from its point of generation to a storage or treatment tank, between hazardous waste storage or treatment tanks to a point of disposal on-site, or to a point of shipment for disposal off-site. Tanks and ancillary equipment are all *components* of a tank system (Section 260.10).

Q: Do tanks used for collecting hazardous-waste spills need to comply with Subpart J?

A: If tanks or sumps serve as part of a secondary containment system to collect or contain releases of hazardous wastes, they are exempt [Sections 264/265.190(b)]

Q: An existing aboveground hazardous waste tank is moved to another location at the same facility. Does it become subject to new tank standards when it is moved? What would the situation be if the tank were underground?

A: For both aboveground and underground tanks, the tank would be classified as a new tank after being moved and reinstalled (51 *FR* 25446, July 14, 1986). The tank would be subject to the requirements for new tank systems. The tank would have to be reinstalled with secondary containment meeting the requirements specified in Sections 264/265.193.

Q: What are the key features of the regulatory program for tank systems?

A: The key features of EPA's regulatory program for hazardous waste tank systems are:

- To maintain the integrity of the primary containment system for both new and existing tank systems

- To require the proper installation of new tank systems
- To outline the installation of secondary containment with leak detection capabilities for new and existing tank systems
- To implement a program for adequate response to releases from tank systems
- To ensure the proper operation, maintenance, and inspection of tank systems
- To implement a program to ensure proper closure and post-closure care for tank systems

General Requirements

Q: Are there exceptions to the general requirements for hazardous waste tank systems?

A: Yes; the following are exceptions to the tank system requirements:

- Tank systems that are used to store or treat hazardous waste that contains no free liquids and that are situated on an impermeable floor (Sections 264/265.190)
- Tank systems used to manage recycled materials in a closed-loop recycling system where secondary materials are returned to the original process and tank systems used are totally enclosed, wastewater treatment, elementary neutralization, or emergency units as defined in Section 260.10
- Limited requirements apply to medium-quantity generators of between 100 and 1,000 kg/mo of hazardous waste [Sections 262.34(d) and 265.201]

Medium-Quantity Generators

Q: What are the requirements for medium-quantity generators accumulating hazardous waste in tank systems?

A: In accordance with Sections 262.34(d) and 265.201, medium-quantity generators (generators of 100 to 1,000 kg/mo) accumulating waste in a tank need to comply only with the following requirements pertaining to tank systems:

- Treatment must not generate any extreme heat, explosions, fire, fumes, mists, dusts, or gases; damage the tank's structural integrity; or threaten human health or the environment in any way

- Hazardous wastes or reagents that may cause corrosion, erosion, or structural failure must not be placed in the tank
- At least 2 feet of freeboard must be maintained in an uncovered tank, unless sufficient containment capacity is supplied
- Continuously fed tanks must have a waste-feed cutoff or by-pass system
- No ignitable, reactive, or incompatible wastes are to be placed in a tank unless these wastes are rendered nonignitable, nonreactive, or compatible
- At least once each operating day the waste-feed cutoff and by-pass systems, monitoring equipment data, and waste level must be inspected
- At least weekly, construction materials and the surrounding area of the tank system must be inspected for possible corrosion, leaks, or visible signs of erosion
- At closure, all hazardous wastes must be removed from the tank, containment system, and discharge control systems

Large-Quantity Generators

Q: What are the requirements for large-quantity generators accumulating waste for fewer than 90 days?

A: Owners and operators of 90-day accumulation tanks, accumulating hazardous waste in accordance with Section 262.34, are required to comply with most of the provisions of Subpart J of Part 265, including:

- A one-time assessment of the integrity of the tank system
- Installation standards for new tanks
- Design standards, including an assessment of corrosion potential
- Secondary containment phase-in provisions
- Closure
- Periodic leak testing, if the tank system does not have secondary containment
- Additional response requirements regarding leaks, including reporting to EPA the extent of any release and requirements for repairing or replacing leaking tanks

However, owners or operators of 90-day accumulation tanks are not required to prepare closure or post-closure plans, prepare contingent closure or post-closure plans, maintain financial responsibility, or conduct waste analysis and trial tests [Section 262.34(a)(1)].

Existing Tank Systems

Q: What are the requirements for existing tanks?

A: Interim status tank systems without secondary containment (they will eventually have to be retrofitted) must have had a written assessment of the tank's integrity completed and on file at the facility by January 12, 1988 (Sections 264/265.191). If a tank contains a solid waste that subsequently becomes a hazardous waste (e.g., newly listed), the assessment must be conducted within 12 months of the listing. The assessment, certified by an independent, qualified, registered professional engineer, must determine the adequacy of the tank's design and ensure that it will not rupture, collapse, or fail based on the following considerations:

- The tank design standards
- Hazardous characteristics of the waste
- Existing corrosion protection
- Documented age of the tank
- Results of a leak test (tank-tightness test), internal inspection, or other integrity test results

Q: What must the subsequent integrity assessments of tank systems without secondary containment include and when must the assessments be conducted?

A: All tank systems without secondary containment (as required under Sections 264/265.193) must regularly undergo a leak test or other integrity test. The leak test or other integrity test must assess the condition of both the tank and its ancillary equipment [Sections 264/265.193(i)].

Nonenterable, underground, interim status tank systems must be leak-tested at least annually [Section 265.192(b)(5)(i)]. Nonenterable, underground, permitted tank systems also must be assessed annually either through the use of a leak test or other integrity test method as approved or required by EPA [Section 264.192(b)(5)(i)].

Tanks that can be entered for inspection, and all ancillary equipment, may be tested by using either a leak test or an internal inspection. With respect to interim status tank systems, leak tests or inspections of other than nonenterable, underground tank systems must be conducted at least annually. With respect to permitted tank systems, the frequency of the leak test or internal inspection for other than nonenterable, underground, permitted tank systems must be sufficient to detect the potential for serious releases before they occur. The frequency is specified in the permit.

Internal inspections must be conducted by an independent, qualified, registered, professional engineer, and the schedule and procedure for the inspection must be adequate to detect obvious cracks, leaks, and corrosion or erosion that may lead to cracks and leaks.

Q: Does EPA specify the techniques to be used when conducting leak tests?

A: No; EPA does not specify the techniques that must be used when conducting leak tests. EPA expects owners and operators to use the most reliable methods available to assess the integrity of their hazardous waste tank systems. Current methods should be capable of detecting leaks of 0.1 gallons per hour (OSWER Directive No. 9483.00-3).

Q: Who may perform and certify the initial tank system integrity assessment and subsequent annual assessments?

A: It is the owner or operator's responsibility to determine that a tank system is not leaking or unfit for use. When conducting the initial assessment of the integrity of tank systems without appropriate secondary containment, an independent, qualified, registered, professional engineer must review and certify the owner or operator's written assessment of the tank system's integrity [Sections 264/265.191(a)].

Q: Where must a record of the initial integrity assessment and any subsequent leak tests or integrity assessments be maintained?

A: The initial certified written assessment of tank system integrity and records of any subsequent tests or inspections must be kept on file at the facility where the tanks are located. In addition, EPA maintains the right to request and inspect these assessments at any reasonable time [Sections 264/265.191(a)].

Q: Do the regulations require leak testing for existing tank systems prior to installation of secondary containment?

A: Yes; the regulations require leak testing in existing tank systems prior to installation of secondary containment. Sections 264/265.193(i) require all existing tank systems to be evaluated for leaks in some manner. Non-enterable, underground tanks must be tested for leaks at least annually.

Secondary Containment

Q: What tanks and ancillary equipment must have secondary containment and when must secondary containment be provided?

A: In accordance with Sections 264/265.193(a), all hazardous waste tanks and ancillary equipment must have secondary containment by specified dates (see Table 13) except for:

- Tanks that are used to store or treat hazardous waste containing no free liquids and that are situated inside a building with an impermeable floor (note that a concrete floor, unless covered by a sealer, is considered permeable)
- Tanks (including sumps) that serve as part of a secondary containment system to collect or contain releases of hazardous wastes
- Aboveground piping (exclusive of flanges, joints, valves, and connections) that is visually inspected for leaks daily
- Welded flanges, joints, and connections that are visually inspected for leaks daily
- Sealless or magnetic coupling pumps that are visually inspected for leaks daily

TABLE 13 Regulatory deadlines for providing secondary containment.

Tank description	Regulatory deadline
All existing tank systems (regardless of age) used to store or treat EPA hazardous waste numbers: F020, F021, F022, F023, F026, or F027.	January 12, 1989
Tank systems used to store or treat a waste that is defined as hazardous after January 12, 1987	Within two years after the date that the waste is listed as hazardous waste (to determine secondary containment deadline, substitute the date handled material becomes a hazardous waste for "January 12, 1987" in the following deadline descriptions)
Existing tank systems of known and documented age	Within two years after January 12, 1987, or when the tank system has reached 15 years of age, whichever comes later
Existing tank systems for which the age cannot be documented	Within 8 years of January 12, 1987, but if the age of the facility is greater than 7 years, secondary containment must be provided by the time the facility reaches 15 years of age or within 2 years of January 12, 1987, whichever comes later
New tank systems after July 14, 1986	Prior to putting the tank system into service

Source: 40 CFR Sections 264/265.193

- Sealless valves
- Pressurized aboveground piping systems with automatic shutoff devices (e.g., excess flow-check valves, flow metering shutdown devices, or loss of pressure-activated shutoff devices) that are visually inspected for leaks daily
- Tanks and ancillary equipment for which a variance from secondary containment requirements has been granted

Q: What is meant by welded flange?

A: For the purposes of Sections 264.193(e) and 265.193(f), EPA interprets *welded flange* to mean weld-neck, lap-joint, slip-on, and socket-weld flanges (53 *FR* 34081, September 2, 1988).

Q: Must ancillary equipment be provided with secondary containment?

A: Yes; all ancillary equipment, unless explicitly exempted, must be provided with secondary containment, in accordance with the schedule in Sections 264/265.193(a). This includes piping used to carry hazardous waste from a process tank or area to a storage or treatment tank and any other equipment used to transfer the hazardous waste to or from other hazardous waste tanks.

Q: What requirements apply to the design, installation, and operation of secondary containment systems?

A: In general, secondary containment systems must be designed, installed, and operated to prevent the migration of any wastes or accumulated liquids out of the system to the soil, groundwater, or surface water during the use of the tank system. In addition, secondary containment systems must be able to detect and collect releases and accumulated liquids until the collected material is removed [Sections 264/265.193(b)(1)].

A leak-detection system must be provided that can detect the failure of either the primary or secondary containment structure or the presence of any release of hazardous waste or accumulated liquid in the secondary containment system within 24 hours [Sections 264/265.193(b)(3)].

Accumulated liquids must be removed within 24 hours or, if the owner or operator can demonstrate to EPA that removal of accumulated liquids cannot be performed within 24 hours, as quickly as possible to prevent harm to human health and the environment [Sections 264/265.193(b) and (c)].

Q: Are there specified secondary containment systems for tanks?

A: Yes; Sections 264/265.193(d) and (e) state that secondary containment for tanks must consist of one or more of the following devices:

- External liner
- Vault
- Double-walled tank
- An equivalent device as approved by EPA

For each of the first three devices, the regulations specify additional design requirements.

Q: What are the requirements for external liner systems?

A: External liner systems must be designed or operated to contain 100 percent of the capacity of the largest tank within its boundary and prevent run-on or infiltration of precipitation into the secondary containment system, unless additional capacity sufficient to contain precipitation from a 25-year, 24-hour rainfall event is provided.

Liners must also be free of gaps and be designed and installed to surround the tank completely, such that both lateral and vertical migration are prevented. If concrete is used in the secondary containment unit, then the regulation for vaults would apply (i.e., impermeable interior coating or lining, water stops, etc., that are compatible with the storage waste must be applied) [Sections 264/265.193(e)(1)].

Q: What are the requirements for vaults?

A: Vaults must be designed or operated to contain 100 percent of the capacity of the largest tank. A vault must also be designed or operated to prevent run-on precipitation from a 25-year, 24-hour rainfall event. In addition, vaults are also required to have chemical resistant water stops at all joints (if any), be provided with an impermeable interior coating or lining that is compatible with the stored waste and will prevent any waste from migrating into the concrete, have a means to prevent the formation and ignition of vapors in the vault if the waste is ignitable or reactive and may form ignitable or explosive vapor, and be provided with an exterior moisture barrier if the vault is subject to hydraulic pressure [Sections 264/265.193(e)(2)].

Q: The vault system must be designed or operated to contain 100 percent of the capacity of the largest tank within its boundary. If the largest tank within the boundary contains nonhazardous waste, must the vault be designed to contain the capacity of the nonhazardous waste tank or the capacity of the largest hazardous waste tank?

A: The regulations are not applicable to tanks containing nonhazardous waste; therefore the vault must be designed to contain 100 percent of the capacity of the largest hazardous waste tank.

Q: What are the requirements for double-walled tanks?

A: Double-walled tanks must be designed as an integral structure so that the outer shell will retain any releases from the inner tank. If constructed of metal, the primary tank interior and the external surface of the outer shell must be cathodically protected and provided with a continuous leak-detection system [Sections 264/265.193(e)(3)].

Q: What are the requirements for ancillary equipment?

A: Secondary containment for ancillary equipment (e.g., trenching, jacketing, or double-walled piping) must satisfy the minimum requirements described above. Unlike the requirements for tanks, no particular methods are specified or required for ancillary equipment [Sections 264/265.193(f)].

Q: Can an owner or operator receive a variance from the secondary containment requirement?

A: Yes; an owner or operator can receive either a technology-based variance or a risk-based variance from the secondary containment requirement [Sections 264/265.193(g)].

Q: What is required for a technology-based variance?

A: To receive a technology-based variance, the owner or operator must demonstrate to EPA that alternative design and operating practices, together with location characteristics, will prevent migration of hazardous waste into ground- or surface water at least as well as secondary containment [Sections 264/265.193(g)(1)].

For further information, consult EPA's *Technical Resource Document for Obtaining Variances from the Secondary Containment Requirements for Hazardous Waste Tank Systems:Volume I—Technology-Based Variance*, February 1987.

Q: What is required for a risk-based variance?

A: To receive a risk-based variance, the owner or operator must demonstrate to EPA that, if release from a hazardous waste tank system occurs, the resulting level of contamination in the environment will not pose a substantial present or future hazard to human health or the environment. To do so, the applicant must demonstrate that, as a result of environmental conditions at the site and the characteristics and concentrations of hazardous constituents present in the hazardous waste in the tank system, no substantial current or future hazard to human health or the environment will result due to either no exposure or an acceptable level of exposure to the hazardous waste. This variance provision is not available to owners or operators of new, underground, hazardous waste tank systems [Sections 264/265.193(g)(2)].

For further information, consult EPA's *Technical Resource Document for Obtaining Variances from Secondary Containment Requirements for Hazardous Waste Tank Systems:Volume II—Risk-Based Variance,* February 1987.

General Operating Requirements

Q: What are the general operating requirements for tank systems?

A: In accordance with Sections 264/265.194, the following requirements are general operating requirements applicable to tank systems:

- Incompatible wastes or reagents must not be placed into the tank system
- No ignitable, reactive, or incompatible wastes are to be placed in a tank unless they are rendered nonignitable, nonreactive, or compatible
- The tank must have spill prevention (e.g., check valves) and overfill prevention controls
- Sufficient freeboard (distance between the top of the tank and the surface of the waste) must be maintained to prevent overtopping by precipitation or wind action

Q: Are there inspection requirements?

A: Yes; pursuant to Sections 264/265.194, the owner or operator must:

- At least once each operating day, inspect above-ground portions of the tank system for corrosion or any release of waste and inspect data from monitoring and leak detection equipment
- Inspect the construction materials and the area around the accessible portion of the tank system, including the secondary containment system, to detect erosion or releases of waste

- Develop and follow a schedule for inspecting overfill controls
- Inspect cathodic protection systems if present; and the proper operation of the cathodic protection system must be confirmed within 6 months of its installation and checked annually thereafter
- Inspect or test as appropriate sources of impressed current, at least every other month
- Document in the operating record of the facility that the required inspections were completed

Q: Would video monitoring of the above-ground portions of a tank system meet the daily inspection requirements?

A: Yes; provided the system would have a level of performance comparable to actual close-up visual inspection of the entire system and the capability of effectively detecting leaks within 24 hours (HMR).

Q: Where an owner cannot inspect the bottom of a tank (e.g., existing flat-bottom tank sitting on a concrete pad), is an inspection of the visible portions of the tank a satisfactory method for detecting leaks and corrosion?

A: Yes; in the case where the tank bottom is obscured from view (e.g., bottom sitting on a concrete pad) such an inspection is not required. However, special efforts should be made to carefully observe for any leakage around the base of the tank, possibly indicating releases from the tank bottom. Furthermore, when secondary containment is provided, the owner or operator must provide a leak detection system capable of detecting any release from the tank bottom (OSWER Directive No. 9483-00-3).

Q: What procedures must an owner or operator follow upon the discovery of a leak or spill?

A: In accordance with Sections 264/265.196, if a leak occurs from a tank system or secondary containment system, the owner or operator must:

- Immediately stop the flow or addition of wastes into the tank system or secondary containment system
- Promptly remove the waste from the tank system or secondary containment system. Removal procedures are as follows:
  - — Releases from the primary tank system—Within 24 hours, the owner or operator must remove as much of the waste as necessary to prevent further release of hazardous waste to the environment. If the owner or operator can demonstrate that it is not

possible to remove the waste within 24 hours, then the wastes must be removed at the earliest practicable time.
— Releases to the secondary containment system—Within 24 hours, or in as timely a manner as possible to prevent harm to human health and the environment, the owner or operator must remove all wastes from the secondary containment system.

- Prevent further migration of the leak or spill to soils or surface water and remove and properly dispose of any visible contamination of the soil or surface water
- Notify EPA; however, releases that are contained within the secondary containment system need not be reported

Q: What are the notification requirements concerning a leak or spill?

A: If a leak or spill of hazardous waste is less than or equal to 1 pound, and if it is immediately contained and cleaned up, the owner or operator does not have to notify EPA [Sections 264/265.196(d)]. Otherwise, the following notification requirements apply:

- Within 24 hours after a leak is detected, an owner or operator must notify EPA
- Within 30 days after a leak is detected, an owner or operator must submit a report to EPA with the following information:
  — Likely migration route of the leaked or spilled hazardous material
  — Characteristics of the surrounding soil (i.e., soil composition, geology, hydrogeology, and climate)
  — Results of any monitoring or sampling conducted in connection with the release (if available). If results are not available until after 30 days, they must be submitted to EPA as soon as they become available.
  — Proximity of the release to downgradient drinking water, surface water, and population areas
  — Description of response actions taken or planned

Q: What must an owner or operator do to return a leaking tank system to service?

A: If the tank system was taken out of service in response to a spill or overflow that did not damage the integrity of the system, the tank system may be returned to service as soon as the released waste is removed. However, if the cause of the release was a leak from the primary tank system, then the tank must be repaired before it can be returned to service [Sections 264/265.196(e)].

Q: What are the requirements for putting ignitable or reactive wastes in tank systems?

A: Ignitable or reactive wastes must not be placed in tank systems except in the following cases:

- The waste is treated, rendered, or mixed before or immediately after placement in the tank system, so that the waste is no longer ignitable or reactive.
- The waste is stored or treated in such a way that it is protected from conditions that may cause the waste to ignite or react.
- The tank system is used only for emergencies.

The owner or operator of a facility where ignitable or reactive waste is stored or treated in tanks must maintain protective distances between the waste management area and public ways and streets (Sections 264/265.198).

Q: Are there special requirements for the management of incompatible wastes?

A: Yes; incompatible wastes or materials must not be placed in the same tank system unless the owner or operator has taken precautions to prevent reactions that generate extreme heat or pressure, produce uncontrolled toxic gases in sufficient quantities to threaten human health or the environment, produce uncontrolled flammable fumes in sufficient quantities to pose a risk of fire, or damage the structural integrity of the device or facility (Sections 264/265.199).

Similarly, hazardous wastes must not be placed in a tank system that has not been decontaminated and that previously held an incompatible waste or material, unless the owner or operator has taken precautions to prevent any reactions.

Closure

Q: What general procedures must be followed by an owner or operator when closing a tank system?

A: The owner or operator must remove or decontaminate all waste residues, contaminated system components (i.e., secondary containment liners or vaults), contaminated soil, and structures and equipment contaminated with hazardous waste. This decontamination requirement includes all components of the tank system, not just the tank. The decontamination requirement also applies explicitly to all soils contaminated by releases from the unit, including saturated soils (Sections 264/265.197).

New Tank Systems

Q: What design considerations must be followed for new tank systems?

A: Owners and operators of new tank systems (including new accumulation tank systems) must obtain a written assessment, reviewed and certified by an independent, registered, professional engineer, attesting that the tank system has sufficient structural integrity and is acceptable for storing or treating hazardous waste [Sections 264/265.192(a)].

Q: What are the requirements for this assessment?

A: This assessment must be presented to EPA at the time of submittal of the Part B permit application, except that generators accumulating hazardous waste in new tank systems must maintain the assessment on file [Sections 264/265.192(a)].

The assessment must include, at a minimum, the following information:

- Design standard(s) according to which tank(s) and ancillary equipment are constructed
- Hazardous characteristics of the waste(s) to be handled
- For new tank systems where the external shell of a metal tank or any external metal component of the system will be in contact with the soil or water, a determination by a corrosion expert regarding the potential for corrosion and appropriate corrosion protection needed to ensure the integrity of the tank system during its use
- For underground tank system components that are likely to be adversely affected by vehicular traffic, a determination of the design and operational measures that will protect the tank system against potential damage
- Considerations made in the tank system design to ensure that tank foundations will maintain the load of a full tank, that tank systems will be anchored to prevent flotation or dislodgment, and that tank systems will withstand frost heave

Q: What installation procedures must be followed for new tank systems?

A: The owner or operator of a new tank system must ensure that proper handling procedures are used to prevent damage to the systems during installation.

Prior to covering, enclosing, or placing a new tank system in use, an independent, qualified, installation inspector or an independent, qualified, registered professional engineer, either of whom is trained and experienced in the proper installation of tank systems, must inspect the system for any weld breaks, punctures, scrapes of protective coatings, cracks, corrosion, or other structural damage, or signs of inadequate construction or installation [Sections 264/265.192(b)].

In addition, Sections 264/265.192(c) through 264/265.192(g), require:

- Backfill material for new underground tanks must be noncorrosive, porous, and homogeneous and must be placed completely around the tank to ensure even support.
- All new tanks and ancillary equipment must be tested for tightness prior to being covered, enclosed, or placed in use. If the tank is found not to be tight, all repairs to remedy leaks in the system must be made prior to being covered, enclosed, or placed into use.
- Ancillary equipment must be supported and protected against damage and stress caused by settlement, vibration, expansion, or contraction.
- Installation of corrosion protection that is field fabricated must be supervised by an independent corrosion expert. If EPA believes that other forms of corrosion protection are necessary in addition to the requirements referred to above, the owner or operator must provide that protection.
- The owner or operator must keep written statements on file at the facility from the qualified, independent, registered professional engineers or installation inspectors who are required to certify the design and proper installation of tank systems.

SURFACE IMPOUNDMENTS

Applicability

Q: What is a surface impoundment?

A: A *surface impoundment* is a facility or part of a facility that is a natural topographic depression, man-made excavation, or diked area formed primarily of earthen materials (although it may be lined with man-made materials) that is designed to hold an accumulation of liquid wastes or wastes containing free liquids and that is not an injection well. Examples of surface impoundments are holding, storage, settling, and aeration pits; ponds; and lagoons (Section 260.10).

For regulatory purposes, including the land-disposal ban, it is important to distinguish the regulatory difference between a tank and a surface impoundment. If all the surrounding earthen material is removed from a unit and the unit maintains its structural integrity, it is a tank. However, if the unit does not maintain its structural integrity without support from surrounding earthen material, it is a surface impoundment (RIL No. 110).

Q: Are all surface impoundments regulated the same?

A: No; there is a distinction between a storage surface impoundment and a disposal surface impoundment. Generally, a storage surface impoundment will be clean closed, whereas a disposal surface impoundment will be closed as a landfill.

Interim Status Units

Q: What are the general operating requirements for interim status impoundments?

A: An impoundment must have at least 2 feet of freeboard (the space between the surface of the waste and the top of the unit), which must be inspected daily to ensure compliance. The surface impoundment, including the dike and surrounding vegetation, must be inspected at least weekly to detect any leaks, deterioration, or failures in the impoundment (Section 265.222).

All earthen dikes must have a protective cover, such as grass, shale, or rock, to minimize wind and water erosion, as well as to preserve the structural integrity of the dike.

Q: What are the waste analysis requirements?

A: In addition to the waste analysis program required by the general standards, whenever a surface impoundment is to be used to chemically treat a waste different from that previously treated or to employ a substantially different method to treat the waste, the owner or operator of the impoundment must conduct waste analyses and treatment tests (bench or pilot scale) or obtain written, documented information on similar treatment of similar waste under similar conditions to show that this treatment will not create any extreme heat, fire, explosion, violent reactions, toxic mists, fumes, gases, dusts, or damage the structural integrity of the impoundment. The documentation or test results must be entered into the facility's operating record (Section 265.225).

Q: Are liners required?

A: Yes; any new expansion or replacement of an interim status surface impoundment unit requires a double-liner system that contains a leachate collection and removal system that is in compliance with the minimum technological requirements of HSWA. Existing interim status surface impoundments must have been retrofitted to comply with the minimum technological requirements by November 8, 1988 (Sections 265.221).

Q: What are the closure requirements for impoundments?

A: At closure, the owner or operator may elect to remove or decontaminate all hazardous waste and waste residues or close as a landfill. Removal or decontamination of the underlying and surrounding contaminated soil must include the removal of any contaminated groundwater (Section 265.228). This is known as a *clean closure* or closure by removal (see Chapter 5).

Q: What if clean closure is not possible?

A: If a clean closure cannot be attained, the closure plan must be modified and approved to close the surface impoundment as a landfill [Section 265.228(c)]. A post-closure permit will be required. When an owner or operator applies for a permit, corrective action provisions will be required [i.e., 3004(u)] (see Chapter 10).

Permitted Units

Q: What are the liner requirements for permitted impoundments?

A: All permitted surface impoundments must have a liner for all portions of the impoundment. The liner must be constructed in a manner to prevent any migration of wastes to the surrounding environment. Any new, expanded or replaced surface impoundment must have a double-liner system that contains a leachate collection and removal system that is in compliance with the minimum technological requirements of HSWA.

Q: Is there a variance from the double-liner requirement?

A: Yes; the double-liner requirement may be waived if the unit is a monofill that handles only foundry wastes that are hazardous solely due to the toxicity characteristic and has at least one liner [Section 264.221(e)].

Q: What are the inspection requirements?

A: The impoundment must be inspected weekly to detect any possible leakage [Section 265.226(b)].

Q: What if leakage is suspected or occurs at a surface impoundment?

A: If the dike leaks or the level of liquid suddenly drops without known cause, the impoundment must be taken out of service, in which case the waste inflow must be stopped, any surface leakage must be contained, and the leak must be immediately stopped. If the leak cannot be stopped, the contents must be removed. If any leakage enters the leak detection-system, EPA must be notified, in writing, within 7 days. The impoundment cannot be placed back in service until all repairs are completed (Section 264.227).

Q: Can ignitable or reactive waste be placed in an impoundment?

A: No; ignitable or reactive wastes may not be placed in an impoundment unless the waste no longer retains the ignitable or reactive characteristic (Section 264.229).

Q: What are the closure requirements for permitted impoundments?

A: If a unit clean closed, it is no longer subject to RCRA. However, if it cannot clean close, it must be closed as a landfill. The clean closure standard of Part 264 requires that all waste and waste residues must be removed or decontaminated (Section 264.228).

WASTE PILES

General Requirements

Q: What is a waste pile?

A: A *waste pile* is a noncontainerized accumulation of solid, nonflowing waste (Section 260.10).

Interim Status Waste Piles

Q: What are the operating requirements for interim status waste piles?

A: According to Sections 265.251 and 265.253, a waste pile with interim status must comply with the following requirements:

- The pile must be protected from possible wind dispersion
- The pile must be located on an impermeable base
- No liquid may be placed on the pile
- The waste pile must be separated from other potentially incompatible materials
- A run-on control and collection system must be installed and able to prevent a flow onto the active portion of the pile from a peak discharge from a 25-year storm, and the pile must be protected from precipitation (i.e., have a roof)
- No ignitable or reactive wastes may be placed on the pile. However, once a waste is treated and is no longer ignitable or reactive, it may be placed the pile.

Q: What are the waste analysis requirements?

A: In addition to the waste analysis requirements of the general standards, the owner or operator must analyze a representative sample of the waste from each incoming shipment before adding the waste to a pile, unless the only wastes that the facility receives are amenable to piling and are compatible with each other. The waste analysis must be capable of differentiating between the types of hazardous wastes that the owner or operator placed on the pile to protect against inadvertent mixing of incompatible wastes (Section 265.252).

Q: What are the closure requirements for interim status waste piles?

A: All contaminated subsoils, liners, equipment, and structures must be removed or decontaminated in accordance with the closure plan. If not all hazardous waste and residues are removed or decontaminated, the pile must be closed as a landfill; otherwise, it may clean close (Section 265.258).

Permitted Waste Piles

Q: What are the operating requirements for permitted waste piles?

A: In accordance with Section 264.251, a permitted waste pile must have the following:

- A liner on a supporting base or foundation that prevents the migration of any waste vertically or horizontally

- A leachate collection-and-removal system, situated above the liner, that is maintained and operated to remove all leachate
- A run-on control system able to collect and control the water from at least a 24-hour, 25-year storm
- Inspections weekly and after any storm

Q: Can ignitable or reactive waste be placed on a pile?

A: No; any ignitable or reactive waste must be rendered nonignitable or nonreactive before being placed on the pile (Section 264.257).

Q: What are the closure requirements for permitted waste piles?

A: The owner or operator must remove or decontaminate all contaminated equipment, structures, liners, and soils in accordance with the closure plan. If the unit cannot clean close, it must be closed as a landfill (Section 264.258).

LAND TREATMENT

General Requirements

Q: What is land treatment?

A: *Land treatment* is the process of using the land or soil as a medium to simultaneously treat and dispose of hazardous waste. This disposal option is used primarily for the listed petroleum wastes (K048-K052).

Q: What are the general operating requirements for land treatment units?

A: In accordance with Sections 264.273 and 265.272, hazardous waste must not be placed on the land unless the waste can be made less hazardous or nonhazardous by biological degradation or chemical reactions in the soil. A treatment demonstration must be conducted prior to the application of wastes to verify that the hazardous constituents will be treated by the unit.

Monitoring, according to a written plan, must be conducted of the soil beneath the treatment area, and the resulting data must be compared with data obtained on the background concentrations of constituents in untreated soils to detect any vertical migration of hazardous wastes or constituents.

Q: At land treatment units, operators often dump the waste to be treated on the ground, and within a few hours or a day, spread it on the land treatment area. Does this dumping of waste constitute storage in a waste pile subject to regulation?

A: This process is typical at many land treatment units. It may not be viewed as storage in a waste pile if the waste is dumped on the actual treatment area and remains for only a limited time prior to spreading. If the waste is dumped in an area other than the treatment area, then it could be regulated as a waste pile or landfill (HMR).

Q: What is the waste analysis requirement for land treatment facilities?

A: Waste analyses must be conducted prior to placing wastes in or on the land to determine the concentration of:

- Any substance in the waste that exceeds the concentrations in Table I of Section 261.4 rendering the waste characteristically toxic
- Hazardous waste constituents
- Arsenic, cadmium, lead, and mercury (if food-chain crops are grown on the land)

Q: Can food-chain crops be grown on land treatment facilities?

A: The growing of food-chain crops is prohibited in a treated area containing arsenic, cadmium, lead, mercury, and other hazardous constituents, unless it is demonstrated that they would not be transferred to the food portion of the crop or occur in concentrations less than in identical groups grown on untreated soil in the same region (Sections 264/265.276).

Q: What are the closure requirements for land treatment facilities?

A: During the closure period, the owner or operator must continue the operating practices that are designed to maximize degradation, transformation, and immobilization of the wastes at the unit. Operating practices designed to maximize treatment include tilling of the soil, control of soil pH and moisture content, and fertilization. These practices must generally be continued throughout the closure period. In addition, the owner or operator must continue those practices that were designed to minimize precipitation run-off from the treatment zone and to control wind dispersion (if needed) during the closure period.

Proper closure of a land treatment unit includes the placement of a vegetative cover that is capable of maintaining growth without extensive maintenance. A vegetative cover consists of any plant material established on the treatment zone to provide protection against wind or water erosion or to aid in the treatment of hazardous constituents (Sections 264/265.280).

LANDFILLS

General Requirements

Q: What are the general requirements for landfills?

A: Landfilling has historically been the preferred means of disposing of hazardous waste. However, Congress took the position that the requirements for land disposal were inadequate to protect human health and the environment, and through HSWA it discouraged land disposal. This includes the land disposal ban as well as other land disposal restrictions (see Chapter 8).

The problems that hazardous waste landfills have presented can be divided into two broad classes, which are addressed in the interim status requirements. The first class includes fires, explosions, production of toxic fumes, and similar problems resulting from the improper management of ignitable, reactive, and incompatible wastes. Owners and operators are required to conduct waste analysis to provide enough information for proper management. Mixing of incompatible wastes is prohibited in landfill cells, and ignitable and reactive wastes may be landfilled only when they are rendered nonignitable or nonreactive.

The second type of problem is the contamination of surface and groundwater. To prevent contamination, there are required operational controls, such as prohibiting the placement of bulk, noncontainerized, liquid, hazardous wastes, nonhazardous liquid waste, or hazardous waste containing free liquids in a landfill. This prevents the formation of hazardous leachate. An exemption on disposing of nonhazardous liquids may be obtained if the only reasonably available disposal method for such liquids is a landfill or unlined surface impoundment. Similar to surface impoundments and waste piles, a new landfill unit (including expansions or replacements) must install two or more liners and a leachate collection system (one above and one between the liner) in compliance with the minimum technological requirements of HSWA.

Other measures incorporated in the interim status regulations are diversions of run-on away from the active portion of the landfill, proper closure (including a cover) and post-closure care to control erosion and the infiltration of precipitation, and crushing or shredding most landfilled containers, so that they cannot later collapse and lead to subsidence and cracking of the cover. In addition, these regulations require the collection of precipitation and other run-off from the landfill to control surface water pollution The segregation of wastes, such as acids, which could mobilize, solubilize, or dissolve other wastes or waste constituents such as heavy metals, is also required.

Q: What are the closure and post-closure requirements for landfills?

A: A final cover must be placed over a landfill at closure. The closure plan must address the functions as well as specify the design of the final cover. It is necessary to place an appropriate cover on a landfill to control the infiltration of moisture that could increase leaching and to prevent erosion or escape of contaminated soil.

There are specific requirements (Sections 264/265.310) regarding the type, depth, permeability, and number of soil layers required for the final cover. The requirements also list a minimum set of technical factors that the owner or operator must consider in addressing the control objectives.

Q: What are these factors?

A: These factors, with regard to cover design characteristics, include cover materials, surface contours, porosity and permeability, thickness, slope, run length of slope, and vegetation type. The cover design should take into account the number of soil compaction layers and the indigenous vegetation. It should avoid or make allowances for deep-rooted vegetation and should prevent water from pooling.The final cover design in many instances can simply be the placement, compaction, grading, sloping, and vegetation of on-site soils, or it could be a complex design, such as a combination of compacted clay or a membrane liner placed over a graded and sloped base and covered by topsoil and vegetation.

INCINERATORS

Applicability

Q: What is an incinerator?

A: An *incinerator* is any enclosed device using controlled flame combustion that neither meets the criteria for classification as a boiler nor is listed as an industrial furnace (Section 260.10).

Q: Are there exemptions for incinerating certain waste streams?

A: Yes; there are specified waste streams exempted from the incinerator requirements, except for the closure requirements [Sections 264/265.340(b)]. For the owner or operator to operate under an exemption, it must be documented that the exempted waste streams would not reasonably be expected to contain any of the Appendix VIII hazardous constituents and would be classified as one of the following wastes:

- A listed hazardous waste solely because it is ignitable, corrosive, or both
- A listed hazardous waste solely because it is reactive for reasons other than its ability to generate toxic gases, vapors, or fumes that will not be burned when other hazardous wastes are present in the combustion zone
- A characteristic hazardous waste solely because it is ignitable, corrosive, or both
- A characteristic hazardous waste solely because it is reactive for reasons other than its ability to generate toxic gases, vapors, or fumes that will not be burned when other hazardous wastes are present in the combustion zone

Interim Status Units

Q: What are the general requirements for interim status incinerators?

A: During the start-up and shutdown phase of the incinerator, the owner or operator must not feed hazardous waste into the unit unless the incinerator is at *steady-state* (normal) conditions of operation, including steady-state operating temperature and air flow (Section 265.345).

Q: Are there inspection requirements?

A: Yes; the incinerator and all associated equipment (e.g., pumps, valves, conveyors) must be inspected at least daily for leaks, spills, fugitive emissions, and all emergency shutdown controls and alarm systems to assure proper operation. Instruments relating to combustion and emission control must be monitored at least every 15 minutes. If needed, corrections must be made immediately to maintain steady-state combustion (Section 265.347).

Q: What are the waste analysis requirements for interim status incinerators?

A: In addition to the waste analysis required under Section 265.13 of the general standards, the owner or operator must sufficiently analyze any hazardous waste that has not been previously burned in the incinerator. The purpose of the additional waste analysis requirements is to establish steady-state operating conditions, which include waste and auxiliary fuel feed and air flow requirements, and to determine the type of pollutants that might be emitted (Section 265.341). At aminimum, the analysis must determine:

- Heating value (Btu) of the waste
- Halogen and sulfur content of the waste
- Concentrations of both lead and mercury in the waste, unless there is written documentation that shows that these elements are not present

Q: What are the closure requirements for interim status incinerators?

A: At closure, all hazardous waste and hazardous waste residues must be removed from the incinerator. Hazardous waste residues include, but are not limited to, ash, scrubber waters, and scrubber sludges (Section 265.351).

Generally, incinerators also have regulated storage units. Thus, appropriate cleanup of these units must be addressed as well.

Permitted Units

Q: What are the general requirements for permitted incinerators?

A: For permitted incinerators, each permit will specify the hazardous waste that is allowed to be incinerated and the operating conditions required for each waste feed. When an owner or operator wishes to burn waste feeds for which operating conditions have not been set in a permit, either a permit modification (see Chapter 9) must be secured, or, if the burn is to be of short duration and for the specific purposes listed in Section 270.62(a), a temporary trial burn permit must be obtained.

Q: What are the required levels of incinerator performance?

A: Under Section 264.343, the performance standards require:

- A minimum destruction and removal efficiency (DRE) of 99.99 percent for organic compounds designated in the permit as the principal organic hazardous constituents (POHCs)
- A minimum destruction and removal efficiency of 99.9999 percent for waste codes F020, F021, F023, F026, and F027

- Removal of 99 percent of hydrogen chloride gas from the incinerator emissions, unless the quantity of hydrogen chloride emitted is less than 4 pounds per hour
- A limit of 180 milligrams of particulate matter per dry standard cubic meter of gas emitted through the stack

Q: What is DRE?

A: EPA's principal measure of incinerator performance is destruction and removal efficiency (DRE). Destruction refers to the combustion of the waste, while removal refers to the cleansing of pollutants from the combustion gases before they are released from the stack. For example, a 99.99 percent DRE (commonly called "four nines DRE") means that one molecule of an organic compound is released to the air for every 10,000 molecules entering the incinerator; a DRE of 99.9999 percent ("six nines") reduces this to one molecule released out of every 1,000,000 molecules.

Q: Are the performance standards designated for all pollutants present in the waste feed?

A: No; selected hazardous compounds, called the *principal organic hazardous constituents* (POHCs), are designated in the permit. POHCs are selected based on their high concentration in the waste feed and their resistance to burning compared with other organic compounds in the wastefeed. If the incinerator achieves the required DRE for POHCs, then the incinerator should achieve the same or better DRE for organic compounds that are easier to incinerate.

Q: How are new incinerators permitted?

A: The permit for new incinerators covers, four phases of operation:

1. A shakedown period, during which the newly constructed incinerator is brought to the level of normal operating conditions in preparation for the trial burn.
2. A trial burn period, during which burns are conducted so that performance can be tested over a range of conditions.
3. A period following the trial burn during which the data from the trial burn is evaluated and the facility may operate under conditions specified by the permitting agency.
4. A final operating period, which continues throughout the life of the permit (10 years or less).

Trial Burns

Q: What is a trial burn?

A: A trial burn is a test of an incinerator's ability to meet all applicable performance standards when burning a waste under a specific set of operating conditions. Before final permits for hazardous waste incinerators can be issued, owners or operators must demonstrate that their incinerators meet performance standards. Data to be used in evaluating an incinerator's performance is generally gathered by conducting a trial burn.

Q: What happens during the trial burn?

A: During the trial burn, the owner or operator measures the waste feed rate, levels of carbon monoxide in the stack emissions, combustion temperature, combustion gas velocity, and other parameters. To make judgments concerning the incinerator's destruction and removal efficiency, the owner or operator must also measure the quantities of the principal organic hazardous constituents emitted from the incinerator. The permitting agency selects one or more POHCs for each waste feed tested. Emissions of particulate matter and hydrogen chloride are also measured during the trial burn, as is the efficiency of hydrogen chloride removal systems if hydrogen chloride emissions exceed 4 pounds per hour [Section 270.62(b)(6)].

Q: What happens if the incinerator does not meet the performance standards during the trial burn?

A: If the incinerator fails to meet these performance standards, the incinerator design or operation must be modified, and the trial burn must be repeated before the permitting process can proceed to the next step.

Thus, in the case of an interim status incinerator, the permitting agency will not prepare a draft permit until the incinerator has been demonstrated to meet performance standards under at least one set of operating conditions. For a new incinerator, the owner or operator will be required to obtain a permit modification before a second trial burn can be conducted.

Q: How are the results of the trial burn evaluated?

A: Within 90 days following the trial burn, the applicant must provide data from the trial burn and analysis of this data for each waste feed incinerated during the trial burn. All data collected by the applicant must be submitted to the permitting agency for evaluation. The 90-day period following the trial burn allows time for analyzing both the samples collected and other pertinent data [Section 270.62(b)(7)].

Q: How are the results of the trial burn reflected in the permit?

A: For each type of waste feed to be burned by the incinerator, the permit specifies a set of operating conditions consistent with those conditions demonstrated during the trial burn to assure compliance with the performance standards. At a minimum, the permit specifies operating conditions for combustion gas flow rate and acceptable variations in the waste feed composition. The operating conditions may allow for normal fluctuations in these parameters that do not affect performance, as demonstrated during the trial burn [Section 270.62(c)].

After review of trial burn data from an interim status facility, the permitting agency prepares a draft permit based on the trial burn results. For a new facility, if the data from the trial burn shows that the operating conditions included in the permit for the final operating period are sufficient, the facility may enter into this phase of operations. Otherwise, the permit will require modification before this phase of operations may begin.

Q: What are the permit requirements for mobile incinerators?

A: Under RCRA, a mobile incinerator must obtain a permit for each site where it intends to operate. This is because a mobile incinerator meets the definition of a new facility. The owner or operator can submit trial burn data from previous operations in lieu of data from the anticipated site.

When applying for a permit for a mobile incinerator, the applicant should use the model permit application and permit developed for the first permitted mobile treatment site for EPA's Office of Research and Development's (ORD) mobile incinerator for the Denny Farm Site in Missouri (OSWER Directive No. 9527-02).

Q: What are the closure requirements for incinerators?

A: At closure, all hazardous waste and hazardous waste residues must be removed from the incinerator. Hazardous waste residues include but are not limited to ash, scrubber waters, and scrubber sludges (Section 264.351).

Generally, incinerators also have regulated storage units. Thus, appropriate closure of these units must be addressed as well.

THERMAL TREATMENT

Applicability

Q: What is thermal treatment?

A: *Thermal treatment* means the treatment of hazardous waste in a device that uses elevated temperature as the primary means to change the chemical, physical, or biological character or composition of the waste (Section 260.10).

Q: How does thermal treatment differ from incineration?

A: Thermal treatment involves subjecting hazardous waste to elevated temperatures to change its chemical, physical, or biological character or composition. Incineration uses flame combustion in a device to degrade thermally (oxidize) hazardous waste. Other forms of thermal treatment include pyrolysis, microwave discharge, wet air oxidation, calcination, and molten salt processes.

Q: What are the operating requirements for thermal treatment units?

A: Owners or operators who thermally treat hazardous wastes (other than incinerators) must operate their units in compliance with most of the same requirements applied to an incinerator (Section 265.373). However, the thermal treatment standards prohibit open burning of hazardous waste except for the detonation of explosives, such as outdated military ordinances (Section 265.382).

Q: Is the open burning of waste explosives the same as the detonation of waste explosives?

A: Although Section 265.382 distinguishes between "open burning" and "detonation" of waste explosives (on the basis of the speed of the chemical reaction), RCRA regulates them the same way. The regulations require a minimum distance from the unit to "property of others," depending on the quantity (pounds) of waste explosives or propellants.

Q: What are the permit requirements for thermal treatment units?

A: Interim status standards are available for these units only. The Part 264 standards for these units are addressed in the Subpart X standards for miscellaneous units.

CHEMICAL, PHYSICAL, AND BIOLOGICAL TREATMENT

General Requirements

Q: What is chemical, physical, and biological treatment?

A: Treatment, although most frequently conducted in tanks, surface impoundments, and land treatment facilities, can also occur in other types of equipment by such processes as centrifugation, reverse osmosis, ion exchange, and filtration (Section 265.400).

Q: What are the operating requirements for chemical, physical, and biological treatment units?

A: Because the processes are frequently waste-specific, EPA has not developed detailed regulations for any particular type of process or equipment. Instead, general requirements have been established to ensure safe containment of hazardous wastes. In most respects, these other treatment methods are very similar to using tanks for treatment; therefore, they are essentially regulated the same way. The requirements that must be met include avoiding equipment or process failure that could pose a hazard, restricting the use of reagents or wastes that could cause equipment or a process to fail, and installation of safety systems in continuous flow operations to stop the waste inflow in case of a malfunction (Section 265.401).

Q: What are the permitting requirements for chemical, physical, and biological treatment units?

A: There are interim status standards available for these units only. The Part 264 standards for these units are addressed in the Subpart X standards for miscellaneous units.

MISCELLANEOUS UNITS

Applicability

Q: What is a miscellaneous unit?

A: A *miscellaneous unit* is a hazardous waste management unit where hazardous waste is treated, stored, or disposed of and that is not a container, tank, surface impoundment, waste pile, land treatment area, landfill, incinerator, boiler, industrial furnace, or underground injection well (Section 260.10).

Q: What is the purpose of miscellaneous units?

A: On December 10, 1987 (52 *FR* 46946), EPA promulgated final operating standards (Part 264) applicable to miscellaneous units. Although EPA previously issued regulations for the major hazardous waste management technologies (e.g., incineration, landfilling), these regulations were not appropriate for some technologies.

Subsequently, EPA promulgated final regulations under Subpart X to cover both existing and future treatment, storage, and disposal technologies that meet the definition of miscellaneous units, thus ensuring that any and all hazardous waste management technologies are regulated.

General Operating Requirements

Q: What are the general operating requirements for miscellaneous units?

A: A miscellaneous unit must be located, designed, constructed, operated, maintained, and closed in a manner that will ensure protection of human health and the environment. Permits for miscellaneous units must contain such conditions as necessary to protect human health and the environment, including, but not limited to (as appropriate) design and operating requirements, detection and monitoring requirements, and requirements for responses to releases of hazardous waste or hazardous constituents from the unit. Permit conditions will incorporate the requirements under Subparts I through O of Part 264 that are appropriate for the miscellaneous unit being permitted.

To meet the performance standards, EPA has determined that the principal hazardous substance migration pathways must be protected. The protection of these pathways (groundwater, surface water, wetlands, air quality, and soil) must be addressed in the permit. However, a permit does not have to specify conditions that protect each of these pathways (52 *FR* 46955, December 10, 1987).

Q: What is required for groundwater and subsurface pathways?

A: The groundwater and subsurface pathway takes into account various factors that could affect the migration of hazardous wastes and constituents. The factors include:

- Volume and physical and chemical characteristics of the waste, including its potential to migrate through soil and liners
- Hydrogeological characteristics of the site
- Patterns of land use in the region
- Potential for damage to food-chain crops, wildlife, vegetation, physical structures, and domestic animals

Q: What is required for the surface water and surface soil pathways?

A: Because most land-based hazardous waste management units are open or only semiclosed, units must be designed to prevent contamination of surface water, surface soil, and wetlands. The design factors must take into account:

- Volume and physical and chemical characteristics of the waste
- Hydrology and surrounding topography of the site
- Effectiveness and reliability of systems and structures that contain, confine, and collect migrating substances
- Proximity of the unit to surface water
- Current and potential uses of nearby surface waters
- Existing quality of surface waters and soils
- Potential for damage to wildlife, crops, vegetation, physical structures, and domestic animals

Q: What is required for the air pathway?

A: Some hazardous waste management units (e.g., thermal treatment and open burning/open detonation units) may present a significant potential for releases into the air. Thus, the permit applicant must address factors to ensure protection of the surrounding air quality. These factors include:

- Volume and physical and chemical characteristics of the waste, including its potential to emit gases, aerosols, and particulates
- Atmospheric, meteorologic, and topographic characteristics
- Existing quality of air
- Operating characteristics of the unit, including the effectiveness and reliability of systems and structures to reduce or prevent emissions of hazardous wastes or constituents to the air
- Potential for damage to wildlife, crops, vegetation, physical structures, and domestic animals

Chapter 8

Land Disposal Restrictions

The Hazardous and Solid Waste Amendments established a strong statutory presumption against land disposal by prohibiting the continued land disposal of hazardous wastes beyond specified dates, unless such disposal is determined by EPA to be protective of human health and the environment. In its enactment of HSWA, Congress stated explicitly: "reliance on land disposal should be minimized or eliminated, and land disposal, particularly landfills and surface impoundments, should be the least favored method for managing hazardous wastes" [RCRA Section 1002(b)(7)].

The statute requires EPA to set levels or methods of treatment, if any, that substantially diminish the toxicity of the waste or substantially reduce the likelihood of migration of hazardous constituents from the waste so that short-term and long-term threats to human health and the environment are minimized. Wastes that meet the treatment standards established by EPA are not prohibited from land disposal.

EPA has codified the land disposal restrictions under 40 CFR Part 268. The applicable treatment standards are contained in Sections 268.41–42.

SCOPE AND APPLICABILITY

Applicability

Q: Will EPA prohibit the land disposal of all hazardous waste?

A: RCRA does not impose an absolute ban on the land disposal of hazardous waste [Section 268.1(c)]. A waste may be excluded from the ban under the following circumstances:

- When wastes meet treatment standards established by EPA in Sections 268.41 and 42
- When a nationwide extension to the effective date is granted because of a lack of available treatment capacity
- When EPA grants a site-specific variance that demonstrates that there will be no migration of hazardous constituents from the disposal unit for as long as the waste remains hazardous
- When an individual extension to an effective date is granted based on the characteristics of a specific waste
- When untreated waste is treated in a surface impoundment that complies with minimum technological requirements (i.e., double liner with leachate collection and removal system) and if the treatment residues that are hazardous are removed within a year of placement in the impoundment

Q: How is land disposal defined?

A: *Land disposal* is defined to include, but not be limited to, any placement of hazardous waste in a landfill, surface impoundment, waste pile, injection well, land treatment facility, salt dome formation, or underground mine or cave (Section 268.2). EPA also considers placement of hazardous wastes in concrete vaults or bunkers intended for disposal purposes as methods of waste management subject to the land disposal restrictions. However, EPA does not consider open burning and detonation to be methods constituting land disposal and has concluded that the land-disposal restrictions program is not applicable (51 *FR* 40580, November 7, 1986).

Q: Do the land disposal restrictions apply to disposal of restricted wastes at commercial land disposal facilities only?

A: No; all restricted wastes must meet the treatment standards (or be granted a "no-migration" variance) before being placed in or on the land regardless of whether the land disposal facility is located on-site or is a commercial facility.

Q: What is restricted waste?

A: *Restricted wastes* are those categories of hazardous waste that are prohibited from land disposal either by regulation or statute regardless of whether the waste is under a nationwide variance, case-by-case variance, or a successful "no-migration" petition (53 *FR* 31208, August 17, 1988). Thus, the universe of restricted wastes includes all wastes for which the statutory land disposal restriction deadline has passed, including those wastes that may currently be land disposed due to an extension or variance or by meeting the applicable treatment standard, as well as those that are currently banned from land disposal.

Q: Is there a difference between a restricted waste and a prohibited waste?

A: Yes; prohibited wastes are a subset of restricted wastes. A "prohibited" waste is a "restricted" waste that is currently banned from land disposal (53 *FR* 31208-9, August 17, 1988). This includes all wastes that do not meet the applicable treatment standards for which the applicable land disposal restriction date has passed, and for which no variances or extensions have been granted.

Q: Who is responsible for determining whether restricted wastes meet the applicable treatment standard prior to land disposal?

A: The owner or operator of the disposal facility has the ultimate responsibility for verifying that only wastes meeting the treatment standards are land disposed. The land disposal facility must maintain documentation to demonstrate that the wastes are in compliance with the applicable treatment standards.

Q: Are there requirements specifically for generators pertaining to the land disposal ban?

A: Yes; according to Section 268.7(a)(1), a generator must:

- Determine whether their waste is restricted.
- Retain a copy of all demonstrations and other waste analyses or documentation for all wastes sent to either on-site or off-site treatment, storage, or disposal units.
- Manage restricted waste must determine whether the wastes meet the applicable treatment standards at the point of generation, or are otherwise exempt from those standards.
- Prepare a notice for each off-site shipment setting out the hazardous

waste identification number, applicable treatment standard or prohibition level, manifest number, and available waste analysis data.

Q: Is the residue resulting from treatment of restricted wastes subject to the treatment standards?

A: Yes; because of the derived-from rule [Section 261.3(c)(2)(i)], all residuals resulting from the treatment of the original listed waste are likewise considered to be listed wastes and, thus, must meet the treatment standards if they are to be disposed of in or on the land (53 *FR* 31142, August 17, 1988).

Q: Must restricted wastes that have been treated to meet promulgated treatment standards still be managed as hazardous wastes under RCRA?

A: Yes; if listed hazardous wastes are treated, under the provisions of Section 261.3(c)(2), they will remain hazardous wastes until or unless they have been delisted in accordance with Section 260.22. Thus, these wastes must continue to be managed as hazardous waste. Characteristic wastes that no longer exhibit a characteristic of hazardous waste have special requirements. These are addressed later in this chapter (see page 246).

Q: How does the land ban affect reclamation/recycling of restricted hazardous wastes?

A: Restricted wastes may continue to be recovered or reclaimed. However, though the reclamation operation itself is exempt from regulation under RCRA, storage of restricted wastes prior to reclamation is still subject to the provisions specified in Section 268.50. Still bottoms and other residues from reclamation of restricted wastes remain subject to the land disposal restrictions if they meet the definition of hazardous waste.

Q: If restricted wastes are excavated and removed (e.g., corrective action), are they subject to restrictions, although they were originally placed in the ground prior to an effective date of a land disposal prohibition?

A: Yes; where restricted wastes land disposed prior to the applicable effective date are removed from the disposal unit, subsequent placement of such wastes in or on the land would be subject to the applicable prohibitions and treatment provisions.

Q: Do the land disposal restrictions apply to the disposal of lab packs?

A: Yes; if a lab pack contains restricted wastes, the entire lab pack is subject to the land disposal restrictions. Thus, a lab pack may not be land disposed after the effective date of an applicable restriction, unless the restricted wastes are removed before land disposal, the restricted wastes in the lab pack meet the treatment standard, or a successful petition demonstration has been made under Section 268.6 (i.e., a "no-migration" petition).

Q: Are small-quantity generators (<100 kg/mo) subject to the land disposal ban?

A: No; these generators are excluded from RCRA Parts 262 through 268, 270, and 124; therefore, they are not subject to the land disposal ban (Section 261.5).

Schedule of Restrictions

Q: What is the scehdule for the land disposal restrictions?

A: Congress set forth a schedule of land disposal restrictions in HSWA. The statute automatically restricted (the hammer provision) wastes under the heading"Solvents" and "California List."

The statute also requires EPA to make determinations on prohibiting land disposal, within the indicated timeframes, for the following:

- At leat one-third of all ranked and listed hazardous wastes by August 8, 1988, (the "First Third")
- At least two-thirds of all ranked and listed hazardous wastes by June 8, 1989, (the "Second Third")
- All remaining ranked and listed hazardous wastes and all hazardous wastes identified by the characteristics by May 8, 1990, (the "Third Third")

EPA promulgated this schedule on May 28, 1986 (51 *FR* 19300), under Sections 268.10, 11, and 12.

Q: What wastes are included in the First Third?

A: The wastes included in the First Third are:

F006, F007, F008, F009, and F019

K001, K004, K008, K011, K013, K014, K015, K016, K017, K018, K020, K021, K022, K024, K030, K031, K035, K036, K037, K044, K045,

K046, K047, K048, K049, K050, K051, K052, K060, K061, K062, K069, K071, K073, K083, K085, K086, K087, K099, K101, K102, K103, K104, and K106

P001, P004, P005, P010, P011, P012, P015, P016, P018, P020, P030, P036, P037, P039, P041, P048, P050, P058, P059, P063,P068, P069, P070, P071, P081, P082, P084, P087, P089, P092, P094,P097, P102, P105, P108, P110, P115, P120, P122, and P123

U007, U009, U010, U012, U016, U018, U019, U022, U029, U031, U036, U037, U041, U043, U044, U046, U050, U051, U053, U061, U063, U064, U066, U067, U074, U077, U078, U086, U089, U103, U105, U108, U115, U122, U124, U129, U130, U133, U134, U137, U151, U154, U155, U157, U158, U159, U171, U177, U180, U185, U188, U192, U200, U209, U210, U211, U219, U220, U221, U223, U226, U227, U228, U237, U238, U248, and U249

Q: What wastes are included in the Second Third?

A: The wastes included in the Second Third are:
F010, F011, F012, and F024

K009, K010, K019, K025, K027, K028, K029, K038, K039, K040, K041, K042, K043, K096, K097, K098, and K105

P002, P003, P007, P008, P014, P026, P027, P029, P040, P043, P044, P049, P054, P057, P060, P072, P074, P085, P098, P104,P106, P107, P111, P112, P113, and P114

U002, U003, U004, U005, U008, U011, U014, U015, U020, U021, U023, U025, U026, U028, U032, U035, U047, U049, U057, U058, U059, U060, U062, U070, U073, U080, U083, U092, U093, U094, U094, U097, U098, U099, U101, U106, U107, U109, U110, U111, U114, U116, U119, U127, U128, U131, U135, U138, U140, U142, U143, U144, U146, U147, U149, U150, U161, U162, U163, U164, U165, U168, U169, U170, U171, U172, U173, U174, U176, U177, U178, U179, U189, U193, U196, U203, U205, U206, U208, U213, U214, U215, U216, U217, U218, U235, U239, and U244

Q: What wastes are included in the Third Third?

A: All remaining listed hazardous wastes and the characteristic wastes are in the Third Third.

Q: What is the schedule based on?

A: The schedule is based on a ranking of the listed hazardous wastes by intrinsic hazard and volume generated. High-volume hazardous wastes with a high intrinsic hazard are scheduled first, and low-volume wastes with a lower intrinsic hazard are scheduled last.

Q: What is the schedule for wastes newly listed after November 8, 1984?

A: EPA must make a land disposal prohibition determination for any newly listed waste after November 8, 1984, within 6 months of the listing.

Q: What if EPA fails to promulgate treatment standards within 6 months of a newly listed waste?

A: There is no automatic prohibition on land disposal of such wastes if EPA fails to meet this deadline.

Q: What if EPA failed to promulgate a treatment standard by the specified date?

A: If EPA failed to promulgate treatment standards by the specified date, the statute bans the placement of all affected wastes in a land disposal unit. However, if EPA fails to set a treatment standard for any waste included in the "First Third" or "Second Third," the waste may be land disposed of in a landfill or surface impoundment until treatment standards are promulgated or until May 8, 1990, provided such facility is in compliance with the minimum technological requirements and the generator certified that such disposal is the only practical alternative to treatment currently available to the generator.

Treatment Standards

Q: How will EPA implement the land disposal prohibitions?

A: By each statutory deadline, EPA promulgated treatment standards for the applicable hazardous wastes (restricted wastes). Wastes that meet these treatment standards may be directly land disposed. Wastes that do not meet these standards must be treated before they are placed in a land disposal unit.

Q: What are the treatment standards based on?

A: The standards, expressed as achievable constituent levels in the waste, are based on best demonstrated available technology (BDAT). The specific BDAT technology upon which the standards are based does not need to be used except when technologies are set as standards, e.g., incineration was established as the treatment standard for halogenated organic compounds.

Q: Is there a provision for changing the treatment standards if new treatment technologies are developed?

A: Yes; if a new technology is shown to be more effective in reducing the concentration of hazardous constituents in the waste (or the waste extract) than the existing technology upon which the treatment standard has been based, EPA can revise the treatment standard. Such a revision would follow the normal regulatory amendment process (i.e., publication in the *Federal Register* and public comment).

Q: Can a generator dilute a restricted waste to meet the treatment standards?

A: No; as stated in Section 268.3, "No generator, transporter, handler, or owner or operator of a treatment, storage, or disposal facility shall in any way dilute a restricted waste or the residual from treatment of a restricted waste as a substitute for adequate treatment to achieve compliance with Subpart D of this part."

Q: Is dilution that occurs as part of a treatment process prohibited?

A: No; treatment that necessarily involves some degree of dilution (such as biological treatment or steam stripping) is acceptable. Also, mixing wastes prior to treatment is not considered dilution. Dilution is prohibited if it is conducted in lieu of adequate treatment for purposes of attaining the applicable treatment standards.

Q: Is solidification considered an acceptable method of treating restricted wastes?

A: Yes; wherever possible, EPA establishes treatment standards as performance standards, rather than requiring a specific treatment method. In such cases, any method (other than inappropriate solidification practices that would be considered dilution to avoid adequate treatment) that can meet the treatment standard is acceptable. When EPA has specified a technology as the treatment standard, the applicable wastes must be treated using the specified technology. Solidification may, nonetheless, be a necessary prerequisite to land disposal to comply with the prohibition against free liquids in landfills (Sections 264/265.314).

Q: What does a generator use to determine compliance with a treatment standard?

A: The regulations specify the method to use for each designated waste. For example, the solvent- and dioxin-containing wastes are required to use the toxicity characteristic leaching procedure (TCLP).

Q: How is the TCLP used?

A: Under the final rules for the solvent- and dioxin-containing wastes, EPA requires that the TCLP be used to determine whether a waste requires treatment. The TCLP is a procedure that generates an extract of the waste and determines the constituents in the extract. However, some restricted wastes, e.g., California wastes, require a total composition analysis instead of an extract analysis. Thus, in some cases, the TCLP is inappropriate.

Documentation

Q: What actions must be taken by a generator to determine the appropriate treatment standard for a restricted hazardous waste?

A: The generator may identify the applicable treatment standard based on waste analysis data, knowledge of the waste, or both. Where this determination is based solely on the generator's knowledge of the waste, the generator is required to maintain all supporting data used to make this determination in the facility files.

Q: What is required if a generator determines that a waste to be shipped off-site is a restricted waste and exceeds the applicable treatment standards (i.e., prohibited from land disposal)?

A: If a generator determines that the waste does not meet the specified treatment levels (and is thus prohibited from land disposal), the generator must notify the designated facility in writing.

Q: What information is required in the notification?

A: According to Section 268.7(a)(1), the notice must contain the:

- EPA hazardous waste number
- Corresponding treatment standard
- Waste analysis data (where applicable)
- Manifest number associated with the shipment of waste

***Q:* What is required if the waste meets the treatment standards and thus is allowed to be land disposed?**

A: If a generator determines that a restricted waste can be land disposed of without further treatment, each shipment of that waste to a designated facility must have a notice and certification stating that the waste meets all applicable treatment standards. The signed certification must state the following:

> *I certify under penalty of law that I have personally examined the waste through analysis and testing or through knowledge of the waste to support this certification that the waste complies with the treatment standards specified in 40 CFR 268.32 or RCRA Section 3004(d). I believe that the information submitted is true, accurate, and complete. I am aware that there are significant penalties for submitting a false certification, including the possibility of a fine or imprisonment.*

***Q:* Is a generator required to notify a recycler that a restricted waste is being shipped for recycling?**

A: Yes; a recycler is a treatment facility; therefore, a generator must notify the recycler according to the requirements in Section 268.7(a)(1).

***Q:* What documentation must an off-site disposal facility that accepts restricted waste maintain?**

A: An off-site disposal facility must maintain:

- Waste analysis data obtained through testing of the waste
- A copy of the notice and certification required by the owner or operator of a treatment facility under Sections 268.7(b)(1) and (2)
- A copy of the notice and certification required by the generator in cases where the wastes meet the treatment standard and can be land disposed without further treatment Section 268.7(a)(2)
- Records of the quantities (and date of placement) for each shipment of hazardous waste placed in the unit under an extension of the effective date (a case-by-case extension or a 2-year extension of the effective date) or a no-migration petition and a copy of the notice under Section 268.7(a)(3)

Variances

***Q:* Are there variances from a particular treatment standard?**

A: Yes; generators or owner and operators of facilities may petition EPA for a variance from a treatment standard. Wastes may be granted a variance due to physical and chemical characteristics that are significantly different from the wastes evaluated by EPA in setting the treatment standards. For example, in some cases, it may not be possible to treat a restricted waste to the applicable treatment standards [Section 268.44(a)].

Q: Must the facility comply with the land disposal restriction rules while a variance from the treatment standard is being considered by EPA?

A: Yes; in accordance with Section 268.44(g), during the petition review process, the applicant is required to comply with all restrictions on land disposal under Part 268, once the effective date for the waste has been reached.

Q: Will wastes still be restricted if there is not enough capacity, nationwide, to treat the wastes to meet the restrictions?

A: No; if there is insufficient capacity of alternative treatment, recovery, or disposal technologies for a particular waste or group of wastes nationwide, EPA can grant a nationwide extension to the effective date of the restriction and has done so. The purpose of the extension is to allow time for the development of capacity. This extension may not exceed 2 years beyond the applicable statutory deadline for the waste.

No-Migration Petition

Q: Are there variances for site specific land disposal of restricted wastes?

A: Yes; a facility owner or operator can petition EPA to allow land disposal of a specific waste at a specific site. The applicant must demonstrate that the waste can be contained safely in a particular type of disposal unit, so that no migration of any hazardous constituents occurs from the unit for as long as the waste remains hazardous. If EPA grants the petition, the waste is no longer prohibited from land disposal in that particular type of unit at that site [Section 268.6(a)].

Q: What must be demonstrated in a "no-migration" petition?

A: The statutory standard for evaluation of "no-migration" petitions requires that the petitioner demonstrate, to a reasonable degree of certainty, that there will be no migration (to air, groundwater, and surface water) of hazardous constituents from the disposal unit or injection zone for as long as the wastes remain hazardous [RCRA Section 3004(d)]. It is expected, however, that a successful demonstration of "no-migration" will be difficult, but not impossible, to achieve. The majority of units obtaining a no-migration variance will probably be land treatment units treating refinery wastes.

Q: Who reviews "no-migration" petitions?

A: EPA is requiring that applicants submit petitions directly to the EPA Administrator. Where possible, EPA intends to process Part B permit applications and "no-migration" petitions concurrently. However, if review of Part B applications or "no-migration" petitions is unduly delayed by concurrent reviews, EPA will process such applications separately. Applications for "no-migration" petitions will be reviewed at EPA headquarters. EPA headquarters will coordinate reviews with appropriate regional and state staff responsible for reviewing Part B applications for the same facility [Section 268.6(c)].

Case-by-Case Extensions

Q: Are there any provisions for those cases where a waste was not granted a national capacity extension and alternative capacity cannot be found by the effective date?

A: Yes; any person who generates or manages a restricted hazardous waste may submit an application to the EPA Administrator for a case-by-case extension of the applicable effective date. The applicant must demonstrate that: (1) He or she has made a good-faith effort to contract with a treatment, storage, or disposal facility; (2) a binding contract has been entered into to construct or otherwise provide alternative capacity; and (3) that this alternative capacity cannot reasonably be made available by the applicable effective date due to circumstances beyond the applicant's control.

Q: Is there a deadline for submitting applications for case-by-case extensions?

A: There is no deadline for submitting these applications to the EPA Administrator. However, case-by-case extensions cannot extend beyond 48 months from the date of the statutory land disposal restriction [Section 268.5(e)].

Storage

Q: Can restricted wastes be stored prior to treatment or disposal?

A: Yes; the storage of hazardous wastes restricted from land disposal is prohibited except where storage is needed to accumulate sufficient quantities to allow for proper recovery, treatment, or disposal [Section 268.50(a)].

Q: How long can a hazardous waste management facility store restricted wastes?

A: Owners and operators of hazardous waste management facilities may store restricted wastes as needed to accumulate sufficient quantities to allow for proper recovery, treatment, or disposal. However, where storage occurs beyond 1 year, the owner or operator bears the burden of proving, in the event of an enforcement action, that such storage is solely for the purpose of accumulating sufficient quantities to allow for proper recovery, treatment, or disposal. For periods less than or equal to 1 year, the burden of demonstrating whether or not a facility is in compliance with the storage provisions lies with EPA [Section 268.50(b)].

Q: Can an owner or operator of a hazardous waste management facility store restricted waste in land-based units?

A: No; under the storage prohibitions, these facilities may store restricted wastes only in containers and tanks, provided that each container or tank is clearly marked to identify its content and the date the hazardous waste entered storage [Section 268.50(a)(2)].

PHASE I—SOLVENT- AND DIOXIN-CONTAINING WASTES

Applicability

Q: What is Phase I?

A: On November 7, 1986 (51 *FR* 40572), EPA promulgated Phase I of the land disposal ban. This rule set out the regulatory framework for implementing the land disposal prohibitions and promulgated treatment standards for specified solvent and dioxin wastes.

Q: What specific wastes are covered by the Phase I rule?

A: The wastes under Phase I, the solvent- and dioxin-containing wastes, are those wastes numbered F001, F002, F003, F004, F005, F020, F021, F022, F023, F026, F027, and F028.

Solvent Wastes

Q: What solvent wastes are covered under the F001–F005 listings?

A: A solvent waste must meet the following criteria to be covered by the solvent listings (i.e., EPA hazardous waste nos. F001, F002, F003, F004, and F005):

- The solvent waste must be defined as hazardous (i.e., spent).
- The waste must be generated from solvents that are used for their "solvent" properties, i.e., to solubilize (dissolve) or mobilize other constituents.
- With respect to spent solvent mixtures/blends, the mixture must contain, before use, a total of 10 percent or more (by volume) of one or more of the solvents listed in F001, F002, F004, or F005.

Q: When does a solvent become spent?

A: A solvent is considered *spent* when it has been used and is no longer fit for use without being regenerated, reclaimed, or otherwise reprocessed.

Q: Are solvent-based paints covered under the F001–F005 solvent listings concerning the Phase I land disposal restrictions?

A: No; the spent solvent listings do not cover manufacturing process wastes containing solvents when the solvents were used as reactants or ingredients in the formulation of commercial chemical products. Therefore, waste solvent-based paints are not within the scope of the F001–F005 spent solvent listings, because the solvents are ingredients in the paint.

Q: What are the treatment requirements for solvent wastes?

A: The treatment levels, to be determined by the TCLP, are specified in Table 14.

Q: How does one determine whether a spent solvent waste should be treated to the levels specified in Table 14 for "wastewater" or "all other spent solvent wastes?"

TABLE 14 Constituents contained in solvent waste extract.

Constituent	Wastewater(mg/L)	All other wastes(mg/L)
Acetone	0.050	0.590
n-Butyl alcohol	5.00	5.00
Carbon disulfide	1.05	4.81
Carbon tetrachloride	0.05	0.96
Chlorobenzene	0.15	0.05
Cresols (cresylic acid)	2.82	0.75
Cyclohexanone	0.125	0.75
1,2-Dichlorobenzene	0.65	0.125
Ethyl acetate	0.05	0.75
Ethylbenzene	0.05	0.053
Ethyl ether	0.05	0.75
Isobutanol	5.00	5.00
Methanol	0.25	0.75
Methylene chloride	0.20	0.96
Methylene chloride (pharmaceutical)	2.7	0.96
Methyl ethyl ketone	0.05	0.75
Methyl isobutyl ketone	0.05	0.33
Nitrobenzene	0.66	0.125
Pyridine	0.12	0.33
Tetrachloroethylene	0.079	0.05
Toluene	1.12	0.33
1,1,1-Trichloroethane	1.05	0.41
Trichloroethylene	0.062	0.091
Trichlorofluoromethane	0.05	0.96
1,1,2-Trichloro-1,2,2-trifluoroethane	1.05	0.96
Xylene	0.05	0.15

Source: 40 CFR 268.41

A: For the purposes of defining applicability of the treatment standards for wastewaters containing F001–F005 spent solvents, wastewaters are defined as solvent-water mixtures containing total organic carbon (TOC) of 1 percent or less. Wastewaters containing greater than 1 percent TOC must meet the treatment standard for "all other spent solvent wastes."

Q: What test method should be used to determine if there is a greater than 1 percent total F001–F005 solvent concentration in a waste?

A: A total organic carbon analysis can be used, or, alternatively, individual tests can be run to determine the concentration of each of the F001–F005 constituents present or expected to be present in the wastes. The individual concentrations are then summed to determine if the 1 percent threshold is exceeded.

Q: What treatment technologies were used as the basis for the treatment standards for F001–F005 solvent-containing wastes?

A: Treatment standards for solvent-containing" wastewater" were based on either a combination of biological treatment, steam stripping, and activated carbon technologies or on the technologies individually. The treatment standards for "all other spent solvent wastes" were based on incineration.

Q: Are spent solvents that are shipped off-site to incinerators or recycling operations subject to the Phase I rule?

A: Yes; hazardous wastes are determined to be restricted from land disposal at the point of generation. Therefore, the generator is required to notify the treatment facility of the appropriate treatment standard, assuming the solvent waste as generated does not meet the treatment standard and is not otherwise exempt from restriction.

Q: Are there any circumstances where dilution of solvent wastes is allowed?

A: Yes; if dilution is a legitimate step in a properly operated treatment process (e.g., if a waste is mixed with other wastes prior to incineration), or if a treatment method includes the addition of reagents to physically or chemically change the waste (and does not merely dilute hazardous constituents into a larger volume of waste so as to lower the constituent concentration), then dilution is allowed.

Q: Are the residuals (still bottoms) from the recycling or distillation of listed solvents subject to the Phase I rule?

A: Yes; residuals from solvent recycling operations that accept restricted wastes are subject and, thus, must meet the applicable treatment standard before being land disposed.

Dioxin Wastes

Q: What are the dioxin containing wastes?

A: The regulations apply to those hazardous wastes identified by the hazardous waste codes F020, F021, F022, F023, F026, F027, and F028.

Q: What are the treatment standards for the dioxin wastes?

A: The treatment standards for the dioxin wastes are outlined in Table 15.

Q:* *What test procedure should be used for detecting 1 ppb of dioxin?

A: One part per billion is the routinely achievable detection limit using test method 8280, *Method of Analysis for Chlorinated Dibenzo-p-Dioxins and -Dibenzofurans*. This procedure is described in SW-846 and Appendix X of 40 CFR Part 261.

PHASE II—THE CALIFORNIA WASTES

Applicability

Q:* *What are the California wastes?

A: The *California List Wastes,* so named because the State of California initiated the restriction of land disposal of certain liquid waste, are shown in Table 16. The associated treatment levels are also specified in the table.

Q:* *Is the TCLP used to determine the concentration levels?

A: No; the concentrations expressed in Table 16 are determined by using a total composition analysis, rather than an extract analysis using the TCLP.

TABLE 15 Constituents contained in dioxin waste extract.

Constituent	Concentration
HxCDD - All Hexachlorodibenzo-*p*-dioxins	<1 ppb
HxCDF - All Hexachlorodibenzofurans	<1 ppb
PeCDD - All Pentachlorodibenzo-*p*-dioxins	<1 ppb
PeCDF - All Pentachlorodibenzofurans	<1 ppb
TCDD - All Tetrachlorodibenzo-*p*-dioxins	<1 ppb
TCDF - All Tetrachlorodibenzofurans	<1 ppb
2,4,5-Trichlorophenol	<0.05 ppm
2,4,6-Trichlorophenol	<0.05 ppm
2,3,4,6-Tetrachlorophenol	<0.10 ppm
Pentachlorophenol	<0.01 ppm

Source: 40 CFR 268.41

TABLE 16 The California list wastes.

Constituent	Level*
Free cyanides	1,000 mg/L
Arsenic	500 mg/L
Cadmium	100 mg/L
Chromium	500 mg/L
Lead	500 mg/L
Mercury	20 mg/L
Nickel	134 mg/L
Selenium	100 mg/L
Thallium	130 mg/L
PCBs	50 ppm
Corrosive wastes	<2.0 pH

*A waste is a California list waste if it is a liquid and it meets or exceeds any of these levels.
Source: 40 CFR Part 268

Q: What criteria must a California waste meet to be subject to the land disposal restriction?

A: For a waste to be subject to Phase II of the land disposal prohibitions, the waste must meet the following criteria:

- It must be listed or identified as a RCRA hazardous waste.
- It must contain a California list constituent at or above the concentrations specified in Table 16.
- Its physical form must be liquid as determined by the paint filter test, Method No. 9095. (However, a California waste [except for halogenated organic compound wastes] can be transformed into a solid and thus be land disposed.)

Wastes Containing HOCs

Q: What are the requirements for California wastes containing halogenated organic compounds?

A: All liquid and nonliquid hazardous waste containing halogenated organic compounds (HOCs) in total concentration greater than or equal to 1,000 mg/kg, except for dilute halogenated compound wastewaters, such as HOC/water mixtures that contain primarily water and that have less than 10,000 mg/l HOCs, must be incinerated in accordance with existing RCRA regulations.

HOC wastewaters need not be incinerated, but they must be treated to the 1,000 mg/l prohibition level.

Q: What criteria are used to determine a halogenated organic compound?

A: For the purposes of the land disposal prohibition, EPA has defined the HOCs that must be included in the calculation as any compound having a carbon-halogen bond; these are listed in Appendix III to part 268. The Appendix III list is contained in Table 17.

TABLE 17 List of halogenated organic compounds.

Volatiles	
Bromodichloromethane	1,1-Dichloroethylene
Bromomethane	*trans*-1,2-Dichloroethene
Carbon tetrachloride	1,2-Dichloropropane
Chlorobenzene	*trans*-1,3-Dichloropropene
2-Chloro-1,3-butadiene	*cis*-1,3-Dichloropropene
Chlorodibromomethane	Iodomethane
Chloroethane	Methylene chloride
2-Chloroethyl vinyl ether	1,1,1,2-Tetrachloroethane
Chloroform	1,1,2,2,-Tetrachloroethane
Chloromethane	Tetrachloroethene
3-Chloropropene	Tribromomethane
1,2-Dibromo-3-chloropropane	1,1,1-Trichloroethane
1,2-Dibromomethane	1,1,2-Trichloroethane
Dibromomethane	Trichloroethene
trans-1,4-Dichloro-2-butene	1,2,3-Trichloropropane
Trichloromonofluoromethane	Vinyl chloride
Dichlorodifluoromethane	
1,1-Dichloroethane	
1,2-Dichloroethane	
Semivolatiles	
Bis(2-chloroethoxy)ethane	Hexachlorobutadiene
Bis(2-chloroethyl)ether	Hexachlorocyclopentadiene
Bis(2-chloroisopropyl)ether	Hexachloroethane
p-Chloroaniline	Hexachloropropene
Chlorobenzilate	Pentachlorobenzene
4,4′-Methylenebis(chloroaniline)	
p-Chloro-*m*-cresol	

TABLE 17 *(Continued)*

2-Chloronapthalene	Pentachloroethane
2-Chlorophenol	Pentachloronitrobenzene
3-Chloropropionitrile	Pentachlorophenol
m-Dichlorobenzene	Pronamide
o-Dichlorobenzene	1,2,4,5-Tetrachlorobenzene
p-Dichlorobenzene	2,3,4,6-Tetrachlorophenol
3,3′-Dichlorobenzidine	1,2,4-Trichlorobenzene
2,4-Dichlorophenol	2,4,5-Trichlorophenol
2,6-Dichlorophenol	2,4,6-Trichlorophenol
Hexachlorobenzene	Tris(2,3-dibromopropyl)phosphate
Organochlorine Pesticides	
Aldrin	Endosulfan I
alpha-BHC	Endosulfan II
beta-BHC	Endrin
delta-BHC	Endrin aldehyde
gamma-BHC	Heptachlor
Chlordane	Heptachlor epoxide
DDD	Isodrin
DDE	Kepone
DDT	Methoxychlor
Dieldrin	Toxaphene
Phenoxyacetic Acid Herbicides	
2,4-Dichlorophenoxyacetic acid	Silvex
2,4,5-T	Polychlorinated Biphenyls
Aroclor 1016	Aroclor 1248
Aroclor 1221	Aroclor 1254
Aroclor 1232	Aroclor 1260
Aroclor 1242	PCBs N.O.S.
Dioxins and Furans	
Hexachlorodibenzo-*p* dioxins	Tetrachlorodibenzo-*p*-dioxins
Hexachlorodibenzofuran	Tetrachlorodibenzofuran
Pentachlorodibenzo-*p*-dioxins	
2,3,7,8-Tetrachlorodibenzo-*p*-dioxin	
Pentachlorodibenzofuran	

Source: 40 CFR Part 268, Appendix III.

PCBs

***Q:* What are the requirements for wastes containing polychlorinated biphenyls?**

A: Because PCBs by themselves are not defined as hazardous waste, a waste containing PCBs at or above 50 ppm will be subject to the land disposal ban, only if the PCBs are either mixed with a listed hazardous waste or exhibit a characteristic of hazardous waste.

Liquid hazardous wastes containing PCBs at concentrations greater than or equal to 50 ppm must be treated in accordance with the existing Toxic Substances Control Act's (TSCA) disposal regulations in 40 CFR Part 761. These regulations require that PCB wastes be either incinerated or burned in a high-efficiency boiler.

THE LISTED AND CHARACTERISTIC WASTES

Listed Hazardous Wastes

***Q:* When were the treatment standards promulgated for the listed and characteristic wastes?**

A: The listed and hazardous wastes were divided into thirds and were promulgated as follows:

The First Third final rule was published on August 17, 1988 (53 FR 31138). This rule set out the conditions under which wastes included in the first one-third of the schedule of restricted hazardous wastes may continue to be land disposed.

The Second Third final rule was published on June 23, 1989 (54 FR 26594), that set forth the conditions which certain Second Third wastes, some First Third wastes, and some newly listed hazardous wastes may continue to be land disposed.

The Third Third final fule was published on June 1, 1990 (55 FR 22520), that set forth conditions under which Third Third wastes may continue to be land disposed.

Characteristic Hazardous Wastes

***Q:* How are the characteristic hazardous wastes regulated?**

A: The characteristic hazardous wastes were addressed in the Third Third rule, which requires the following:

- Generators must determine whether listed hazardous wastes also exhibit characteristics of a hazardous waste.
- Wastes that are listed and also exhibit a characteristic (i.e., those wastes that carry more than one waste code) must be treated to meet the treatment standard for each characteristic, unless the characteristic constituent or property is specifically addressed in the treatment standard for the listed waste [Section 268.9(b)].
- Disposal of a waste that exhibits a characteristic at the point of disposal is prohibited unless the treatment standard for that characteristic component is above the characteristic level [Section 268.9(c)].

Q: If a characteristic waste no longer exhibits a characteristic, isn't the waste no longer hazardous and thus, not subject to the land disposal restrictions?

A: EPA stated (June 1, 1990, 55 FR 22651) that it believes that the statute provides EPA ample authority to determine whether additional treatment beyond removal of the characteristic is necessary for particular types of wastes to achieve the goals of the statute. Thus, in some cases, additional treatment beyond removing the characteristic is required; whereas in other cases, removal of the characteristic itself has been deemed sufficient.

Q: If a generator's characteristic waste has been treated to meet the treatment standard before being sent off-site to Subtitle D facility, what is required?

A: According to Section 268.9(d)/notification need not be sent to the facility. However, a notification and the certification [268.7(b)(5)(i)] must be sent to the EPA Regional Administrator or authorized state. The notification must include:

- Name and address of the Subtitle D facility
- Description of the waste at generation, hazardous waste code, and treatability group
- Applicable treatment standards

Chapter 9

Permits and Interim Status

Owners or operators of hazardous waste treatment, storage, and disposal facilities (TSDs) are required to obtain a permit to operate those facilities legally. Permits stipulate the required administrative and technical performance standards for these facilities in the form of permit conditions. Because all hazardous waste management facilities are required to obtain a permit, with some exceptions, it is the key to implementing Subtitle C of RCRA.

Congress established the interim status provisions for existing facilities. A facility with interim status is considered to be operating under a permit until such time as EPA takes final administrative action on the facility's permit application. The permit procedures and the interim status regulations are contained in 40 CFR Parts 124 and 270.

SCOPE AND APPLICABILITY

Applicability

Q: Are all owners and operators of hazardous waste management facilities required to obtain a permit?

A: No; pursuant to Section 270.1(c)(2), the following persons or processes are not required to obtain either a permit or interim status to operate under RCRA:

- Facilities that are state-approved to exclusively handle small-quantity generator waste (less than 100 kg/mo)
- Facilities that meet the definition of a totally enclosed facility in Section 260.10
- A generator accumulating hazardous waste in compliance with Section 262.34
- A farmer disposing of waste pesticides from his or her own use in compliance with Section 262.51
- An elementary neutralization unit or a wastewater treatment unit, as defined in Section 260.10
- A person engaged in the immediate treatment or containment of a discharge of hazardous waste
- A transporter storing manifested waste at a transfer facility, in compliance with Section 262.30
- The addition of absorbent material to waste or vice versa, provided it occurs at the time the waste is first placed in the container

Q: Is it the particular process or the unit that is permitted?

A: It is not the process, but the type of unit that requires a permit. Thus, solidification or fixation is not permitted per se, but the processing tank itself requires a permit.

Q: If a facility has numerous units, do they all have to be permitted at the same time even if they are different types?

A: No; a permit may be issued or denied for one or more units at a facility without simultaneously issuing or denying a permit to all the units at a facility. The interim status of any unit for which a permit has not been issued or denied does not affect the issuance or denial of a permit for any other unit at a facility [Section 270.1(c)(4)].

Q: Section 102(2)(c) of the National Environmental Policy Act (NEPA) requires all "major Federal actions significantly affecting the environment" to prepare an environmental impact statement (EIS). Because the issuance of a permit by RCRA does involve "Federal action," does the issuance of a RCRA permit require an EIS?

A: No; EPA has determined that RCRA permits are not subject to the environmental impact statement provisions of Section 102(2)(c) of NEPA [40 CFR 124.9(b)(6)]. EPA has asserted (45 *FR* 33173, May 19, 1980) that RCRA permit requirements are the functional equivalent of an EIS. Thus, an EIS is not required.

INTERIM STATUS

Applicability

Q: What is interim status?

A: Interim status is the statutorily conferred authorization for a hazardous waste management facility to operate pending issuance or denial of a RCRA permit.

Q: What are the requirements for obtaining interim status?

A: The requirements to obtain interim status include establishing that a facility was *in existence* on the effective date of a regulatory or statutory change, filing a Section 3010 notification (for an EPA identification number), and filing a Part A permit application (Section 270.70).

For example, a facility may be managing a solid waste. If EPA lists that particular waste as a hazardous waste, that facility is then eligible for interim status (unless the facility has a permit for which a modification is required) provided the owner or operator files a Section 3010 notification and a Part A permit application by the date specified in the rule. Thus, any facility in existence on the effective date of a new regulation is eligible for interim status, provided that interim status, was not previously terminated for that facility [Section 270.1(b)].

Q: A waste management facility becomes regulated under RCRA because of a revision to the regulations. When must the owner or operator file a Part A?

A: Section 270.10(e) states that the EPA Administrator will specify in the preamble to the *Federal Register* containing the revision when the facility must submit a Part A permit application to obtain interim status.

Q: Does a facility always follow the Part 265 standards if it is operating or has operated under interim status?

A: No; a facility is subject to the Part 265 standards only until receiving its final permit. On the day the final permit is issued, the facility is subject to the conditions specified in its permit, which are based on the Part 264 standards.

Q: If a facility's interim status is terminated, must the owner or operator of the facility still meet the Part 265 interim status standards for closure, post-closure, and financial responsibility?

A: Yes; a facility that has had its interim status terminated must meet the Part 265 standards, including those for closure, post-closure, and financial responsibility. A technical amendment to the interim status standards, which was published in the November 21, 1984 *Federal Register* (49 *FR* 46094), clarified that interim status standards are applicable to facilities whose interim status is terminated until their closure and post-closure requirements are fulfilled.

Q: Can a new facility obtain interim status and, thus, commence construction and/or operation?

A: No; an owner or operator of a new facility must submit both the Part A and Part B of the permit application at least 180 days before the planned initiation of *physical construction*. It is important to note that physical construction cannot commence until a permit is issued [Section 270.1(b)].

Operation during Interim Status

Q: Are there requirements beyond the Part 265 standards for interim status facilities?

A: Yes; beyond the Part 265 operating standards, there are specific regulatory constraints contained in Sections 270.71 and 270.72, including:

- A facility may not treat, store, or dispose of any hazardous waste not specified in the facility's Part A permit application.
- A facility may not employ processes not specified in the Part A permit application.
- A facility may not exceed the design capacities that were specified in the Part A permit application.
- A facility may not make any changes to the facility during interim status that constitute reconstruction of that facility. *Reconstruction* means a capital investment exceeding 50 percent of the capital cost of a comparable new hazardous waste management facility. However, changes that are prohibited under this provision do not include changes to treat or store in containers or tanks hazardous wastes subject to the land disposal prohibitions imposed by Part 268, provided that such changes are made solely for the purpose of complying with Part 268.

Q: Can any of these changes occur?

A: Yes; these changes (except reconstruction) during interim status may be allowed. New wastes can be handled simply by submitting a revised

Part A permit application. The process and design capacity changes require a revised Part A permit application, written justification for a change, and state or federal approval [Sections 270.72(b) and (c)].

In addition, a change of ownership or operational control of a facility is allowed if the facility submits a revised Part A at least 90 days prior to the change. The previous owner or operator must comply with the financial requirements until the new owner or operator can demonstrate compliance with the financial requirements to EPA. The new owner or operator must comply with the financial requirements within 6 months of the change [Section 270.72(d)].

Q: A facility operator with interim status for container storage closed all storage units. Can the operator now build a treatment unit under interim status or is a permit required?

A: Once a facility has been granted interim status, the facility will retain its interim status until either (1) the final disposition of a permit application by that facility has been made, (2) the interim status is terminated per Section 270.10(e)(5) or under Section 3008(h), or (3) the facility loses interim status under 270.73(c)–(f).

If the owner or operator of the facility in question wanted to build and operate a treatment unit after all the container storage units had been closed, the owner or operator would have to comply with the following RCRA regulations. First, the cost of the construction of the treatment unit could not exceed 50 percent of the cost of building a container storage area similar to the one for which interim status was originally granted according to Section 270.72(e). Second, the owner or operator would have to submit a modified Part A and have the activity approved by the director [Section 270.72(c)]. If the cost to build a new treatment unit exceeds the 50 percent reconstruction threshold, then the owner or operator would need a RCRA permit prior to starting construction of the new treatment unit.

Q: An interim status facility has a surface impoundment for storing hazardous waste. This facility wants to build another impoundment for a new product line that will produce a hazardous waste that was not designated on the facility's Part A application. Would building such a storage surface impoundment for accepting a generated hazardous waste new to the facility be considered an increase in design capacity or a process change?

A: Adding a new storage surface impoundment would be an increase in design capacity. This would not be considered a process change, because the process is not changing; the new unit is also a storage surface impoundment. An increase in design capacity requires the owner or operator to submit a revised Part A application, which includes a justification for the change, and to obtain approval from the regional administrator or state director [Section 270.72(b)]. Also, the owner or operator must comply with Section 270.72(e) concerning reconstruction of the facility.

Q: At the time an interim status facility has its Part B application called, surface impoundments are being rebuilt with clay liners. Does this constitute an increase in design capacity or a change in process under Sections 270.72(b) and (c)? If so, can EPA refuse to allow the change under interim status and require the surface impoundments to meet the Part 264 standards?

A: No; if the capacity of the surface impoundments is not enlarged and no new units are being added, improvements to the surface impoundments are a permissible change under interim status as long as the reconstruction provision of Section 270.72(e) is not violated. This is not a change in process. Rebuilt surface impoundments will be treated as existing units for purposes of compliance with Part 264 standards. At the time these existing units are permitted, however, only the existing portions, i.e., the land surface area upon which wastes are placed prior to permit issuance, will be exempt from Part 264 requirements to install liners and leachate collection systems [OSWER Directive No. 9431.01(84)].

Q: An owner or operator of an interim status hazardous waste management facility is considering expanding the facility to incorporate new technologies. Does the 50 percent restriction apply to each expansion at the facility or to total expansion costs over the interim status period?

A: The 50 percent of capital cost pertains to the total expansion costs over the interim status period, not to each individual expansion. Expansion costs would include the cost of the land and construction, but not design and engineering costs (RIL No. 98).

Q: An interim status tank storage facility plans to upgrade its tanks to meet the new secondary containment standards of Section 265.193. Does upgrading a tank to meet secondary containment requirements constitute a "change during interim status" under Section 270.72?

A: Yes; upgrading a tank to meet the hazardous waste tank secondary containment requirements does constitute a change subject to Section 270.72. According to Section 270.72(c), an owner or operator who wishes to make a change at an interim status facility must submit the revised Part A application and the justification for the change prior to making the change.

In general, Section 270.72(e) does not allow a change under interim status where costs exceed 50 percent of the capital cost of construction of a comparable new facility. However, Section 270.72(e) contains an exception to this prohibition for tanks that must be retrofitted to comply with Section 265.193. Therefore, the cost of retrofitting a tank to comply with Section 265.193 would be allowed to exceed 50 percent of the cost of constructing a new tank facility. Retrofitting to meet the secondary containment standards of Part 265 Subpart J is still considered to be a change requiring submittal of a revised Part A application and justification.

OPERATING PERMITS

Scope

***Q:* What are the administrative procedures involved in obtaining a permit?**

A: The administrative procedures in the permitting process, regulated under Part 124 and depicted in Figure 3, include the application process, completeness review, draft permit or denial, public comment period, and final permit decision.

Permit Application Process

***Q:* What is the permit application?**

A: Owners and operators of hazardous waste management facilities are required to submit a permit application that covers aspects of the design, operation, and maintenance of the facility, as outlined in Part 270. The permit application is divided into two parts, A and B.

***Q:* How is the issuance of a permit handled in relationship to HSWA and state authorization?**

A: Permits are issued by EPA, an authorized state, or both. Currently, under HSWA, until states receive specific HSWA authorization, all permits are handled as a joint issuance, provided the state has the authority to implement pre-HSWA provisions (see Chapter 12). A joint permit may be issued in two ways. There can be one complete permit with signatures of both the state director and the EPA regional administrator on the same document. The other option is to issue two incomplete permits, one signed by EPA and one signed by the state. In either situation, signatures by EPA and the state are necessary to provide the facility with authority

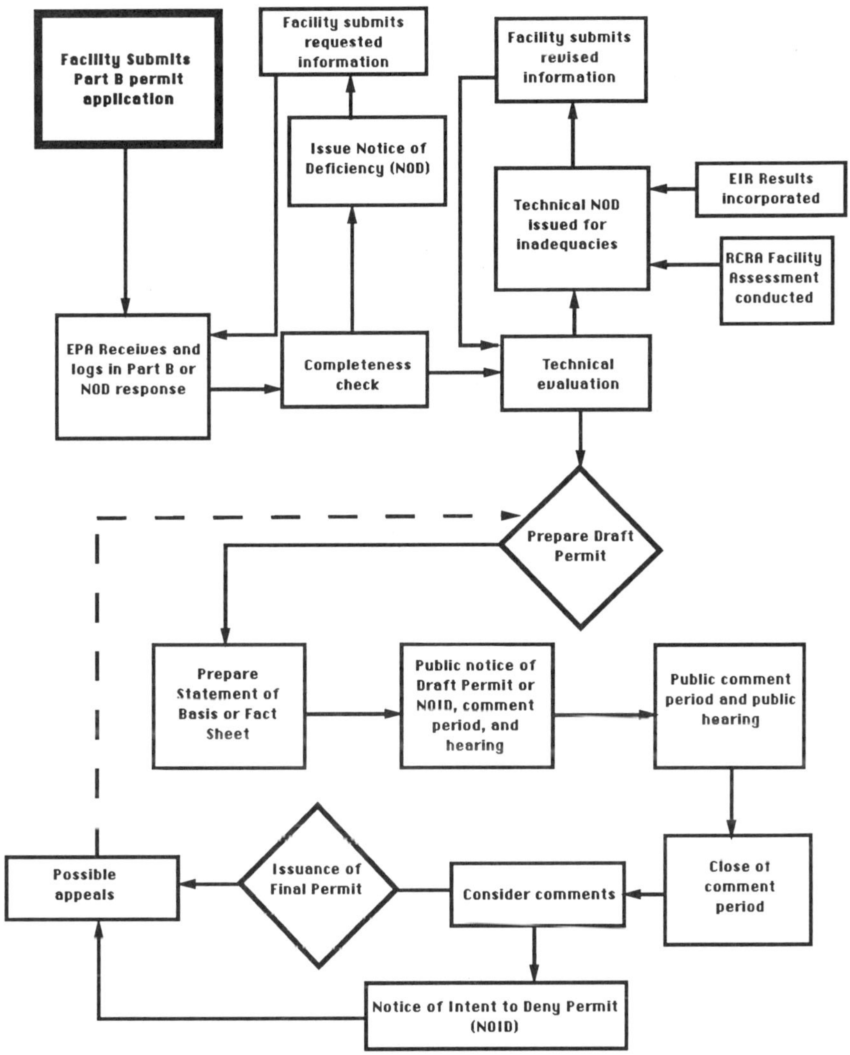

Figure 3. Permit Process

Source: 40 CFR Part 124

to operate under RCRA. If a single permit is issued, it is important to have a clear understanding of which provisions stem from federal authorities and which stem from state authorities. This distinction will enable a permittee to determine the appropriate authority to approach when appealing a given permit condition. If a state has no authorization, EPA issues the permit.

***Q:* If an operator of a facility submits a permit application, does the owner of the facility have to sign it also?**

A: Yes; if the operator of a facility submits the permit application, the owner of the facility must also sign the permit application; if the owner fails to sign the application, then a RCRA permit cannot be issued (OSWER Directive No. 9523.01).

***Q:* A new facility will have a separate owner and operator. Though both parties will sign the permit, is one party chosen as the "permittee," i.e., which party is liable during the operating life, closure, and post-closure?**

A: Both the owner and operator are the "permittees" on the permit; however, it is common for the operator to assume responsibility for meeting permit conditions. Both the owner and operator are liable during the facility's operating life. Both the owner and operator are liable during closure/post-closure of the facility, unless the closure/post-closure plans specify that the owner of the facility is becoming the operator as well as the owner. This action would be accompanied by a permit modification and relieve the original operator from liability (under RCRA) during the closure/post-closure period.

Confidentiality

***Q:* Can a permit applicant claim any of the information contained in an application confidential?**

A: Yes; at the time of submittal, applicants for a RCRA permit may assert a claim of business confidentiality for proprietary information included in the application. General EPA regulations governing claims of confidentiality are found in 40 CFR Part 2. Specific provisions for claims of confidentiality submitted with permit applications are found in Section 270.12.

***Q:* What is the definition of business information?**

A: Under 40 CFR Section 2.201, *business information* is defined as ". . . any information which pertains to the interest of any business, which was developed or acquired by that business, and (except where the context otherwise requires) which is possessed by EPA in recorded form."

***Q:* What limitations are placed on confidentiality assertions?**

A: Applicants should list their requests for confidential treatment to such material that, if released, is likely to cause substantial harm to the competitive position of the irrespective companies. Claims of confidentiality should not be asserted for information that is reasonably obtainable without the applicant's consent (for example, standard engineering designs). A claim of confidentiality for the entire permit application is likely not justifiable, and such claims may be rejected.

Q: When must the confidentiality assertion occur?

A: Claims of confidentiality must be asserted when the permit application is submitted. If no claim is asserted at that time, EPA can make the information available to the public without further notice to the applicant.

Q: How is a claim of confidentiality asserted?

A: To assert a claim, the applicant must attach a cover sheet to the information, or stamp or type a notice on each page of the information, or otherwise identify the confidential portions of the application. Words such as "trade secret," "confidential business information," "proprietary," or "company confidential" should be used. Such words of warning should be placed as hollow-letter stamps across the center of each applicable page. The notice should also state whether the applicant desires confidential treatment only until a certain date or a certain event.

Q: What if the public requests information concerning a facility's application?

A: Prior to releasing any information for which a claim of confidentiality has been made, EPA will give the applicant an opportunity to substantiate its claim and will then determine whether the information warrants confidential treatment. If considered confidential by EPA, information will not be released to the public under the Freedom of Information Act.

Part A Application

Q: What is the Part A permit application?

A: The Part A is an application for interim status for existing facilities and the initial permitting step for new facilities. The Part A permit application is a standardized form (EPA Form No. 3510-1) that requires general information about a facility.

Q: What information is required for the Part A?

A: The required information includes:

- Name of the owner or operator and the facility address, including longitudinal and latitudinal coordinates
- The activities to be conducted that require a RCRA permit
- The facility's design capacity
- Description of the processes for treating, storing, or disposing of hazardous wastes
- A detailed description of hazardous wastes to be handled
- Description of the facility's processes generating hazardous waste
- A scale drawing of the facility, including photographs
- A topographic map
- A listing of all federal environmental permits obtained or applied for by the facility

Q: What happens if an applicant submits a Part A with deficient information?

A:. EPA must issue a notice of deficiency (NOD) to any owner or operator of an existing hazardous waste management facility who submits a deficient Part A permit application. The owner or operator has 30 days after being notified to explain or amend the application before being subject to EPA enforcement. After such notification and opportunity for response, EPA will determine if that facility qualifies for interim status (OSWER Directive No. 9528:50-1A).

Part B Application

Q: What is the Part B permit application?

A: There is no specified form for the Part B. The Part B permit application is a document prepared by the facility owner or operator that addresses each point of concern specified in Section 270.22.

Q: What is the required format for the Part B permit application?

A: The actual application format is left to the discretion of the applicant. However, EPA suggests, in its *Permit Applicant's Guidance Manual for Land Disposal, Treatment, and Storage Facilities,* a general format for the submission of a Part B permit application as follows:

Part I - General Information Requirements

- A copy of the Part A permit application
- General description of the facility
- The process codes (from Part A) that identify the type of units for which the permit is required
- Chemical and physical analysis of hazardous waste to be handled and a copy of the waste analysis plan
- Security description for the active portion of the facility
- General inspection schedule and description of the procedures, including the specific requirements for particular units
- Preparedness and prevention documentation
- Contingency plan documentation
- Preventative procedures, structures, and equipment documentation for control of unloading hazards, waste run-off, water-supply contamination, effects of equipment failure and power outages, and undue personnel exposure to wastes
- Prevention of accidental ignition or reaction documentation, including specific requirements for particular unit types
- Facility traffic documentation
- Facility location documentation
- Personnel training program documentation
- Closure plan documentation, including specific requirements for particular unit types
- Post-closure plan, when applicable, including specific requirements for particular unit types
- Documentation for deed notice
- Closure cost estimate and documentation of financial assurance mechanism
- Documentation of financial assurance for liability
- Topographical map showing contours with 0.5- to 2.0-meter intervals, map scale, map date, 100-year floodplain area, surface waters including intermittent streams, surrounding land uses, wind rose, north orientation, legal boundaries of facility, access control, injection and withdrawal wells, buildings and other structures, utility areas, barriers for drainage or flood control, and location of operating units, including equipment cleaning areas. Each hazardous waste management unit should be shown on the map with a unique identification code and the associated process code from the Part A.

Part II - Specific Information Requirements

The information contained in this section depends on the particular type of unit for which the permit is required. Refer to Subparts I, J, K, L, M,

N, O, and X of Part 264 for the appropriate specific information requirements.

Part III - Additional Information Requirements

- Summary of groundwater monitoring data obtained during the interim status period
- Identification of aquifers beneath the facility
- Delineation of waste management area and point of compliance for groundwater monitoring on topographic map
- Description of any existing plume of contamination in the groundwater
- Detailed groundwater monitoring program description
- Detection monitoring program description, if applicable
- Compliance monitoring program description, if applicable

Part IV - Information Requirements for Solid Waste Management Units

- Location of the solid waste management units (SWMUs) on topographic map
- Designation of the type of SWMU (i.e., impoundment, pile)
- General dimensions and structural descriptions of the SWMUs, including any available drawings
- Dates of operation for the SWMU
- An inventory of all wastes that have been managed at the SWMU
- All available information pertaining to any release of hazardous waste or hazardous constituents from such SWMUs

***Q:** Are there any regulatory options available to a facility owner or operator of an interim status facility that must construct or install equipment required for Part 264?*

A: Yes; permit compliance schedules can be used to allow facilities to construct or install equipment that is not required under Part 265, but is required under Part 264. A permit compliance schedule, which is an enforceable component of a permit, must allow for public notice and comment and allow the applicant additional time only when legitimately needed (OSWER Directive No. 9524.01).

***Q:** Can permit compliance schedules be used to allow facilities to develop information required under Part 270?*

A: No; the permit compliance schedule provision is only for equipment and not information.

Q: Can an applicant submit information along with the Part B, for potential expansions to the facility and obtain a permit for those expansions when there is no definite expansion date?

A: Yes; the applicant, however, must submit information at the same level of detail as if construction were to begin immediately upon receipt of a RCRA permit or at a later date, consistent with a schedule of compliance specified in the permit. The Part B application must be in such detail that the permit writer can draft an enforceable permit and so that there can be meaningful public participation and review of the proposed facility and permit conditions. In other words, the applicant must fully satisfy all the information requirements of a Part B application and the Part 264 standards for a new facility. In addition, when the applicant does finally decide to undertake the expansion, the plans and specifications contained in the permit must conform exactly. The applicant should also be advised of any relevant regulations regarding the procedures for expanding the capacity of a permitted facility [OSWER Directive No. 9431.01(84)].

Q: Must a facility's Part B permit application submission include a closure date if the facility has no plans for closure in the near future?

A: Yes; Section 270.14(b)(13) requires a closure plan to be submitted with the Part B application, and the estimated closure date is part of that closure plan [Section 264.112(a)(4)]. If the estimated closure date changes, the closure plan must be amended as a modification to the permit [Section 270.42(g)].

Permit Review

Q: What occurs during the permit review process?

A: When EPA receives a RCRA permit application, it reviews the application separately for administrative and technical completeness. The completeness review generally takes 60 days for existing facilities and 30 days for new ones.

Q: What if the application is deemed incomplete?

A: If the application is incomplete, EPA will request the missing information through a Notice of Deficiency (NOD) letter. This letter details the information needed to complete the application and specifies the date for submission of this data. EPA can issue a warning letter to accompany the NOD requiring submission of the necessary information within a specified additional period.

Q: What occurs if the permit applicant fails to supply the requested information?

A: If an applicant fails or refuses to correct the deficiencies in the application, the permit may be denied and appropriate enforcement actions may be taken under statutory provisions, including RCRA Section 3008 (OSWER Directive No. 9521.01).

Q: What happens if all the required information has been submitted?

A: If all the required information has been submitted, the permit application is declared *complete*. This declaration does not constitute approval or acceptance of the permit, it only acknowledges a completed application.

A determination that an application is complete is not necessarily a determination that the application is free of deficiencies. During the detailed review of the application and drafting of the permit conditions, it may become necessary to clarify, modify, or supplement provisions previously submitted before progressing to a draft permit or a decision to deny.

Q: Can EPA deny a permit during the review process?

A: Yes; if during the review process, EPA decides to deny the permit, a Notice of Intent to Deny must be sent to the applicant. The applicant may appeal the denial decision.

Q: What is the purpose of the technical review?

A: The purpose of the technical review is to determine whether a specific facility should be granted a permit; in other words, to determine if the facility has satisfied all the siting, design, and operation criteria and the closure, post-closure, and financial requirements.

Q: What are the steps of the technical review?

A: The steps of the technical review include:

- Preliminary review
- Site visit

- Verification of accuracy
- Compliance assessment

Q: What is the preliminary review?

A: The preliminary review involves primarily a secondary completeness check to ensure that the applicant has responded adequately to any previous notices of deficiency.

Q: What is involved in the verification of accuracy?

A: The verification of accuracy involves a check on the reasonableness of data and the accuracy of computations. It also requires a professional assessment of information, assumptions, and methodology presented to identify weaknesses that may require applicant response.

Q: What is the compliance assessment?

A: The compliance assessment, conducted concurrently with the accuracy verification, is the process to ensure that the permitted operations will be in compliance with Part 264.

Q: What is the next step after the technical evaluation?

A: After the technical evaluation, the permit writer recommends whether a draft permit should be prepared or a permit should be denied.

Exposure Information Report

Q: What is an exposure information report?

A: The goals of the exposure information report (EIR) are to identify human exposures to past releases from subject units and potential exposures from future releases, so that they may be mitigated by specific permit conditions.

The Hazardous and Solid Waste Amendments (Section 3019) require all permit applications for landfills or surface impoundments submitted after August 7, 1985, to be accompanied by an exposure information report.

Q: What will EPA use the information for?

A: EPA intends to use the information to establish permit conditions to mitigate exposure or potential exposure, even when there is no current exposure.

Q: What information is required for an exposure information report?

A: Per Section 270.10(j), at a minimum, the owner or operator must submit information addressing the following:

- Reasonably foreseeable potential releases from both normal operations and accidents at the unit, including releases associated with transportation to or from the unit
- The potential pathways of human exposure to hazardous wastes or constituents resulting from the releases described above
- The potential magnitude and nature of the human exposure resulting form such releases

Q: What are the potential pathways that must be addressed in the exposure information report?

A: The pathways of concern (OSWER Directive No. 9523.00-2A) are:

- Groundwater
- Surface water
- Air
- Soil
- Subsurface gas
- Transportation
- Management practices

Q: Is submission of exposure information a condition for permit issuance?

A: No; Section 270.10(c) states that an application for a Part B permit is not considered incomplete if the owner or operator fails to submit the exposure information described in Section 270.10(j). Failure to submit exposure information is a separate violation of Section 3019 of RCRA.

Draft Permit

Q: What is a draft permit?

A: Upon completion of the technical review, EPA tentatively decides whether to issue or deny a RCRA permit. If the tentative decision is to issue the permit, EPA prepares a draft permit. EPA must also prepare a fact sheet or a statement of basis for the public, which explains in simple language each condition included in the draft permit and the reasons for each condition. The draft permit, prepared primarily for public review, contains tentative conditions.

Conditions of All Permits

Q: Are there conditions applicable to all RCRA permits?

A: Yes; under Section 270.30, the following conditions apply to all RCRA permits and are incorporated into the permits either expressly or by reference. These requirements include:

- Compliance with all conditions of the permit
- Proper operation and maintenance, including the operation of backup or auxiliary units when necessary to achieve compliance
- Halting production when necessary to ensure compliance, and taking all reasonable steps to minimize releases to the environment
- Providing all relevant information requests by EPA regarding facility operation, including the reporting of any noncompliance that may endanger health or the environment within 24 hours of the time the owner or operator becomes aware of such conditions (note: a notice of anticipated noncompliance does not suspend or negate any permit condition)
- Allowing authorized representatives to inspect the facility upon presentation of credentials
- Maintaining and reporting all records and monitoring information necessary to document and verify compliance, including such data as continuous-strip chart recordings from monitoring instruments
- Reporting all planned changes and anticipated periods of noncompliance

Q: Can EPA set permit conditions that go beyond the Part 264 requirements?

A: Yes; Section 3005(c)(3) of RCRA gives the permit-issuing agency broad authority concerning permit conditions. This section reads: "Each permit issued under this Section (3005) shall contain such terms and conditions as the Administrator (or State) determines necessary to protect human health and the environment."

This provision allows the issuing agency to impose permit conditions beyond those contained in the Subtitle C regulations, as may be necessary to protect human health and the environment from risks posed by a particular facility. For example, this authority could be used if a facility were to adversely affect vulnerable groundwater because of a lack of adequate location standards.

Q: What is the permit shield provision?

A: The regulations provide that compliance with a RCRA permit during its term constitutes compliance with the statute (except for the land disposal restriction regulations). The purpose of the shield provision is to protect permittees who are in full compliance with their permit, yet may not be in full compliance with other provisions under RCRA. The shield provision protects the permittee from enforcement action brought on by EPA and states, as well as civil actions from citizen groups. However, this provision does not provide a defense to an imminent hazard action under Section 7003 of RCRA (45 *FR* 33428, May 19, 1980).

Q: Are permit schedules of compliance available for draft permits?

A: Yes; when writing a RCRA permit, EPA may specify a schedule for compliance (Section 270.33) at the time of permit issuance. A compliance schedule allows an existing facility to operate while the permittee upgrades the operations to meet all the regulatory requirements. In its decision regarding a schedule for compliance, EPA also considers such factors as availability of any materials required to upgrade the facility, construction time, and the time required to contract for such services.

Public Notice

Q: When is public notice required?

A: In accordance with Section 124.10(a), public notice must be given whenever:

- A permit application has been tentatively denied
- A draft permit has been prepared
- An informal public hearing has been scheduled
- An appeal has been granted by EPA regarding a final permit decision

Q: What are the requirements for public notice?

A: EPA must issue a public notice identifying the applicant and the facility and state where copies of the draft permit and other related information may be obtained (e.g., statement of basis or fact sheet). The notice must be circulated in local newspapers for major permits and mailed to various agencies and parties expressing interest.

Public notice provides interested people a minimum of 45 days to comment on the draft permit. If written opposition to EPA's intent to issue a permit and a request for a hearing are received during the comment period, a public hearing may be held. Notification of the hearing is issued at least 30 days prior to the scheduled date, and the public comment period is extended until the close of the public hearing.

Q: What is the statement of basis and fact sheet?

A: These documents, which must accompany the public notice of the draft permit, briefly set forth the principal facts and the significant factual, legal, methodological, and policy questions considered in preparing the draft permit.

The fact sheet must be prepared for all "major" facilities (i.e., land disposal facilities and incinerators) and for those facilities that EPA finds are "the subject of widespread public interest or raise major issues" (Section 124.8).

Q: What specific information is contained in these documents?

A: These documents include:

- A description of the activity that requires the draft permit
- The type and quantity of wastes that are proposed to be or are being treated, stored, or disposed of
- Reasons why any requested variances or alternatives to required standards do or do not appear justified
- A description of the procedures for reaching a final decision on the draft permit, including:
    - —The extent of the public comment period
    - —Procedures for requesting a hearing
    - —Any other procedures in which the public can participate
- Name and telephone number of a person to contact

Permit Denial

Q: What if EPA decides to deny the permit?

A: If EPA tentatively decides to deny a RCRA permit, a notice of intent to deny a permit is prepared. This notice is considered a type of draft permit and follows the same procedures as any draft permit. These procedures include preparation of a statement of basis or fact sheet containing reasons supporting the tentative decision to deny the permit, public notices of the denial, acceptance of comments, a possible hearing, preparation of a final decision, and possible receipt of a request for appeal.

Q: For what reasons can EPA deny a permit?

A: A permit may be denied for the following reasons:

- The facility cannot meet the requirements of the standards set forth in 40 CFR Part 264
- Activities at the facility would endanger human health or the environment
- An applicant is believed to have not fully disclosed all relevant facts in the application or during the RCRA permit-issuance process
- An applicant has misrepresented relevant facts
- The application did not fully meet the requirements of Part 270 (e.g., did not have the signatures of both the owner and the operator)

Q: What procedural recourse or appeal process is available to an owner or operator whose permit application is denied?

A: An owner or operator wanting to appeal a permit *denial* would follow the procedures in Section 124.19, which addresses recourse for permit denial. This section contains procedures for informal hearings. Briefly, the owner or operator has a 30-day period in which to request a review by serving notice to the regional administrator.

Final Permit

Q: What is the next step after the draft permit?

A: After the close of the public comment period, which includes the public hearing period, EPA must either grant or deny the permit application.

In either case, the applicant and those submitting questions and those requesting notification must be notified, including information regarding appeal procedures.

Final RCRA permits become effective 30 days after the date of the public notice of decision to issue a final permit, unless a later date is specified in the permit or the conditions of Section 124.15(b) are met. At the time the final RCRA permit is issued, EPA also issues a response to any significant public comments received and indicates any provisions of the draft permit that have been changed and the reasons for the changes. The response to comments becomes part of the administrative record.

Q: What does a permit consist of?

A: The actual permit consists of written approval of the completed permit application (Parts A and B). The permit requires the applicant to adhere to all statements made in the application and includes conditions with which the applicant must comply.

Q: Can a person appeal the contents of a final RCRA permit?

A: Yes; persons who submitted comments on the draft RCRA permit or participated in any public hearing are allowed 30 days after the final permit decision to file a notice of appeal and a petition for review with the EPA Administrator in Washington, D.C., who will review and then grant or deny the petition within a reasonable time. If the Administrator decides to conduct a review, the parties are given the opportunity to file briefs in support of their positions. Within the 30-day period, the Administrator may, on his or her own motion, decide to review the decision to grant or deny a hearing. The Administrator then notifies the parties and schedules a hearing. On review, the Administrator has several options regarding the final decision. It may be summarily affirmed without opinion, modified, set aside, or remanded for further proceedings. This petition for review is a prerequisite for judicial review of the Administrator's final decision. Details of appeals of RCRA permits are found in Section 124.19.

Q: What is the duration of a RCRA permit?

A: A permit may be issued for any length of time up to 10 years. All land disposal permits must be reviewed every 5 years and modified if necessary. However, a permit can be terminated at any time if noncompliance with any condition of the permit occurs, any false information was included in the application, or the facility's activity endangers human health or the environment.

Q: A permitted facility closes its container storage area, its only RCRA-regulated unit. The owner now wishes to terminate the permit, because the facility no longer has any active units and is not subject to the post-closure care requirements of Section 264.117. The facility has complied with all the permit conditions and has disclosed all relevant facts for the permit. On what basis may EPA terminate the facility's permit?

A: Normally the owner or operator of a facility that has closed all its RCRA units and has no post-closure care requirements would allow the permit to expire. Although the owner or operator is still subject to the Part 264 standards, there are no hazardous waste management activities to regulate. The owner or operator's financial responsibilities should end when EPA receives certification of closure.

If a facility owner or operator wishes to terminate its permit before the termination date in the permit, it should request a permit modification under Section 270.41. According to Section 270.50, the maximum permit duration is 10 years, but a permit may cover a shorter period. In this situation, EPA could modify the permit so that it would expire earlier.

Permit Modifications

Q: Once a permit is issued, can it be modified?

A: Yes; EPA has stated that permits must be viewed as living documents that can be modified to allow facilities to make technological improvements, comply with new environmental standards, respond to changing waste streams, and generally improve waste management practices. Because permits are usually written for 10 years of operation, the facility or the permit writer cannot anticipate all or even most of the administrative, technical, or operational changes required over the permit term for the facility to maintain an up-to-date operation. Therefore, permit modifications are inevitable. In fact, EPA estimates that many permits may have to be modified two or three times a year (53 *FR* 37913, September 28, 1988).

Q: How are modifications handled?

A: Previously, there were two types of permit modifications, major and minor. Any modification that was not classified as a minor modification was considered a major modification. Major modifications required public notice and comment, as well as the issuance of a draft permit, which was a lengthy process.

On September 28, 1988 (53 *FR* 37912), EPA revised the regulations governing permit modifications contained in Sections 270.41 and 270.42 to incorporate a process that better accommodates the different types of modifications. The revisions provide both owners and operators and EPA more flexibility to change specified permit conditions, expand public notification and participation opportunities, and allow for expedited approval if no public concern exists regarding a proposed change.

Q: What is the basic structure of the new permit-modification program?

A: The program establishes procedures that apply to changes that facility owners and operators may want to make at their facilities. EPA has categorized selected permit modifications into three classes and established administrative procedures for approving modifications in each of these classes. It is important to note that this rule addresses only modifications requested by a permittee. It does not change the procedures for modifications sought solely by the regulatory agency.

Q: What are the three classes of permit modifications?

A: Class 1 and 2 permit modifications do not substantially alter existing permit conditions or significantly affect the overall operation of the facility, whereas Class 3 modifications do. Appendix I to Section 270.42 (see Appendix F of this book) contains a list of specific modifications and assigns them to Class 1, 2, or 3.

Q: What type of permit modification is required for newly listed or identified wastes?

A: Section 270.42(g) provides permittees with a special procedure for modifying permits when wastes that they are already managing are newly listed or identified by EPA as hazardous. Under this provision, the permittee must submit a Class 1 modification request at the time the waste becomes subject to the new listing or identification, that is, on or before the effective date of the rule listing or identifying the waste.

If the changes at the facility constitute a Class 2 or 3 modification, the permittee must submit, in addition to the above Class 1 request, a complete permit modification request within 180 days of the effective date. Until a final decision is made on the modification request, the permittee must comply with Part 265 standards. In addition, where new wastes or units are added to a facility's permit under this approach, they would not count against the 25 percent expansion limit for Class 2 modifications. Finally, for land disposal units, the owner or operator is required to certify compliance with all applicable groundwater monitoring and financial responsibility requirements within 1 year of the effective date. Owners or operators who fail to make this certification will lose authorization to operate their units.

Q: What if an owner or operator wants to modify a permit and the modification does not have a predesignated class number?

A: EPA has established procedures [Section 270.42(d)] for a permittee wishing to make a permit modification not included in Appendix I to Section 270.42 (see Appendix F of this book) to submit a Class 3 modification request or alternatively ask EPA for a determination that Class 1 or 2 modification procedures should apply. In making this determination, EPA will consider the similarity of the requested modification to listed modifications and will also apply the general definitions of Class 1, 2, and 3 modifications.

Class 1 Modifications

Q: What is a Class 1 modification?

A: Class 1 modifications cover changes that are necessary to correct minor errors in the permit, to upgrade plans and records maintained by the facility, or to make routine changes to the facility or its operation. They do not substantially alter the permit conditions or significantly affect the overall operation of the facility. Generally, these modifications include the correction of typographical errors; necessary updating of names, addresses, or phone numbers identified in the permit or its supporting documents; updating of sampling and analytical methods to conform with revised EPA guidance or regulations; updating of certain types of schedules identified in the permit; replacement of equipment with functionally equivalent equipment; and replacement of damaged groundwater monitoring wells.

Q: How are Class 1 modifications implemented?

A: Section 270.42(a) specifies the approval procedures for Class 1 modifications. There are two categories of Class 1 modifications: those that do not require prior EPA approval and those that do.

Q: What are the requirements for Class 1 modifications that do not require prior EPA approval?

A: Under the procedures, the permittee may, at any time, put into effect any Class 1 modification that does not require prior EPA approval. The permittee is required to notify EPA by certified mail or by any other means that establishes proof of delivery within 7 calendar days of making the change. The notice must specify the change being made to the permit conditions or documents referenced in the permit and explain briefly why it was necessary.

The permittee is also required to notify by mail, persons on the *facility mailing list* within 90 days of making the modification.

Q: What is the facility mailing list?

A: EPA is required under 40 CFR 124.10(c)(viii) to compile and maintain a mailing list for each RCRA permitted facility. The list must include all persons who have asked in writing to be on the list (for example, in response to public solicitations from EPA). Also, it generally includes both local residents in the vicinity of the facility and statewide organizations that have expressed interest in receiving such information on permit modifications.

Q: Who is required to develop the facility mailing list?

A: EPA is required to develop the facility mailing list. The permittee is responsible for obtaining from EPA a complete facility mailing list and for updating it by contacting EPA periodically. However, it is also the permitting agency's responsibility to periodically inform the facility of new additions to the list. The facility owner or operator will not be held responsible for failure to notify persons recently added to the EPA list when the owner or operator has made a reasonable effort to keep its list current (53 *FR* 37915, September 28, 1988).

Q: What are the procedures for Class 1 modifications that require prior EPA approval?

A: The approval procedure is analogous to the former minor modification procedure. That is, a Class 1 permit modification requiring prior EPA approval (see Appendix F) may be made only with the prior written approval of EPA. In addition, upon approval of such request, the permittee must notify persons on the facility mailing list within 90 calendar days after approval of the request.

Q: Is there an established time frame for EPA decisions regarding Class 1 modification requests?

A: No; there are no time frame requirements for EPA action concerning a decision. However, Section 270.42(a)(3) allows a permittee to elect to follow the Class 2 process (instead of the Class 1 procedures). The Class 2 process will assure that a decision will be made on the modification request within established time frames (usually 90 to 120 days).

Q:* *Can the public request a review of a permittee's Class 1 modification?

A: Yes; although the permittee may make most Class 1 modifications without EPA approval or prior public notice, under Section 270.42(a)(iii) the public may ask the permitting agency to review any Class 1 modification.

Q:* *What occurs if EPA denies a Class 1 modification request?

A: In the event such a review is conducted, if EPA denies a Class 1 modification request, EPA shall notify the permittee in writing of this ruling, and the permittee is required to comply with the original permit conditions.

Class 2 Modifications

Q:* *What is a Class 2 modification?

A: Class 2 modifications cover changes that are necessary to enable a permittee to respond, in a timely manner, to:

- Common variations in the types and quantities of the wastes managed under the facility permit
- Technological advancements
- Regulatory changes, where such changes can be implemented without substantially altering the design specifications or management practices prescribed by the permit

Class 2 modifications include increases of 25 percent or less in a facility's non-landbased treatment or storage capacity, authorizations to treat or store new wastes that do not require different unit design or management practices, and modifications to improve the design of hazardous waste-management units or improve management practices.

Q:* *How is a Class 2 modification obtained?

A: Under Section 270.42(b)(1), a permittee who wishes to make a Class 2 modification is required to submit to EPA a modification request describing the exact change to be made to the permit conditions. The permittee must also submit supporting documents that identify the modification as a Class 2 modification, explain why the modification is needed, and provide the applicable information required by Sections 270.13 through 270.21, 270.62, and 270.63.

Q: Do Class 2 permit modification requests require public notification?

A: Yes; under Section 270.42(b)(2), the permittee must notify persons on the facility mailing list and appropriate units of state and local government and publish a notice in a local newspaper regarding the modification request.

Q: What information is to be included in the public notification?

A: Section 270.42(b)(2) specifies the information required in the notice:

- Announcement of a 60-day comment period, during which interested persons may submit written comments to the permitting agency.
- Announcement of the date, time, and place for an informational public meeting.
- Name and telephone number of the permittee's contact person, whom the public can contact for information on the request.
- Name and telephone number of an agency contact person, whom the public could contact for information about the permit, the modification request, applicable regulatory requirements, permit modification procedures, and the permittee's compliance history.
- Information on viewing copies of the modification request and any supporting documents.
- A statement that the permittee's compliance history during the life of the permit is available from the agency's contact person. Section 270.42(b)(2) also requires the permittee to submit to the permitting agency evidence that this notice was published in a local newspaper and mailed to persons on the facility mailing list.

The permittee must also make a copy of the permit modification request and supporting documents accessible to the public in the vicinity of the permitted facility (for example, at a public library, local government agency, or location under control of the owner).

Q: Is an informational public meeting required?

A: Yes; the permittee is required to hold an informational public meeting, which is open to all members of the public, no fewer than 15 days after the start of the comment period, and at least 15 days before the end of the comment period. An official transcript of the statements made at the meeting is not required, and EPA is not obligated to attend the meeting or respond to comments made at the meeting (53 *FR* 37916, September 28, 1988).

***Q:* What is meant by the "permittee's compliance history"?**

A: The regulation does not specifically define what would constitute a "compliance history"; however, it should be designed to give the public a sense of the way the facility has been operated during the permit term. For example, the compliance history could be a summary list of permit violations, dates that the violations occurred, and whether these violations have been corrected. It would not include any instances where the allegations were dismissed and would not contain confidential inspection reports or other confidential items not found in the public record (e.g., sensitive information pertaining to a pending enforcement action).

***Q:* Is there a specified time frame for EPA to respond to a Class 2 permit modification request?**

A: Yes; under Section 270.42(b)(6)(i), EPA must make one of the following five decisions within 90 days of receiving the modification request:

- Approve the request with or without changes
- Deny the request
- Determine that the modification request must follow the procedures for Class 3 modifications
- Approve the request, with or without changes, as a temporary authorization having a term of up to 180 days
- Notify the permittee that it will make a decision on the request within 30 days

If the permitting agency notifies the permittee of a 30-day extension for a decision (or if it fails to make any of the decisions), it must, by the 120th day after receiving the modification request, make one of the following decisions:

- Approve the request, with or without changes
- Deny the request
- Determine that the modification request must follow the procedures for Class 3 modifications
- Approve the request as a temporary authorization for up to 180 days

Q: Can the deadline for action be extended?

A: Yes; Section 270.42(b)(6)(vii) allows EPA to extend indefinitely or for a specified length of time, the deadlines for action on a Class 2 request after obtaining the written consent of the permittee. This option may be useful where EPA requests additional information from the permittee or when the permittee wishes to conduct additional public meetings.

Q: What occurs if EPA fails to make one of the specified decisions by the 120th day?

A: If EPA fails to make one of the four decisions listed above by the 120th day, the activities described in the modification request, as submitted, are authorized for a period of 180 days as an "automatic authorization" without EPA action. However, at any time during the term of the automatic authorization, EPA may approve or deny the permit modification request. If EPA does so, this action will terminate the automatic authorization. If EPA has not acted on the modification request within 250 days of receipt of the modification request (i.e., 50 days before the end of the automatic authorization), under Section 270.42(b)(6)(iv) the permittee must notify persons on the facility mailing list within 7 days and make a reasonable effort to notify other persons who submitted written comments that the automatic authorization will become permanent unless EPA acts to approve or deny it.

Q: What happens if EPA does not approve or deny the modification during the automatic authorization?

A: If EPA fails to approve or deny the modification request during the term of the automatic authorization, the activities described in the modification request become permanently authorized without EPA action on the day after the end of the term of the automatic authorization. However, if the owner or operator fails to notify the public when EPA has not acted on an automatic authorization 50 days before its termination date, the clock on the automatic authorization will be suspended. The permanent authorization will not go into effect until 50 days after the public is notified. Until the permanent authorization becomes effective, EPA may approve or deny the modification request at any time. In addition, the owner or operator will be subject to potential enforcement action. This permanent authorization lasts for the life of the permit unless modified later by the permittee (Section 270.42) or EPA (Section 270.41). This procedure for automatic authorization is commonly referred to as the *default* provision.

Class 3 Modifications

Q: What is a Class 3 modification?

A: Class 3 modifications cover changes that substantially alter the facility or its operations. Generally, they include increases in the facility's land-based treatment, storage, or disposal capacity; increases of more than 25 percent in the facility's non-landbased treatment or storage capacity; authorization to treat, store, or dispose of wastes not listed in the permit, which requires changes in unit design or management practices; substantial changes to landfill, surface impoundment, and waste pile liner and leachate collection/detection systems; and substantial changes to the groundwater monitoring systems or incinerator operating conditions.

Q: What are the application procedures for a Class 3 modification?

A: The first steps in the application procedures for Class 3 modifications are similar to the procedures for Class 2. Under Section 270.42(c)(1), the permittee must submit a modification request to EPA indicating the change to be made to the permit; identifying the change as a Class 3 modification; explaining why the modification is needed; and providing applicable information required by Sections 270.13 through 270.21, 270.62, and 270.63. As with Class 2 modifications, the permittee is encouraged to consult with EPA before submitting the modification request.

Q: What are the requirements for notifying the public?

A: Section 270.42(c)(2) requires the permittee to notify persons on the facility mailing list and local and state agencies concerning the modification request. This notice must occur not more than 7 days before the date of submission nor more than 7 days after the date of submission. The notice must contain the same information as the Class 2 notification, including an announcement of a public informational meeting. The meeting must be held no fewer than 15 days after the notice and no fewer than 15 days before the end of the comment period.

Q: What occurs after the public comment period?

A: After the conclusion of the 60-day comment period, the permitting agency then initiates the permit issuance procedures of 40 CFR Part 124 for the Class 3 modification. Thus, the permitting agency will prepare a draft permit modification, publish a notice, allow a 45-day public comment period on the draft permit modification, hold a public hearing on the modification if requested, and issue or deny the permit modification.

Temporary Authorizations

***Q:** If an owner or operator needs a modification immediately, is a temporary authorization available?*

A: Yes; EPA can grant a temporary authorization, without prior public notice and comment, to conduct activities necessary to respond promptly to changing conditions [Section 270.42(e)].

***Q:** What activities can be issued under a temporary authorization?*

A: An EPA-issued temporary authorization may be obtained for activities that are necessary to:

- Facilitate timely implementation of closure or corrective action activities
- Allow treatment or storage in tanks or containers of restricted wastes in accordance with Part 268
- Avoid disrupting ongoing waste management activities at the permittee's facility
- Enable the permittee to respond to changes in the types or quantities of wastes being managed under the facility's permit
- Carry out other changes to protect human health and the environment

Temporary authorizations can be granted for any Class 2 modification that meets these criteria or for a Class 3 modification that is necessary to:

- Implement corrective action or closure activities
- Allow treatment or storage in tanks or containers of restricted waste
- Provide improved management or treatment of a waste already listed in the permit where necessary to avoid disruption of ongoing waste management, allow the permittee to respond to changes in waste quantities, or carry out other changes to protect human health and the environment.

***Q:** What is the duration of a temporary authorization?*

A: A temporary authorization will be valid for a period of up to 180 days. The term of the temporary authorization will begin at the time of its approval by EPA or at some specified effective date shortly after the time of approval. The authorized activities must be completed at the end of the authorization.

SPECIAL PERMITS

Permits-by-Rule

Q: What is the permit-by-rule program?

A: To avoid duplicative permitting, EPA has established a permit-by-rule program. A permit-by-rule is essentially an amendment to an existing federal environmental permit stating that a facility or activity is deemed to have a RCRA permit if it meets specified requirements. Those eligible for a permit-by-rule are:

- Publicly owned treatment works (POTW) that have a permit under the National Pollutant Discharge Elimination System (NPDES) under the Clean Water Act (CWA)
- Persons with ocean dumping permits under the Marine Protection, Research, and Sanctuaries Act (MPRSA)
- Permitted Underground Injection Control (UIC) facilities under the Safe Drinking Water Act (SDWA)

The specified requirements for obtaining a permit-by-rule for the above activities are contained in Section 270.60.

Q: Does a wastewater treatment unit or elementary neutralization unit (as defined in Section 260.10) qualify for a permit-by-rule?

A: No; contrary to many publications that have stated this, these units are excluded from the permitting requirements under Section 270.1(1)(c)(1) and are not eligible for a permit-by-rule.

Post-Closure Permits

Q: What facilities must have a post-closure permit?

A: According to Section 270.1(c), owners or operators of surface impoundments, landfills, land treatment units, and waste piles that received wastes after July 26, 1982, or that certified closure after January 26, 1983, must have a post-closure permit, unless the unit satisfied the requirements for clean closure.

Q: If a facility has a final RCRA operating permit, is a separate post-closure permit required?

A: No; if a facility currently has an operating permit, there is no separate permit for post-closure care. The post-closure care, if applicable, should exist as a permit condition. A separate post-closure permit is intended for interim status units that close without an operating permit. Thus, if a surface impoundment loses its interim status or voluntarily closes under interim status, EPA can call in the facility's Part B for a post-closure permit. Failure to submit a Part B on time or the submission of incomplete information is grounds for the termination of a facility's interim status [Section 270.10(e)(5)].

Q: A facility has a surface impoundment. The facility stopped receiving waste on January 25, 1983. However, the facility did not have certification of closure until September 10, 1984. Is this facility required to have a post-closure permit?

A: Yes; permits covering the post-closure care period are required for all disposal units that close after January 26,1983 [Section 270.1(c)]. Units are considered closed once certification of closure is received, not when the unit stops receiving waste (50 *FR* 28712, July 15, 1985).

Q: How long is the post-closure care period?

A: The post-closure care period is 30 years, but this can be extended or reduced as appropriate by EPA (Sections 264/265.117).

Q: What information is required for a post-closure permit?

A: The information required for post-closure permits is less than the standard operating permit. At a minimum, an owner or operator must submit the following information:

- The post-closure plan
- A copy of the post-closure inspection schedule
- Location information, including a delineation of the floodplain area
- Documentation of the notice in the deed
- Cost estimate for post-closure
- A copy of the financial mechanism to be used
- Exposure information
- Groundwater data and a demonstration of compliance with Subpart F of Part 264
- Information pertaining to the 3004(u) corrective action provision
- Demonstration of financial responsibility for corrective action, if required

Emergency Permits

Q: What is an emergency permit?

A: If EPA finds that an imminent and substantial endangerment to human health or the environment exists, a temporary emergency permit may be issued (Section 270.61).

Q: What can be permitted under these procedures?

A: In accordance with Section 270.61(a), an emergency permit may be issued to:

- A nonpermitted facility to allow treatment, storage, or disposal of hazardous waste
- A permitted facility to allow treatment, storage, or disposal of wastes not covered by an effective permit

Q: How is the permit issued?

A: The permit may be oral or written. If oral, a written emergency permit must be sent to the facility within 5 days [Section 270.61(b)(1)].

Q: What are the conditions of an emergency permit?

A: In accordance with Section 270.61(b), the conditions of an emergency permit are:

- Shall not exceed 90 days
- Shall specify the types of wastes to be received and their methods of handling (e.g., disposal or treatment)
- Can be terminated at any time by the EPA
- Shall be accompanied by a public notice
- Shall incorporate, to the greatest extent possible, all applicable standards in Parts 264 and 266

RD&D Permits

Q: What are RD&D permits?

A: The EPA Administrator is authorized to issue research, development, and demonstration (RD&D) permits for innovative and experimental treatment technologies or processes for which permit standards have not been established under Part 264.

Q: What are the terms and requirements for an RD&D permit?

A: The EPA must establish permit terms and conditions for the RD&D activities as necessary to protect human health and the environment (Section 270.65). EPA is required to address construction (if appropriate), limit operation for not longer than 1 year, and place limitations on the waste that may be received to those types and quantities of wastes deemed necessary to conduct the RD&D activities. The permit must include the financial responsibility requirements currently in EPA's regulations and other such requirements as necessary to protect human health and the environment. Other possible requirements include, but are not limited to, provisions regarding monitoring, operation, closure, remedial action, testing, and information reporting. EPA may decide not to permit an RD&D project if it determines that the project, even with restrictive permit terms and conditions, may threaten human health or the environment [Section 270.65(a)].

Q: Are field demonstrations allowed?

A: RD&D activities must be conducted in units or devices made primarily from nonearthen materials. Thus, in situ treatment is not authorized under this provision.

Q: Are there RD&D activities that do not require a permit?

A: Yes; lab and bench-scale testing to determine the characteristics, composition, and compatability of waste [Section 261.4 (d)]; on-site treatment by generators (Section 262.34); and facilities meeting the definition of a wastewater treatment that add a tank (Section 260.10) are examples of activities not required to obtain an RD&D permit.

Q: What are the information requirements for an RD&D permit application?

A: An RD&D permit application must include the following information:

- Facility description
- Waste analysis plan
- Procedures for handling incompatible wastes
- Procedures to inspect testing, safety, and monitoring equipment
- Procedures for remedial actions
- Personnel qualifications
- Closure cost estimate and closure plan
- Financial assurance for closure and liability

Chapter 10

Corrective Action

The primary objective of the RCRA corrective action program is to clean up releases of hazardous waste or hazardous constituents from RCRA hazardous waste management facilities. Prior to the Hazardous and Solid Waste Amendments (HSWA), EPA had limited authority in requiring corrective action at interim status facilities. EPA had to pursue lengthy legal action or attempt to issue the facility a permit because a permitted facility has stricter requirements concerning corrective action.

However, HSWA established three new corrective action programs for interim status, hazardous waste management facilities that greatly strengthened EPA's authority (permitted facilities are governed under Part 264.100). The corrective action programs are: *Prior Releases* [3004(u)], *Beyond Facility Boundaries* [3004(v)], and *Interim Status Corrective Action Orders* [3008(h)].

PRIOR RELEASES

Scope and Applicability

Q: What is the program for corrective action for prior releases?

A: The prior-release program, also known as the Section 3004(u) program, requires corrective action for releases of hazardous waste or hazardous constituents from any solid (includes hazardous waste) waste management unit (SWMU) or hazardous waste unit at a RCRA hazardous waste management facility (Section 264.101).

Q: How is the 3004(u) program implemented?

A: The program is implemented through the permitting process, that is, an owner or operator must implement this program when applying for an operating or post-closure permit.

Q: What is meant by facility?

A: The term *facility* means the entire facility as defined in Section 260.10. Thus, even if a permit is being sought for a unit only, the owner or operator must address possible releases from all waste management units at the facility under this program.

Q: What is meant by the term release?

A: The term *release* means the CERCLA definition of release contained in Section 101(22), which reads:

> *Release* means any spilling, leaking, pumping, pouring, emitting, emptying, discharging, injecting, escaping, leaching, dumping, or disposing into the environment, but excludes (A) any release which results in exposure to persons solely within a workplace, with respect to a claim which such persons may assert against the employer of such persons, (B) emissions from the engine exhaust of a motor vehicle, rolling stock, aircraft, vessel, or pipeline pumping station engine, (C) release of source, byproduct, or special nuclear material from a nuclear incident, as those terms are defined in the Atomic Energy Act of 1954, if such release is subject to requirements with respect to financial protection established by the Nuclear Regulatory Commission under Section 170 of such Act, or, for the purposes of Section 104 of this title or any other response action, any release of source byproduct, or special nuclear material from any processing site designated under Section 102(a)(1) or 302(a) of the Uranium Mill Tailings Radiation Control Act of 1978, and (D) the normal application of fertilizer.

Q: What is meant by hazardous waste and hazardous constituent?

A: The terms *hazardous waste* and *hazardous constituent* mean any waste identified as a RCRA hazardous waste (listed or characteristic), as well as any Part 261, Appendix VIII hazardous constituent. Thus, a nonhazardous waste can release a hazardous constituent and trigger corrective action.

Q: What is meant by solid waste management unit?

A: The term *solid waste management unit* means any discernible waste management unit at a RCRA facility from which hazardous waste or hazardous constituents might migrate, irrespective of whether the unit was intended for the management of solid and/or hazardous waste. Thus, if a facility is seeking a RCRA permit and has nonhazardous waste management units (or units that accepted waste prior to November 19, 1980) on-site, the owner or operator must address those units for possible corrective action. Even though some units are currently exempt from permit standards (e.g., wastewater treatment units), they are considered SWMUs under this provision.

According to OSWER Directive No. 9502.00-6c, the solid waste management unit definition includes:

- Containers; tanks; surface impoundment; container storage areas; waste piles; land treatment units; landfills; incinerators; underground injection wells; and other physical, chemical, and biological units, including units defined as *regulated units*
- Recycling units, wastewater treatment units, and other units that EPA has generally exempted from standards applicable to waste management units
- Areas associated with production processes at facilities that have become contaminated by routine, systematic, and deliberate discharges of waste or constituents.

Q: Are spills or leaks considered solid waste management units?

A: Onetime spills of hazardous waste or constituents are subject to 3004(u) only if the spill occurred from a solid waste management unit. A spill that cannot be linked to a discernible solid waste management unit is not itself a SWMU. Likewise, leakage from product storage and other types of releases associated with production processes would not be considered a SWMU, unless those releases were routine, systemic, and deliberate (50 *FR* 28712, July 15, 1985).

Q: What is meant by routine, systemic, and deliberate discharges?

A: According to OSWER Directive No. 9502.00-6c, routine and systemic releases constitute, in effect, management of wastes; the area at which this activity has taken place can thus reasonably be considered a solid waste management unit. Deliberate does not require a showing that the owner or operator knowingly caused a release of hazardous waste or constituents. Rather, the term *deliberate* was included to indicate EPA's intention not to exercise its Section 3004(u) authority against one-time, accidental spills that cannot be linked to a SWMU. An example of this type of release would be an accidental spill from a truck at a RCRA facility.

Implementation

Q: How is the 3004(u) provision initiated?

A: The 3004(u) provision is initiated when an owner or operator applies for a permit. The permit can be an operating permit or a post-closure permit. Section 270.14(c) requires specified information concerning solid waste management units to be included in the Part B permit application.

Q: A facility received a permit prior to the effective date of HSWA. This facility is now seeking a permit modification to handle new wastes. Is the facility required to address corrective action under 3004(u) during the modification process?

A: No; Section 3004(u) states that corrective action for a facility shall be required as a condition of each permit issued after November 8, 1984. Because a permit modification is not equivalent under Section 270.41 to the issuance of a permit, a facility that is seeking a modification to a permit issued prior to November 8, 1984, is not required to address the corrective action requirements of 3004(u). A facility permit being reviewed for reissuance, however, is subject to the 3004(u) corrective action provisions.

Q: Is an owner or operator seeking a preconstruction permit for a new RCRA treatment, storage, or disposal facility subject to corrective action under Section 3004(u) of RCRA?

A: Yes; Section 3004(u) states that corrective action is required "for all releases of hazardous waste or constituents from any solid waste management unit at a treatment, storage, or disposal facility seeking a permit . . ." under Subtitle C of RCRA, ". . . regardless of the time at which wastes was placed in such unit. . . ." Therefore, any solid waste management unit located on a site that is involved in a permit application is subject to corrective action (Section 264.101), even if there has never been any previous authorization for hazardous waste activity at the site.

Q: Are there any situations in which an interim status facility could avoid corrective action requirements under 3004(u)?

A: Yes; a facility that is not required to obtain a permit under Section 3005(c) of RCRA will not have to meet Section 3004(u) (e.g., a surface impoundment that clean closes). However, if EPA found a release of hazardous waste or hazardous constituents from hazardous or solid waste, it could order corrective action under the interim status corrective action order authority in Section 3008(h). Section 3008(h) orders may be issued both before and after closure.

Q: What information is required concerning SWMUs?

A: The owner or operator must submit the following information in the Part B permit application per Section 270.14(c):

- The location of the unit or units on the required topographical map
- Designation of type of unit (e.g., landfill, impoundment)
- General dimensions and structural description (supply any available drawings)
- The dates of operation
- Specification of all wastes that have been managed at each unit

Q: How is corrective action under the 3004(u) program implemented?

A: The 3004(u) program consists primarily of four steps: the RCRA Facility Assessment (RFA), the RCRA Facility Investigation (RFI), the Corrective Measures Study (CMS), and Corrective Measures Implementation (CMI), as outlined in Figure 4.

RCRA Facility Assessments

Q: What is involved in an RFA?

A: According to OSWER Directive No. 9502.00-6c, and RFA, previously known as the preliminary assessment/site investigation (see Figure 5), will be conducted for each facility seeking a RCRA permit. The major objectives of the RFA are to:

- Identify solid waste management units and collect existing information on releases
- Identify releases or suspected releases needing further investigation

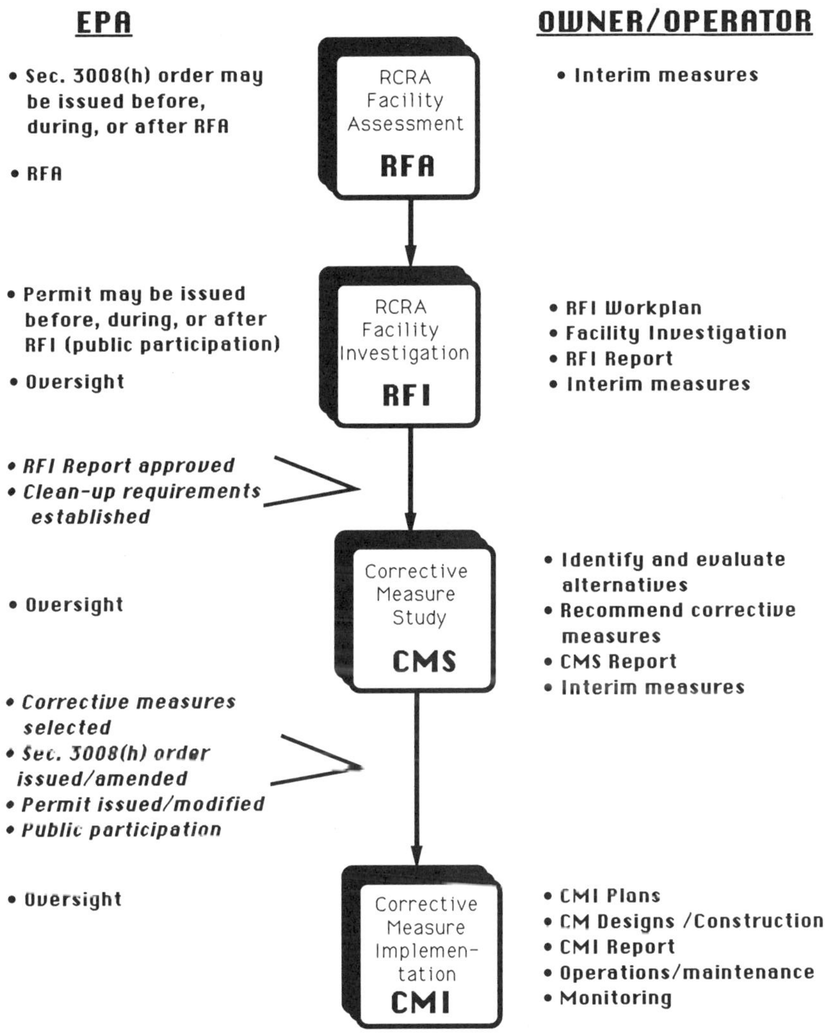

Figure 4. Corrective Action Process

Source: OWPE RCRA Corrective Action Handbook (EPA)

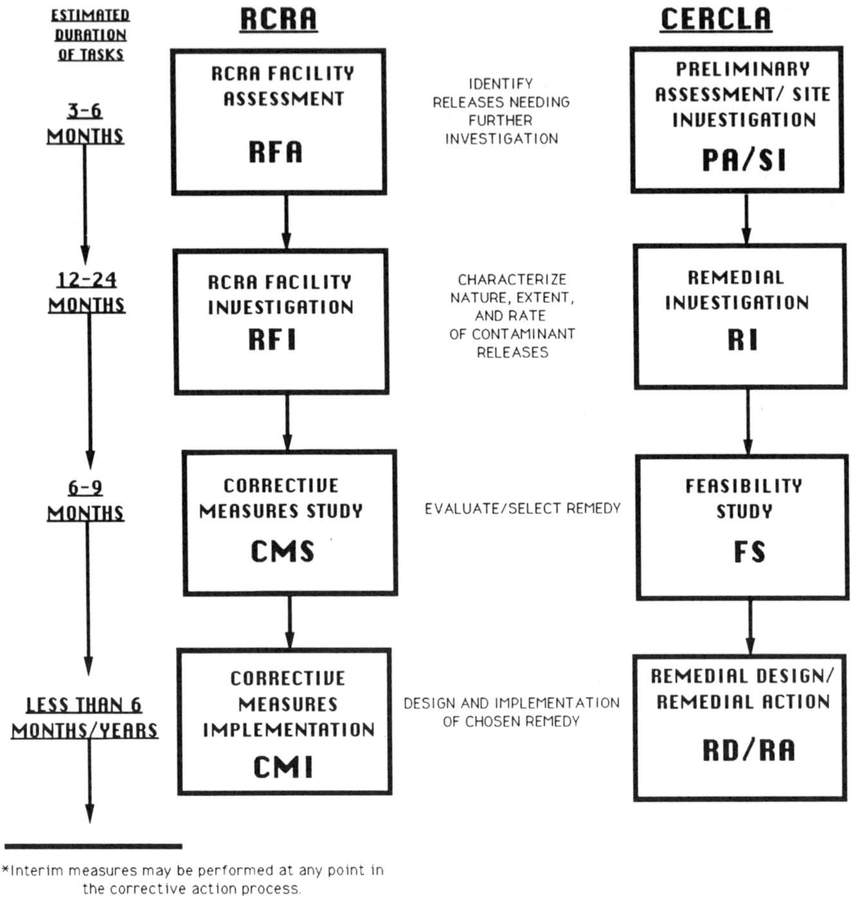

Figure 5. Comparison of CERCLA and RCRA Corrective Action Process
Source: OWPE RCRA Corrective Action Handbook (EPA)

The RFA begins with a preliminary review of pertinent information concerning the facility. If necessary, the review is followed by a visual site inspection to verify information obtained in the preliminary reviews and to gather information needed to develop a sampling plan. If necessary, a sampling visit is subsequently conducted to obtain samples for evaluation of any releases.

Q: Who is responsible for conducting an RFA?

A: As the program is currently set up, EPA is responsible for conducting RFAs. Typically, EPA uses contractors to assist it in conducting these investigations, but EPA retains overall responsibility. However, in some cases, the facility owner or operator is requested to conduct certain sampling activities (OSWER Directive No. 9502.00-5).

Q: What happens after an RFA?

A: The findings of the RFA will result in one or more of the following actions:

- No further action is required, because there is no evidence of a release.
- A RCRA facility investigation by the owner or operator will be required where the information collected indicates a release exists and further information is required.
- Interim corrective measures will be required by the owner or operator where EPA determines that expedited action is necessary to protect human health or the environment.

RCRA Facility Investigations

Q: What is involved during an RFI?

A: The RFI, which is analogous to the remedial investigation in the CERCLA program, may initially involve verification of suspected releases. If releases are confirmed, further characterization of such releases may be necessary. This characterization includes identification of the type and concentration of hazardous waste or constituents released, the rate and direction at which the releases are migrating, and the distance over which releases have migrated.

The RFI also includes characterization requirements to assess the need for a corrective measures study and data required for that study.

Corrective Measures Study

Q: What is the purpose of a CMS?

A: If the need for corrective measures is identified during the RFI process, the owner or operator is required to conduct a corrective measures study. The purpose of the CMS is to identify and recommend specific corrective measures that will adequately correct the release. The CMS process is analogous to the feasibility study under the CERCLA program.

Q: What is typically included in a corrective measures study?

A: A corrective measures study typically includes:

- Evaluation of performance, reliability, ease of implementation, and potential impacts of the remedy, including safety impacts, cross-media impacts, and control of exposure to any residual contamination
- Assessment of the effectiveness of potential remedies in achieving adequate control of sources and cleanup of hazardous waste and constituents released from SWMUs
- Assessment of the time required to begin and complete the remedy
- Estimation of the costs of remedy implementation
- Assessment of the institutional requirements, such as state or local permit requirements, or other environmental or public health requirements which may substantially affect implementation of the remedy

Q: What cleanup levels are used?

A: Because each site is different, EPA has not established national uniform standards for cleanup, per se. EPA-recommended limits or factors are generally used as target cleanup levels. These factors are health-based levels (e.g., RfDs and CPFs) that use the risk range of 10^{-4} to 10^{-6}. Site-specific factors are used to determine the final cleanup levels. These factors include groundwater use (i.e., drinking water), depth of contamination, mobility and persistence of contaminants, and potential receptors.

Q: What occurs after completion of the CMS?

A: Upon completion and receipt of the corrective measures study, EPA evaluates its adequacy. If the plan is deficient, EPA either modifies the plan or requires the owner or operator to make the appropriate modifications.

Corrective Measures Implementation

Q: What is the purpose of the corrective measures implementation?

A: The purpose of the CMI program is to design, construct, operate, maintain, and monitor the performance of selected corrective measures.

Q: Must all of these steps be completed before a permit is issued?

A: No; these provisions can be implemented through compliance schedules contained in the facility's permit. On June 22, 1987 (52 *FR* 2344), EPA promulgated amendments to allow the information related to detailed corrective action planning, required by Sections 270.14(c)(7), (c)(8)(iii), and (c)(8)(iv), to be developed at EPA's discretion after the issuance of a permit through the use of compliance schedules.

The owner or operator will be required to obtain advanced written authorization from EPA waiving the submittal of the corrective action information if issued in compliance schedules. This waiver applies only to the design of the corrective action program and not to information required to assess the need or extent of corrective action.

INTERIM STATUS CORRECTIVE ACTION ORDERS

Scope and Applicability

Q: What is an interim status corrective action order?

A: Authorized by Section 3008(h) of HSWA, EPA can require corrective actions or other measures necessary to protect human health and the environment whenever there is or has been a release of hazardous wastes or hazardous constituents from an interim status facility. Such orders may also revoke or suspend a facility's interim status and/or assess penalties of up to $25,000 per day for noncompliance with previous corrective action orders.

Q: Do the terms* release *and* facility *mean the same as in Section 3004(u)?

A: Yes; these terms are parallel.

Q: What is the difference between 3008(h) and 3004(u)?

A: The authority under Section 3008(h) is not confined to addressing releases from solid waste management units as it is under 3004(u). Thus, one-time spills and other types of contamination at facilities can be addressed under Section 3008(h). In addition, 3004(u) is initiated only when an owner or operator is applying for a permit, whereas 3008(h) has no such limitations for interim status facilities.

Q: An interim status container storage facility has two solid waste management units (SWMUs) on site. If the facility was closed before receiving a final permit, could EPA enforce interim status corrective action measures against the facility to clean up the SMWUs?

A: Yes; the authority to enforce corrective action measures at an interim status facility is not necessarily tied to closure at hazardous waste management units at the facility in question. Facilities with closed units may remain in interim status and are potentially subject to an enforcement action, pursuant to Section 3008(h).

Q: The owner of a surface impoundment has managed hazardous wastes in the impoundment without interim status or a RCRA permit. A release of hazardous waste from the impoundment has contaminated surrounding soil and groundwater. Upon discovery of the contamination, EPA intends to issue a corrective action order under Section 3008(h) of RCRA. Given that the owner never had interim status, can the corrective action order be issued?

A: Yes; Section 3008(h) authorizes the EPA Administrator to issue corrective action orders to address releases of hazardous wastes into the environment from facilities authorized to operate under interim status [Section 3005(e)]. This authority extends to include those facilities that should have had interim status, but failed to notify EPA under Section 3010 of RCRA or failed to submit a Part A application (OSWER Directive No. 9901.1)

BEYOND FACILITY BOUNDARIES

Scope and Applicability

Q: What are the requirements for corrective action beyond facility boundaries?

A: Owners and operators of hazardous waste management units are required to institute corrective action beyond the facility's boundary where necessary to protect human health and the environment, unless the owner or operator is denied access to adjacent property, despite the best efforts of the owner or operator (52 *FR* 45790, December 1, 1987).

Q: How does EPA define best efforts?

A: In determining *best efforts,* EPA will consider case-by-case circumstances; however, at a minimum, the effort should include a certified letter to the adjacent property owner (52 *FR* 45790, December 1, 1987).

Q: What if off-site access is denied?

A: Even if permission from the adjacent landowner is denied (despite the best efforts of the owner or operator), the facility is not necessarily relieved of its responsibility to undertake corrective measures to address releases that have migrated beyond the facility boundary. EPA can require the facility to implement on-site corrective measures in an attempt to clean up releases beyond its boundary. Any corrective measures will be based on their feasibility and appropriateness based on a case-by-case specific basis, considering hydrogeologic conditions and other relevant factors (52 *FR* 45790, December 1, 1988).

Chapter 11

Enforcement

The purpose of any enforcement program, including RCRA's, is to compel compliance with a set of regulations or statutory provisions. The RCRA enforcement program has many tools, both legal and administrative, to help or force compliance with the regulations. The effectiveness of the hazardous waste management regulatory program under RCRA depends on whether or not the regulated community complies with the requirements. Close monitoring of facility activities and legal action when noncompliance exists are necessary. Monitoring allows EPA to determine which facilities are out of compliance and to evaluate the effectiveness of enforcement programs.

FACILITY INSPECTIONS

Scope and Applicability

Q: What initiates an enforcement action?

A: There are two situations that may give rise to the initiation of a RCRA enforcement action. The first is when a facility is found through a compliance inspection to be out of compliance with applicable Subtitle C regulations. The second is when a facility is found from a referral or citizen's complaint to present a substantial threat to human health or the environment.

Q: What is a compliance inspection?

A: A compliance inspection is the primary enforcement mechanism used to detect and verify violations of RCRA. Establishments are selected for inspection either under a neutral administrative inspection scheme or "for cause" (i.e., where probable violations of the act are observed or brought to the attention of EPA through, for example, an employee's complaint or a competitor's tip).

Q: What authority does EPA have for conducting inspections?

A: Section 3007 of RCRA authorizes EPA employees and their representatives (including EPA contractors) to enter at reasonable times the establishment of any person where hazardous wastes are or have been generated; transported from; or treated, stored, or disposed. This authorization includes the allowance to inspect the facility, obtain samples, and obtain and copy records and information related to hazardous wastes (OSWER Directive No. 9938.0).

It is important to note that the authority to inspect a facility is not limited to compliance inspections, but also for the purposes of assisting EPA in developing regulations.

Q: Are EPA representatives in any way limited in what they can inspect and sample?

A: Pursuant to Section 3007, authorized officers, employees and representatives, including authorized contractors, are allowed to enter any portion of a facility that is being used or has been used to generate or manage *hazardous wastes*. Such persons may inspect and obtain samples of hazardous wastes and inspect containers and labeling of such wastes. The inspection must be for the purpose of developing regulations or enforcing provisions of RCRA. The specific objective of the inspection does not have to be written in any form, but the inspection must strictly deal with the generation, management, or transportation of hazardous waste (OSWER Directive No. 9938.0).

Q: What occurs during an inspection?

A: The main purpose of any inspection is to determine whether the facility is operating in accordance with the terms of its permit or in compliance with the interim status standards.

An inspection typically consists of the following steps:

- Before visiting the facility, the inspector reviews the facility's permit or other records to identify any problems that may be encountered.
- The inspector enters the facility, shows indentification, and describes the nature of the inspection. An opening conference is held with the owner or operator to describe the information and samples to be gathered.
- The facility is inspected. The inspection includes examination of facility records, possible collection of samples, and observation of the facility, including any hazardous waste management operations. The inspector will also observe all associated activities, such as unloading of wastes, lab work, and safety procedures. The inspector may use field notebooks, checklists, and photographs to document the visit.
- The inspector holds a closing conference with the owner or operator to respond to questions about the inspection and provide additional information.
- The inspector prepares a report summarizing the results of the inspection, including the results of sampling. Violations are documented in the report. Inspections usually last between a day and a week.

Q: How often is a facility inspected by EPA?

A: RCRA requires that all federal- or state-operated facilities be inspected at least annually and that all other hazardous waste management facilities be inspected at least once every 2 years. Inspections are scheduled by states and EPA regional offices according to criteria that ensure greater attention to facilities of greater concern. Inspections may also be conducted at any time based on suspicion that a violation is occurring.

Q: What types of facilities can be inspected?

A: The scope of a RCRA inspection extends to any establishment or other place where hazardous wastes are or have been generated, stored, treated, disposed of, or transported from. In addition, inspectors have access to and may copy all records relating to the hazardous waste of any person who generates, treats, stores, disposes of, or transports any hazardous waste or who has handled such waste in the past. An inspection may include taking samples of hazardous waste or obtaining samples of any containers or labeling for such waste.

Q: What about confidentiality of inspections?

A: Pursuant to Section 3007(b) of RCRA, any records, reports, or other information obtained as a result of an inspection are available to the public, unless a claim of confidentiality is asserted under EPA's business confidentiality regulations contained in 40 CFR Part 2.

ENFORCEMENT ACTIONS

Administrative Actions

Q: What enforcement actions are available under RCRA?

A: There are several enforcement options available, including:

- Administrative actions
- Civil actions
- Criminal actions

Q: What is an administrative action?

A: An administrative action is a nonjudicial enforcement action taken by EPA or a state under its own authority. These actions require less preparation than a lawsuit (civil action). Two types of administrative actions exist, informal actions and administrative orders. Both constitute an enforcement response outside the court system.

Informal Action

Q: What is an informal action?

A: An informal administrative action is any communication from EPA or a state that notifies a facility that is not in compliance with a specified provision of the regulations, such as a warning letter. If the owner or operator does not take steps to comply within a certain time, a more formal action can be taken. For more serious violations, EPA or the state can use a formal warning letter setting out specific actions to be taken to bring the facility back into compliance. A formal warning also sets out the enforcement actions that will follow if the facility fails to take the required steps.

Q: What is a warning letter?

A: A warning letter is a letter issued by EPA that advises a company or person that a violation of RCRA has been detected. Although issuance of a warning letter is not specifically authorized by RCRA, the letter can be a useful enforcement tool.

Q: What information is contained in a warning letter?

A: A warning letter generally contains the following information:

- Identification, citation, and explanation of the violation
- A deadline for achieving full compliance with the appropriate regulatory or statutory requirements
- A statement indicating that continued noncompliance beyond a particular date would generally result in the issuance of a Section 3008 compliance order or other enforcement action, including the assessment of civil penalties of up to $25,000 per day per violation
- The name and telephone number of an EPA contact person

Q: How is a warning letter issued?

A: EPA issues a warning letter by certified mail, return receipt requested. In addition, a copy of the letter is placed in the case file.

Administrative Orders

Q: What is an administrative order?

A: When a violation is detected that is more severe than those requiring an informal action, EPA or the state can issue an administrative order. An administrative order, issued directly under the authority of RCRA, imposes enforceable legal requirements. Orders can be used to force a facility to comply with specific regulations; to initiate corrective actions; to conduct monitoring, testing, and analysis; or to address a threat to human health or the environment. Four types of orders can be issued under RCRA: compliance orders, corrective action orders, Section 3013 orders, and Section 7003 orders.

Q: What is a compliance order?

A: Section 3008(a) of RCRA authorizes the use of an order requiring any person who is not complying with a requirement thereunder to take steps to come into compliance. A compliance order may require immediate compliance or may set out a timetable to be followed to move toward compliance.

Q: What penalties are available with a compliance order?

A: A compliance order can include a penalty of up to $25,000 for each day of noncompliance and can suspend or revoke the facility's permit or interim status.

Q: Can the person who was served the order request a hearing?

A: Yes; when EPA issues a compliance order, the person to whom the order is issued can request a hearing on any factual provisions of the order. If no hearing is requested, the order will become final 30 days after it is issued.

Q: What is a corrective action order?

A: Section 3008(h) of HSWA authorizes the use of an order requiring corrective action at an interim status facility when there has been a release of hazardous waste or constituents into the environment. These orders can be issued to require corrective action, regardless of when waste was placed in a unit (see Chapter 10).

Q: What penalties are available pursuant to a corrective action order?

A: Corrective action orders can suspend interim status and impose penalties of up to $25,000 for each day of noncompliance with the order.

Q: What is a Section 3013(a) order?

A: In addition to compliance inspections, RCRA compliance monitoring activities may also involve, under certain circumstances, the use of Section 3013(a) administrative orders. Such orders, which are issued to owners or operators of hazardous waste facilities or sites, require that reasonable testing, analysis, and monitoring be conducted with respect to a facility or site to ascertain the nature and extent of a situation that may present a substantial hazard to human health or the environment.

Q: What is required of a person who is served a Section 3013(a) order?

A: A Section 3013(a) order requires the person to whom the order was issued to submit to EPA within 30 days from the issuance of the order a proposal for carrying out the required monitoring, testing, analysis, and reporting. The EPA Administrator may, after providing the person an opportunity to confer with EPA, require that the person carry out the proposal, as well as make any modifications in the proposal as the EPA Administrator deems reasonable to ascertain the nature and extent of the hazard.

Q: What occurs if a person fails or refuses to comply with a Section 3013(a) order?

A: EPA may commence a civil judicial action against any person who fails or refuses to comply with a Section 3013(a) order. Such an action is brought in the United States district court in which the defendant is located, resides, or is doing business. The court may not only require compliance with the order, but may also assess a civil penalty of not more than $5,000 for each day during which such failure or refusal occurs.

Q: What is a Section 7003 order?

A: In any situation where an "imminent and substantial endangerment to health or the human environment" is caused by the handling of hazardous or solid wastes, EPA can issue a Section 7003 order compelling any action necessary to prevent such threat.

Q: What evidence is required to support a Section 7003 order?

A: Evidence possessed to support the issuance of a RCRA Section 7003 order must show that the "handling, storage, treatment, transportation, or disposal of any solid or hazardous waste *may present* an imminent and substantial endangerment to health or the environment."

Q: What is meant by "may present"?

A: The words "may present" indicate that Congress established a standard of proof that does not require a certainty. The evidence is not required to demonstrate that an imminent and substantial endangerment to public health or the environment definitely exists. Instead, an order may be issued if there is sound reason to believe that such an endangerment may exist (OSWER Directive No. 9940.2).

Q: Can a Section 7003 order be used to compel past contributors at a site?

A: Yes; prior to HSWA, in any situation in which an imminent and substantial endangerment to human health or the environment was caused by the handling of hazardous wastes, the responsible agency could order those contributing to the problem to take steps to rectify the situation. Although this language implied that only those currently contributing to the endangerment could be ordered to clean up the problem, Section 7003 was almost always interpreted to include past contributors as well. HSWA validated this interpretation by allowing EPA to bring actions against past or present generators, transporters, or owners or operators of a site. Violation of a Section 7003 order can result in fines of up to $5,000 per day.

Q: When can the Section 3008(h) authority be used? Can a Section 3013 order support the Section 3008(h) action?

A: Section 3008(h) authorizes EPA to require corrective action or any other response necessary to protect human health or the environment when a release of hazardous waste is identified at an interim status hazardous waste treatment, storage, or disposal facility.

Section 3008(h) provides: "Whenever on the basis of any information the Administrator determines that there is or has been a release of hazardous waste into the environment . . ." Actual sampling data is not necessary to show a release.

However, to exercise the interim status corrective action authority, EPA must first have information that there is or has been a release at the facility. Additional sources that may provide information on releases include: inspection reports, RCRA facility assessments, RCRA Part A and Part B permit applications, responses to Section 3007 information requests, information obtained through Section 3013 orders, notifications required by Section 103 of CERCLA, information gathering activities conducted under Section 104 of CERCLA, or citizens' complaints corroborated by supporting information.

A Section 3013 order may be used in some instances in which EPA does not have adequate information that there is or has been a release. Section 3013 provides that EPA may compel monitoring, testing, and analysis if the presence of hazardous waste at a facility or site at which hazardous waste has been treated, stored or disposed of may present a substantial hazard to human health or the environment.

Q: Is a RCRA compliance order issued to the owner of a facility or its operator? Who is responsible for complying with the order?

A: EPA has always held that both the owner and the operator are equally responsible for compliance with the permit issued to a facility. Section

3005(a) of RCRA requires "each person owning or operating" a treatment, storage, or disposal facility to obtain a permit. For example, the permit regulations require both owner and operator to sign the permit application, according to Section 270.10(b). The permit will be issued to both the owner and operator. Preamble discussions in the May 19, 1980, *Federal Register* confirm this concept of dual responsibility at 45 *FR* 33169 and 45 *FR* 33295. Both discussions specifically reference situations where the operator may be different from the landowner or facility owner.

EPA considers both the owner and operator of a facility to be responsible for regulatory compliance. For this reason, EPA may initiate an enforcement action against either the owner, the operator, or both. Normally, the compliance order is issued to the person responsible for the daily operations at the facility, because this person is most likely to be in the position to correct the problems. If the operator is unable or unwilling to rectify the problems, then EPA may issue a separate compliance order to the owner.

Civil Actions

Q: Are there other enforcement responses available to EPA beyond administrative actions?

A: Yes; in addition to administrative enforcement responses, EPA may initiate civil judicial actions under RCRA's federal enforcement and imminent hazard provisions.

Q: What is a civil action?

A: A civil action is a formal lawsuit, filed in court against an individual or facility that either has failed to comply with some regulatory requirement or administrative order or has contributed to a threat to human health or the environment. Civil actions are generally reserved for situations that present repeated or significant violations or serious threats to the environment. The U.S. Department of Justice represents EPA in any civil or criminal prosecution or defense.

Q: When is a civil action more appropriate?

A: Such civil actions may be used to compel compliance with the act's statutory and regulatory requirements, as well as to assess civil penalties in cases of noncompliance. The choice of whether to pursue an administrative or a judicial remedy must be made on a case-by-case basis. Generally, however, where one or more of the following factors are present, a civil judicial action is preferred to the issuance of an administrative order:

- Where a person has failed to comply with an administrative order
- Where a person's conduct must be immediately stopped to prevent irreparable injury, loss, or damage to human health or the environment
- Where long-term compliance by a person needs to be compelled
- Where notoriety of the action is necessary to deter others similarly situated from violating the requirements of the act

Q: What types of civil actions are available?

A: RCRA provides authority for filing four types of civil actions: compliance action, corrective action, monitoring and analysis action, and an imminent hazard Action.

Q: What is a compliance action?

A: EPA or an authorized state can file a suit to force a person to comply with applicable RCRA regulations. The court can also impose a penalty of up to $25,000 per day for noncompliance.

Q: What is a corrective action?

A: In a situation in which there has been a release of hazardous waste from a facility, EPA or a state can sue to have the court order the facility to correct the problem and take any necessary response measures. The court can also suspend or revoke a facility's interim status as a part of its order.

Q: What is a monitoring and analysis action?

A: If EPA or a state has issued a monitoring and analysis order under Section 3013 of RCRA, and the facility to whom the order was issued fails to comply, EPA or a state can sue to get a court to require compliance with the order. In this type of case, the court can levy a penalty of up to $5,000 for each day of noncompliance with the order.

Q: What is an imminent hazard action?

A: As with a Section 7003 administrative order, when any facility or person contributed or is contributing to an imminent hazard to human health or the environment, EPA or a state can sue the person or facility and ask the court to require that person or facility to take action to remove the hazard or remedy any problem. If an agency has first issued an administrative order, the court can also impose a penalty of up to $5,000 for each day of noncompliance with the order.

Q: Can these civil actions be used together?

A: Yes; frequently, several of the civil action authorities will be used together in the same lawsuit. This is likely to happen when a facility has been issued an administrative order for violating a regulatory requirement, has ignored that order, and is in continued noncompliance. In this circumstance, a lawsuit can be filed that seeks penalties for violating the original requirement, penalties for violating the order, and a judge's order requiring future compliance with the requirement and the administrative order.

Q: What does EPA use to determine the amount of a fine under RCRA?

A: EPA uses the *RCRA Civil Penalty Policy* for assessing administrative penalties under RCRA. The purpose of the policy is to assure that RCRA civil penalties are assessed in a fair and consistent manner, that penalties are appropriate for the violation committed, that economic incentives for noncompliance are eliminated, that persons are deterred from committing violations, and that compliance is achieved (OSWER Directive No. 9900.1).

Criminal Actions

Q: What is a criminal action?

A: Section 3008 of RCRA identifies seven acts that constitute a criminal action. These are:

- Transportation of hazardous waste to a nonpermitted or noninterim status facility
- Treatment, storage, or disposal of hazardous waste without a permit or in violation of a condition of a permit
- Omission of required information from a label or manifest
- Generation, storage, treatment, or disposal of hazardous waste without compliance with RCRA's recordkeeping and reporting requirements
- Transportation of hazardous waste without a manifest

- Export of hazardous waste without the consent of the receiving country
- Knowing transportation, treatment, storage, disposal, or export of any hazardous waste in such a way that another person is placed in imminent danger of death or serious bodily injury

Q: What are the penalties associated with criminal acts?

A: The first six criminal acts carry a penalty of up to $50,000 per day or up to 5 years in jail. The seventh criminal act carries a maximum penalty of $250,000 or 15 years imprisonment for an individual or a $1 million fine for a corporation.

EPA Action in Authorized States

Q: Can EPA take enforcement action in authorized states?

A: Yes; although states with authorized programs have the primary responsibility for ensuring compliance with the RCRA program requirements, Section 3008 of RCRA specifically provides EPA with the authority to take enforcement action in authorized states under certain conditions. It is EPA's policy to take enforcement actions in authorized states when the state asks EPA to do so or fails to take timely and appropriate action.

Q: When will EPA take action?

A: If the state has failed to issue an order or complete a referral within 90 days after the discovery of a *high-priority violator,* the EPA regional office may also choose to assess a penalty against a high-priority violator if the state's action failed to include one. The memorandum of agreement (MOA) or grant agreement (GA) between EPA and each state should set out the mechanism by which notice will be provided. The EPA regional office may need to conduct its own case-development inspection and prepare additional documentation before initiating an action. If the state has made reasonable progress in returning the facility to compliance or in processing an enforcement action, EPA may delay a response.

Q: What is a high priority violator?

A: As outlined in OSWER Directive No. 9900.0-1A, a *high priority violator* is a hazardous waste handler who:

- Has caused actual exposure or a substantial likelihood of exposure to hazardous waste or constituents
- Is a chronic or recalcitrant violator
- Deviates from conditions of a permit, order, or decree by not meeting the requirements in a timely manner, and/or by failing to perform work as required by terms of permits, orders, or decrees
- Substantially deviates from RCRA statutory or regulatory requirements

Citizen Suits

***Q:* Can a person bring action against the United States or any person who is in violation of Subtitle C of RCRA?**

A: Yes; Section 7002 of RCRA authorizes any person to commence a civil action (suit) on his or her own behalf against:

- Any person or government agency alleged to be in violation of any permit, standard, regulation, condition, requirement, or order that has become effective under the act
- Any past or present generator, transporter, or owner or operator of a facility who has contributed to or is contributing to a condition that may present an imminent and substantial endangerment to human health or the environment
- The EPA Administrator where there is alleged a failure of the EPA Administrator to perform any act or duty under RCRA that is not a discretionary action

***Q:* Are there limitations to the citizen suit provisions?**

A: Yes; the right of citizens to bring suits under Section 7002 is limited to certain situations, including:

- No suit may be brought if the EPA Administrator or a state has commenced, and is "diligently prosecuting" an enforcement action against the alleged violator.
- Suits are not allowed to be used to impede the issuance of a permit or the siting of a facility (except by state and local governments).
- Transporters are protected from citizen suits in response to incidents arising after delivery of waste.
- The facility is actively engaged in a removal action under CERCLA (Superfund).

Q: What are the requirements for filing a citizens' suit?

A: Before filing a suit under this provision, the person must give a 60-day notice to EPA, the state in which the alleged violation occurs, and to the alleged violator.

If a citizen is suing a past or present generator presenting a substantial endangerment under Section 7002(a)(1)(B), a 90-day notice must be given to EPA, the state in which the alleged violation occurs, and to the alleged violator.

Q: What is required for the notification of intent?

A: There is no prescribed format for the notification. A simple, concise letter (certified) stating the alleged violation, the alleged violator, the location of the alleged violator, and the intent to file suit is sufficient.

The notice to EPA should be sent to:

The Administrator (A-100)
U.S. EPA
401 M Street, SW
Washington, DC 20460

Q: What are the requirements for filing the suit?

A: The suit must be brought in the district court for the district in which the alleged violation or endangerment occurred. If a person is bringing suit against the EPA Administrator, the suit may be brought in either the district court in which the alleged violation occurred or the District Court of the District of Columbia. In addition, whenever action is brought against a past or present contributing generator, the plaintiff must serve a copy of the suit to the Attorney General of the United States and the EPA Administrator.

Q: Can legal costs be recovered if a person files a citizen suit?

A: Yes; the court, in issuing any final order in any action brought under Section 7002, may award costs of litigation (including reasonable attorney and expert witness fees) to the prevailing or substantially prevailing party, whenever the court determines such an award is appropriate.

Chapter 12

State Authorization

It is the intent of RCRA to have the nation's hazardous waste management program administered by the states with only minimal oversight from the federal government. The program elements and the process for states to obtain responsibility for managing the RCRA Subtitle C program are outlined in Part 271. This process involves the development of a state's hazardous waste program and subsequent approval by EPA.

Because the federal RCRA program is constantly changing, states are required to update their program to be consistent with the federal program.

STATE AUTHORIZATION PROCESS

Introduction

Q: How is the federal hazardous waste program implemented by a state?

A: Because EPA's hazardous waste regulations were developed in stages, the states have been given the opportunity to implement a phased approach to the Subtitle C program, which allows the state to eventually operate the program. Previous to HSWA, a state could obtain *interim authorization* (Phase I) for a program that was *substantially equivalent* to the federal program. Interim authorization allowed a state to administer Parts 260, 261, 262, 263, and 265. A state could also obtain authorization to administer specific permitting programs (Phase II) for various units.

For example, a state that had Phase II component A could write permits for tanks and container storage units only; component B was for incinerators, and component C was for land-disposal units.

Q: Has HSWA changed the phased approach?

A: Yes; under HSWA, states with only interim authorization as of January 31, 1986, had their program reverted to full EPA control. Since that date, states have had to seek final (base) authorization for the complete program (the complete Subtitle C program as it was before the enactment of HSWA). A phased approach is no longer available for pre-HSWA requirements. Any state that had final authorization prior to HSWA must obtain authorization specifically for HSWA. Until a state has HSWA authorization, EPA will administer the HSWA provisions, and the state, if authorized, will administer the pre-HSWA requirements.

Q: Can a state's hazardous waste management program be different from the federal program?

A: Yes; a state with authorization may be *broader in scope* or more *stringent* than EPA [Section 271.121(i)].

Q: What is the difference between state program elements that are broader in scope and those that are more stringent than federal requirements?

A: According to program implementation guidance (PIG) 84-1, a state program that is broader in scope than the federal program either: (1) expands the size of the regulated community; or (2) incorporates program elements that do not have a federal counterpart. Examples of requirements that are broader in scope are permits for federally exempt wastewater treatment units, special licenses for transporters, and listing of wastes that are not listed federally.

A state program requirement that is more stringent has a direct federal program counterpart. Examples of more stringent requirements are requiring generators to submit an annual, rather than a biennial report; shorter durations for permits; and stricter management standards for permitted or interim status tanks and containers (e.g., more rigorous performance standards for secondary containment).

The distinction between broader and more stringent state requirements is significant, because EPA may enforce a more stringent state requirement, but not state requirement that is broader in scope. Section 3008(a)(2) of RCRA allows EPA to enforce any provision of an authorized state's approved program. More stringent state requirements fall into this category. State provisions that are broader in scope are not part of the federally approved RCRA program, according to Section 271.1(i), and are therefore not enforceable by EPA (see also PIG 82-3).

Q: Can an unauthorized state operate its own hazardous waste program?

A: Yes; persons will be subject to both the RCRA federal program and the state's hazardous waste program.

Required Information

Q: What information is required from a state when applying for authorization?

A: In accordance with Section 271.5, any state that seeks authorization for its base hazardous waste program must submit an application to the EPA Administrator that consists of the following elements:

- A letter from the governor requesting program approval
- Copies of all applicable state statutes and regulations, including those governing state administrative procedures
- A description of the program
- An attorney general's statement
- A memorandum of agreement

Q: What is required in the program description?

A: As the name implies, the program description submitted to EPA must detail the state hazardous waste program (Section 271.6). It must include descriptions of:

- The scope, structure, coverage, and processes of the state program
- The state agency or agencies that will have responsibility for running the program
- The state-level staff who will carry out the program
- The state's compliance tracking and enforcement program
- The state's manifest system

If a state chooses to develop a program that is more stringent or extensive than the one required by federal law, the description should address those parts of the program that go beyond what is required under Subtitle C.

In addition, the program description must include an estimate of:

- Costs involved in running the program and an itemization of the sources and amounts of funding available to support the program's operation

- The number of generators, transporters, and on-site and off-site disposal facilities (along with a brief description of the types of facilities and an indication of the permit status of these facilities)
- The annual quantities of hazardous wastes generated within the state; transported into and out of the state; and stored, treated, or disposed of within the state (if available).

Q: What is required in the attorney general's statement?

A: Any state that wants to assume the responsibility for Subtitle C must demonstrate to EPA that the laws of the state provide adequate authority to carry out all aspects of the state program. This demonstration comes in the form of a statement written by the state's attorney general. The statement includes references to the statutes, regulations, and judicial decisions that the state will rely on in administering and enforcing its program (Section 271.7).

Q: What is the memorandum of agreement?

A: Although a state with an authorized program assumes primary responsibility for administering Subtitle C, EPA still retains some responsibilities and oversight powers in relation to the state's execution of its program. The memorandum of agreement (MOA) between the state director and the EPA regional administrator outlines the nature of these responsibilities and oversight powers and the level of coordination between the state and EPA in operating the program. As outlined in Section 271.8(b), the MOA includes provisions for:

- Specification of the frequency and contents of reports that the state must submit to EPA
- Coordination of compliance monitoring activities between the state and EPA
- Joint processing of permits for those facilities that require a permit from both the state and EPA under different programs
- Specification of those types of permit applications that will be sent to the EPA regional administrator for review and comment

Q: When state programs are authorized, does EPA cease to be involved in reporting, permitting, and enforcement requirements?

A: No; EPA's role is not completely eliminated, even though the primary responsibility for the program remains with the state. The regulated community would report to the state in accordance with state requirements.

The state, in turn, would report annually to EPA, and more frequently in certain other instances (noncompliance, for example).

In general, the state would administer its own standards. Although EPA will review and comment on some draft permits and permit applications, the state will be the principle permitting authority.

Section 3008 of RCRA allows EPA to enforce any standard, including standards of authorized states. Thus, under certain circumstances (such as when the state fails to take action), EPA may directly enforce such standards.

Q: What conditions of a state program must be satisfied to obtain approval?

A: The state program must be found to be:

- Equivalent and no less stringent than the federal program
- Consistent with other state programs
- Adequately enforceable
- Providing adequate public notice and hearing in the permit process

Q: What is meant by equivalent and no less stringent?

A: The state program must adopt regulatory and statutory requirements that are at least equivalent to and no less stringent than those implemented and enforced under the Subtitle C program at the federal level. This does not mean that the state program cannot differ from the federal program. Indeed, the state's program can be more stringent and extensive than the federal program [Sections 3006(b) and 3009].

Q: What is meant by consistency?

A: The state program must be consistent with the federal program and other authorized state programs. EPA focuses its review of consistency on those provisions of a state program that may interfere with the proper operation of the national regulatory scheme developed under RCRA. Accordingly, if a state program unreasonably restricts, impedes, or operates as a ban on the free movement of hazardous waste across state borders or does not meet the federal manifest requirements, it is deemed inconsistent and cannot be approved. In addition, any aspect of state law or of the state program that has no basis in human health or environmental protection and that acts as a prohibition on the treatment, storage, or disposal of hazardous waste in the state may be deemed inconsistent and therefore not approvable [Section 3006(b)].

Q: What is required of an enforcement program?

A: RCRA requires that state programs contain adequate authority to enforce all the requirements developed under Subtitle C. In assessing enforceability, EPA focuses on the inspection, enforcement remedy, and penalty authorities contained in the program. The state program must also provide for public participation in the enforcement process [Section 3006(b)].

Q: What is required for public notice and hearing in the permit process?

A: Under Section 7004(b) all state programs must provide for public notification prior to the issuance of permits. Furthermore, the program must require that both a public comment period (at least 45 days) and an informal public hearing be held if a request for such a hearing is made during the comment period.

Q: What can the public do to ensure that hazardous waste facilities comply with the regulations in an authorized state?

A: An approved state program must provide for public participation in the permit-issuing process, in the reporting of violations, and in court enforcement actions [40 CFR Sections 123.9(d) and 123.128(f)(2)].

In addition, if citizens believe that RCRA regulations are being violated and that the state or EPA is not adequately enforcing the regulations, they have the right to bring "citizen suits," under Section 7002 of RCRA.

Program Approval

Q: Is there opportunity for public input concerning authorization of a state's program?

A: Yes; once the state has completed its application it must inform the public about its decision to seek approval by issuing a notice. The notice must be distributed, and the public given ample opportunity to review the application's contents. A public hearing may be held if there is enough interest expressed [Section 271.20(a)].

Q: What occurs if the original application for a state program is significantly modified during the comment period?

A: If the application is significantly modified as a result of information received during the public comment period, the state must provide for an additional comment period, at which time public feedback on the modifications is taken. After the application has been fully scrutinized by the public and modified accordingly, it can be submitted to the EPA Administrator for review [Section 271.20(b)].

Q: What is the EPA review and approval process?

A: After the state has submitted a complete application, the EPA Administrator can proceed to determine whether or not the state's program should be authorized. In making this determination, the Administrator must adhere to the following schedule outlined in Section 271.20(d):

> *Tentative Determination*—Within 90 days from the receipt of the complete application, the Administrator must tentatively approve or disapprove the state's application. The tentative determination is then published in the *Federal Register*.
>
> *Public Input*—The public is given 30 days to comment on the state's application and the Administrator's tentative determination. If sufficient interest is expressed, a public hearing will be held within this period.
>
> *Final Determination*—Within 90 days of the appearance of the tentative determination in the *Federal Register*, the Administrator must consider any comments submitted and decide whether or not to approve the state's program. This final determination is then published in the *Federal Register*.

Revising State Programs

Q: Are state programs ever revised?

A: Yes; as federal and state statutory or regulatory authority relating to RCRA is modified or supplemented, it often is necessary to revise a state's program accordingly. Such revisions can be initiated by the state or required as a result of changes in the federal program (Section 271.21).

Q: What is required if a state wants to revise its program?

A: If a state decides to revise its program, it must notify EPA and submit a modified program description, MOA, and any other documentation EPA deems necessary. The revisions become effective upon approval of EPA [Section 271.21(b)].

Q: What is required if a state must change its program?

A: State programs must also be revised in response to changes in the federal program. Because state programs must be as stringent as the federal program, changes will be required. If the state is able to modify its program without passing a statutory amendment, the program must be revised within 1 year of the date of federal promulgation (*Federal Register* publication date). However, if a statutory amendment is required, the state is given 2 years to revise its program [Section 271.21(b)].

Q: How does HSWA affect state program revision?

A: Although authorized states still have 1 or, in some cases, 2 years to modify their programs, the federal government can implement and enforce the HSWA provisions in an authorized state until the state receives administrative approval to do so (Section 271.25).

In certain circumstances states are not required to have their program reauthorized before being able to enforce new federal requirements resulting from HSWA. Any state that has final authorization for the pre-HSWA program may submit to EPA evidence that its program contains requirements that are substantially equivalent to any requirement created by HSWA. Such a state may request interim authorization for HSWA.

Q: Can a state program revert to EPA control?

A: Yes; authorized state programs are continually subject to review. If EPA finds that a state's program no longer complies with the appropriate regulatory requirements, EPA may withdraw program approval.

Q: What would be grounds for reverting a state's program to EPA?

A: A states's authorization may be withdrawn if a state fails to:

- Inspect and monitor activities subject to regulation
- Comply with terms of the MOA
- Take appropriate enforcement action
- Issue permits that conform to the regulatory requirements

Q: Can a state voluntarily revert its program to EPA control?

A: Yes; in some cases (e.g., when there is a lack of sufficient resources), states with approved RCRA programs may voluntarily transfer them to EPA.

Q: What is required of a state that wants to transfer a program to EPA?

A: The state must give the EPA Administrator 180 days notice and submit a plan for the orderly transfer of all relevant program information necessary for EPA to administer the program, e.g., permits, permit files.

Q: Is a state required to have a delisting program?

A: No; a state is not required to have a delisting program to obtain EPA authorization for its hazardous waste regulatory program. However, if a state does have provisions to delist a hazardous waste, the program must conform to the federal delisting program (OSWER Directive No. 9433.00-1).

Q: Is a state required to regulate radioactive mixed waste?

A: Yes; for a state to obtain and maintain authorization for its hazardous waste regulatory program, the state must have the authority to regulate the hazardous components of a *radioactive mixed waste* (51 *FR* 24504, July 3, 1986).

Q: What is the RCRA authorization status for the states?

A: The status of each state's authorization as it pertains to RCRA is outlined in Table 18.

TABLE 18 State authorization status.

State	No authorization	Base authorization	*HSWA authorization	Mixed waste authorization
Alabama		X		
Alaska	X			
Arizona				
Arkansas		X		X
California	X			
Colorado		X	X	X
Connecticut		X		
Delaware		X		
District of Columbia		X		
Florida		X		
Georgia		X	X	X
Hawaii	X			
Idaho		X	X	X
Illinois		X	X	X
Indiana		X		
Iowa	X			
Kansas		X		
Kentucky		X		
Louisiana		X		
Maine		X		

TABLE 18 *(cont.)*

State	No authorization	Base authorization	*HSWA authorization	Mixed waste authorization
Maryland		X		
Massachusetts		X		
Michigan		X		
Minnesota		X	X	X
Mississippi		X		X
Missouri		X		
Montana		X		
Nebraska		X		X
Nevada		X		
New Hampshire		X		
New Jersey		X		
New Mexico		X		X
New York		X		X
North Carolina		X		X
North Dakota		X		X
Ohio		X		X
Oklahoma		X		X
Oregon		X		X
Pennsylvania		X		
Puerto Rico	X			
Rhode Island		X		
South Carolina		X		X
South Dakota		X		
Tennessee		X		X
Texas			X	
Utah		X	X	X
Vermont		X		
Virginia		X		
Washington		X		X
West Virginia		X		
Wisconsin		X		
Wyoming	X			

Source: 40 CFR Part 272

*Authorization for the corrective action provisions.

APPENDIXES

Appendix A

Regulatory Definitions

RCRA DEFINITIONS

Above-ground Tank means a device meeting the definition of tank in Section 260.10 and that is situated in such a way that the entire surface area of the tank is completely above the plane of the adjacent surrounding surface and the entire surface area of the tank (including the tank bottom) is able to be visually inspected.

Active Life of a facility means the period from the initial receipt of hazardous waste at the facility until the Regional Administrator receives certification of final closure.

Active Portion means that portion of a facility where treatment, storage, or disposal operations are being or have been conducted after the effective date of Part 261 of this chapter and that is not a closed portion. (The effective date is November 19, 1981.)

Ancillary Equipment means any device, including, but not limited to, such devices as piping, fittings, flanges, valves, and pumps, that is used to distribute, meter, or control the flow of hazardous waste from its point of generation to a storage or treatment tank, between hazardous waste storage and treatment tanks to a point of disposal on-site, or to a point of shipment for disposal off-site.

Application means the EPA standard national forms for applying for a permit, including any additions, revisions or modifications to the forms; or forms approved by EPA for use in approved states, including any approved modifications or revisions. Application also includes the information required by the director under 270.14 through 270.29 (contents of Part B of the RCRA application).

Approved Program **or** ***Approved State*** means a state that has been approved or authorized by EPA under Part 271.

Aquifer means a geologic formation, group of formations, or part of a formation capable of yielding a significant amount of groundwater to wells or springs.

Boiler means an enclosed device using controlled flame combustion and having the following characteristics:

(1) (i)The unit must have physical provisions for recovering and exporting thermal energy in the form of steam, heated fluids, or heated gases.

(ii) The unit's combustion chamber and primary energy-recovery section must be of integral design. To be of integral design, the combustion chamber and the primary energy-recovery section (such as waterwalls and superheaters) must be physically formed into one manufactured or assembled unit. A unit in which the combustion chamber and the primary energy-recovery section are joined only by ducts or connections carrying flue gas is not integrally designed; however, secondary energy recovery equipment (such as economizers or air preheaters) need not be physically formed into the same unit as the combustion chamber and the primary energy-recovery section. The following units are not precluded from being boilers solely because they are not of integral design: process heaters (units that transfer energy directly to a process stream) and fluidized bed combustion units.

(iii) While in operation, the unit must maintain a thermal energy recovery efficiency of at least 60 percent, calculated in terms of the recovered energy compared with the thermal value of the fuel.

(iv) The unit must export and utilize at least 75 percent of the recovered energy, calculated on an annual basis. In this calculation, no credit shall be given for recovered heat used internally in the same unit. (Examples of internal use are the preheating of fuel or combustion air, and the driving of induced or forced draft fans or feedwater pumps.)

(2) The unit is one that the regional administrator has determined, on a case-by-case basis, to be a boiler, after considering the standards in Part 260.32.

Certification means a statement of professional opinion based upon knowledge and belief.

Closed Portion means that portion of a facility that an owner or operator has closed in accordance with the approved facility closure plan and all applicable closure requirements.

Component means either the tank or ancillary equipment of a tank system.

Confined Aquifer means an aquifer bounded above and below by impermeable beds or by beds of distinctly lower permeability than that of the aquifer itself; an aquifer containing confined groundwater.

Container means any portable device in which a material is stored, transported, treated, disposed of, or otherwise handled.

Contingency Plan means a document setting out an organized, planned, and coordinated course of action to be followed in case of a fire, explosion, or release of hazardous waste or hazardous waste constituents that could threaten human health or the environment.

Corrosion Expert means a person who, by reason of knowledge of the physical sciences and the principles of engineering and mathematics, acquired by a professional education and related practical experience, is qualified to engage in the practice of corrosion control on buried or submerged metal piping systems and metal tanks. Such a person must be certified as being qualified by the National Association of Corrosion Engineers (NACE) or be a registered professional engineer who has certification or licensing that includes education and experience in corrosion control on buried or submerged metal piping systems and metal tanks.

Designated Facility means a hazardous waste treatment, storage, or disposal facility that has received an EPA permit (or a facility with interim status) in accordance with the requirements of Parts 270 and 124 of 40 CFR, a permit from a state authorized in accordance with Part 271 of 40 CFR, or that is regulated under Section 261.6(c)(2) or Subpart F of Part 266 (precious metal recovery facilities) of 40 CFR, and that has been designated on the manifest by the generator pursuant to Section 262.20.

Dike means an embankment or ridge of either natural or man-made materials used to prevent the movement of liquids, sludges, solids, or other materials.

Director means the regional administrator or the state director, as the context requires, or an authorized representative. When there is no approved state program, and there is an EPA-administered program, director means the regional administrator. When there is an approved state program, director normally means the state director. In some circumstances, however, EPA retains the authority to take certain actions even when there is an approved state program. In such cases, the term director means the regional administrator and not the state director.

Discharge or ***Hazardous Waste Discharge*** means the accidental or intentional spilling, leaking, pumping, pouring, emitting, emptying, or dumping of hazardous waste into or on any land or water.

Disposal Facility means a facility or part of a facility at which hazardous waste is intentionally placed in or on any land or water, and at which waste will remain after closure.

Draft Permit means a document prepared under Section 124.6 indicating the director's tentative decision to issue or deny; modify, revoke, and reissue; terminate; or reissue a permit. A notice of intent to terminate a permit and a notice of intent to deny a permit, as discussed in Section 124.5, are types of draft permits. A denial of a request for modification, revocation and reissuance, or termination, as discussed in Section 124.5, are types of draft permits. A denial of a request for modification, revocation and reissuance, or termination, as discussed in Section 124.5, is not a draft permit. A proposed permit is not a draft permit.

Elementary Neutralization Unit means a device that:

(1) Is used for neutralizing wastes that are hazardous wastes only because they exhibit the corrosivity characteristic defined in Section 261.22 of 40 CFR or are listed in Subpart D of Part 261 of 40 CFR only for this reason

(2) Meets the definition of tank, container, transport vehicle, or vessel in Section 260.10 of 40 CFR.

Emergency Permit means a RCRA permit issued in accordance with 270.61.

EPA Hazardous Waste Number means the number assigned by EPA to each hazardous waste listed in Part 261, Subpart D, of 40 CFR and to each characteristic identified in Part 261, Subpart C, of 40 CFR.

EPA Identification Number means the number assigned by EPA to each generator; transporter; and treatment, storage, or disposal facility.

Equivalent Method means any testing or analytical method approved by the Administrator under Sections 260.20 and 260.21.

Existing Hazardous Waste Management Facility or ***Existing Facility*** means a facility that was in operation or for which construction commenced on or before November 19, 1980. A facility has commenced construction if:

(1) The owner or operator has obtained the federal, state, and local approvals or permits necessary to begin physical construction; and either

(2) (i) A continuous on-site, physical construction program has begun; or (ii) the owner or operator has entered into contractual obligations, which cannot be canceled or modified without substantial loss, for physical construction of the facility to be completed within a reasonable time.

Existing Portion means that land surface area of an existing waste management unit, included in the original Part A permit application, on which wastes have been placed prior to the issuance of a permit.

Existing Tank System or ***Existing Component*** means a tank system or component that is used for the storage or treatment of hazardous waste and that is in operation or for which installation has commenced on or prior to July 14, 1986. Installation will be considered to have commenced if the owner or operator has obtained all federal, state, and local approvals or permits necessary to begin physical construction of the site or installation of the tank system and if either (1) a continuous on-site physical construction or installation program has begun, or (2) the owner or operator has entered into contractual obligations, which cannot be canceled or modified without substantial loss, for physical construction of the site or installation of the tank system to be completed within a reasonable time.

Facility means all contiguous land, structures, other appurtenances, and improvements on the land used for treating, storing, or disposing of hazardous waste. A facility may consist of several treatment, storage, or disposal operational units (e.g., one or more landfills, surface impoundments, or combinations of them).

Federal, State, and Local Approvals or Permits Necessary to Begin Physical Construction means permits and approvals required under federal, state or local hazardous waste control statutes, regulations, or ordinances.

Final Authorization means approval by EPA of a state program that has met the requirements of Section 3006(b) of RCRA and the applicable requirements of Part 271, Subpart A.

Final Closure means the closure of all hazardous waste management units at the facility in accordance with all applicable closure requirements, so that hazardous waste management activities under Parts 264 and 265 of 40 CFR are no longer conducted at the facility, unless subject to the provisions in Section 262.34.

Food-Chain Crops means tobacco, crops grown for human consumption, and crops grown for feed for animals whose products are consumed by humans.

Free Liquids means liquids that readily separate from the solid portion of a waste under ambient temperature and pressure. (This is determined by the paint filter test.)

Freeboard means the vertical distance between the top of a tank or surface impoundment dike and the surface of the waste contained therein.

Generator means any person, by site, whose act or process produces hazardous waste identified or listed in Part 261 or whose act first causes a hazardous waste to become subject to a regulation.

Groundwater means water below the land surface in a zone of saturation.

Hazardous Waste means a hazardous waste as defined in Section 261.3 of 40 CFR.

Hazardous Waste Constituent means a constituent that caused the Administrator to list the hazardous waste in Part 261, Subpart D, of 400 CFR or a constituent listed in table 1 of Part 261.24 of 40 CFR.

Hazardous Waste Management Unit is a contiguous area of land on or in which hazardous waste is placed, or the largest area in which there is significant likelihood of mixing hazardous waste constituents in the same area. Examples of hazardous waste management units include a surface impoundment, a waste pile, a land treatment area, a landfill cell, an incinerator, a tank and its associated piping and underlying containment system, and a container storage area. A container alone does not constitute a unit; the unit includes containers and the land or pad upon which they are placed.

In Operation refers to a facility that is treating, storing, or disposing of hazardous waste.

Inactive Portion means that portion of a facility that is not operated after the effective date of Part 261 of 40 CFR. (See also "active portion" and "closed portion.")

Incinerator means any enclosed device using controlled flame combustion that neither meets the criteria for classification as a boiler nor is listed as an industrial furnace in Section 260.10 of 40 CFR.

Incompatible Waste means a hazardous waste that is unsuitable for:

(1) Placement in a particular device or facility, because it may cause corrosion or decay of containment materials (e.g., container inner liners or tank walls)

(2) Commingling with another waste or material under uncontrolled conditions because the commingling might produce heat or pressure, fire or pressure, fire or explosion, violent reaction, toxic dusts, mists, fumes, or gases, or flammable fumes or gases.

Individual Generation Site means the contiguous site at or on which one or more hazardous wastes are generated. An individual generation site, such as a large manufacturing plant, may have one or more sources of hazardous waste, but is considered a single or individual generation site if the site or property is contiguous.

Industrial Furnace means any of the following enclosed devices that are integral components of manufacturing processes and that use controlled flame devices to accomplish recovery of materials or energy:

(1) Cement kilns
(2) Lime kilns
(3) Aggregate kilns
(4) Phosphate kilns
(5) Coke ovens
(6) Blast furnaces

(7) Smelting, melting, and refining furnaces (including pyrometallurgical devices, such as cupolas, reverberator furnaces, sintering machines, roasters, and foundry furnaces)
(8) Titanium dioxide chloride process oxidation reactors
(9) Methane reforming furnaces
(10) Pulping liquor recovery furnaces
(11) Combustion devices used in the recovery of sulfur values from spent sulfuric acid
(12) Such other devices as the Administrator may, after notice and comment, add to this list on the basis of one more of the following factors:
(i) The design and use of the device primarily to accomplish recovery of material products
(ii) The use of the device to burn or reduce raw materials to make a material product
(iii) The use of the device to burn or reduce secondary materials as effective substitutes for raw materials, in processes using raw materials as principal feedstocks
(iv) The use of the device to burn or reduce secondary materials as ingredients in an industrial process to make a material product
(v) The use of the device in common industrial practice to produce a material product
(vi) Other factors, as appropriate

Inground Tank means a device meeting the definition of tank in Section 260.10, whereby a portion of the tank wall is situated to any degree within the ground, thereby preventing visual inspection of that external surface area of the tank that is in the ground.

Injection Well means a well into which fluids are injected.

Inner Liner means a continuous layer of material placed inside a tank or container that protects the construction materials of the tank or container from the contained waste or reagents used to treat the waste.

Installation Inspector means a person who, by reason of knowledge of the physical sciences and the principles of engineering, acquired by a professional education and related practical experience, is qualified to supervise the installation of tank systems.

International Shipment means the transportation of hazardous waste into or out of the jurisdiction of the United States.

Landfill means a disposal facility or part of a facility where hazardous waste is placed in or on land and is not a land treatment facility, a surface impoundment, or an injection well.

Landfill Cell means a discrete volume of a hazardous waste landfill that uses a liner to provide isolation of wastes from adjacent cells or wastes. Examples of landfill cells are trenches and pits.

Land Treatment Facility means a facility or part of a facility at which hazardous waste is applied onto or incorporated into the soil surface; such facilities are disposal facilities if the waste will remain after closure.

Leachate means any liquid, including any suspended components in the liquid, that has percolated through or drained from hazardous waste.

Leak-Detection System means a system that is capable of detecting the failure of either the primary or secondary containment structure or the presence of a release of hazardous waste or accumulated liquid in the secondary containment structure. Such a system must employ operational controls (e.g., daily visual inspections for releases into the secondary containment system of aboveground tanks) or consist of an interstitial monitoring device designed to detect continuously and automatically the failure of the primary or secondary containment structure or the presence of a release of hazardous waste into the secondary containment structure.

Liner means a continuous layer of natural or man-made materials, beneath or on the sides of a surface impoundment, landfill, or landfill cells, that restricts the downward or lateral escape of hazardous waste, hazardous waste constituents, or leachate.

Management or *Hazardous Waste Management* means the systematic control of the collection, source separation, storage, transportation, processing, treatment, recovery, and disposal of hazardous waste.

Manifest means the shipping document EPA Form 8700-22 and, if necessary, EPA Form 8700-22A, originated and signed by the generator in accordance with the instructions included in the Appendix to Part 262.

Manifest Document Number means the U.S. EPA 12-digit identification number assigned to the generator, plus a unique 5-digit document number assigned to the manifest by the generator for recording and reporting purposes.

Mining Overburden Returned to the Mine Site means any material overlying an economic mineral deposit that is removed to gain access to that deposit and is then used for reclamation of a surface mine.

Movement means that hazardous waste transported to a facility in an individual vehicle.

New Hazardous Waste Management Facility or *New Facility* means a facility that began operation or for which construction commenced after October 21, 1976. (See also "Existing hazardous waste management facility.")

New Tank System or *New Tank Component* means a tank system or component that will be used for the storage or treatment of hazardous waste and for which installation commenced after July 14, 1986; except, however, for purpose of Sections 264.193(g)(2) and 265.193(g)(2), a new tank system is one for which construction commenced after July 14, 1986. (See also "Existing tank system.")

Onground Tank means a device meeting the definition of tank in Section 260.10 and that is situated in such a way that the bottom of the tank is on the same level as the adjacent surrounding surface, so that the external tank bottom cannot be visually inspected.

On-Site means the same or geographically contiguous property that may be divided by public or private right-of-way, provided the entrance and exit between the properties is at a crossroads intersection, and access is by crossing, as opposed to going along, the right-of-way. Noncontiguous properties owned by the same person, but connected by a right-of-way controlled by that person and to which the public does not have access, is also considered on-site property.

Open Burning means the combustion of any material without the following characteristics:

(1) Control of combustion air to maintain adequate temperature for efficient combustion
(2) Containment of the combustion reaction in an enclosed device to provide sufficient residence time and mixing for complete combustion
(3) Control of emission of the gaseous combustion products (see also "incineration" and "thermal treatment").

Operator means the person responsible for the overall operation of a facility.

Owner means the person who owns a facility or part of a facility.

Partial Closure means the closure of a hazardous waste management unit in accordance with the applicable closure requirements of Parts 264 or 265 of 40 CFR at a facility that contains other active hazardous waste management units. For example, partial closure may include the closure of a tank (including its associated piping and underlying containment systems), a landfill cell, surface impoundment, waste pile, or other hazardous waste management unit, while other parts of the same facility continue in operation in the future.

Permit means an authorization, license, or equivalent control document issued by EPA or an approved state to implement the requirements of this part and Parts 271 and 124. Permit includes permit-by-rule (270.60) and emergency permit (270.61). Permit does not include RCRA interim status (Subpart G of this part) or any permit that has not yet been the subject of final agency action, such as a draft permit or a proposed permit.

Permit-by-Rule means a provision of these regulations stating that a facility or activity is deemed to have a RCRA permit if it meets the requirements of the provision.

Person means an individual, trust, firm, joint stock company, federal agency, corporation (including a government corporation), partnership, association, state, municipality, commission, political subdivision of a state, or any interstate body.

Personnel or ***Facility Personnel*** means all persons who work at or oversee the operations of a hazardous waste facility, and whose actions or failure to act may result in noncompliance with the requirements of Parts 264 or 265 of 40 CFR.

Physical Construction means excavation, movement of earth, erection of forms or structures, or similar activity to prepare an HWM facility to accept hazardous waste.

Pile means any noncontainerized accumulation of solid, nonflowing hazardous waste that is used for treatment or storage.

Point Source means any discernible, confined, and discrete conveyance, including, but not limited to, any pipe, ditch, channel, tunnel, conduit, well, discrete fissure, container, rolling stock, concentrated animal-feeding operation, or vessel or other floating craft, from which pollutants are or may be discharged. This term does not include return flows from irrigated agriculture.

Publicly Owned Treatment Works or ***POTW*** means any device or system used in the treatment (including recycling and reclamation) of municipal sewage or industrial wastes of a liquid nature that is owned by a "state" or "municipality" (as defined by Section 502(4) of the Clean Water Act). This definition includes sewers, pipes, or other conveyances only if they convey wastewater to a POTW providing treatment.

Regional Administrator means the regional administrator for the EPA region in which the facility is located, or a designee.

Representative Sample means a sample of a universe or whole (e.g., waste pile, lagoon, groundwater) that can be expected to exhibit the average properties of the universe or whole.

Run-Off means any rainwater, leachate, or other liquid that drains over land from any part of a facility.

Run-On means any rainwater, leachate, or other liquid that drains over land onto any part of a facility.

Saturated Zone or ***Zone of Saturation*** means that part of the earth's crust in which all voids are filled with water.

Schedule of Compliance means a schedule of remedial measures included in a permit, including an enforceable sequence of interim requirements (for example, actions, operations, or milestone events) leading to compliance with the act and regulations.

Site means the land or water area where any facility or activity is physically located or conducted, including adjacent land used in connection with the facility or activity.

Sludge means any solid, semisolid, or liquid waste generated from a municipal, commercial, or industrial wastewater treatment plant, water supply treatment plant, or air pollution control facility exclusive of the treated effluent from a wastewater treatment plant.

Solid Waste means a solid waste as defined in Section 261.2 of 40 CFR.

Solid Waste Management Unit means any discernible waste management unit at a RCRA facility from which hazardous waste or constituents might migrate, irrespective of whether the unit was intended for the management of solid and/or hazardous waste.

The solid waste management unit definition includes:

(a) Containers, tanks, surface impoundments, container storage areas, waste piles, land treatment units, landfills, incinerators, underground injection wells, and other physical, chemical and biological units, including those units defined as regulated units under RCRA;

(b) Recycling units, wastewater treatment units, and other units that EPA has generally exempted from standards applicable to hazardous waste management units; and

(c) Areas associated with production processes at facilities that have become contaminated by routine, systemic, and deliberate discharges of waste or constituents.

State means any of the 50 states, the District of Columbia, the Commonwealth of Puerto Rico, the U.S. Virgin Islands, Guam, American Samoa, and the Commonwealth of the Northern Mariana Islands.

Sump means any pit or reservoir that meets the definition of tank and those troughs/trenches connected to it that serve to collect hazardous waste for transport to hazardous waste storage, treatment, or disposal facilities.

Surface Impoundment or ***Impoundment*** means a facility or part of a facility that is a natural topographic depression, man-made excavation, or diked area formed primarily of earthen materials (although it may be lined with man-made materials), that is designed to hold an accumulation of liquid wastes or wastes containing free liquids, and that is not an injection well. Examples of surface impoundments are holding, storage, settling, and aeration pits, ponds, and lagoons.

Tank means a stationary device, designed to contain an accumulation of hazardous waste, that is constructed primarily of nonearthen materials (e.g., wood, concrete, steel, plastic) that provide structural support.

Tank System means a hazardous waste storage or treatment tank and its associated ancillary equipment and containment system.

Thermal Treatment means the treatment of hazardous waste in a device that uses elevated temperatures as the primary means to change the chemical, physical, or biological character or composition of the hazardous waste. Examples of thermal treatment processes are incineration, molten salt, pyrolysis, calcination, wet air oxidation, and microwave discharge. (See also "incinerator" and "open burning.")

Totally Enclosed Treatment Facility means a facility for the treatment of hazardous waste that is directly connected to an industrial production process and that is constructed and operated in a manner that prevents the release of any hazardous waste or any constituent thereof into the environment during treatment. An example is a pipe in which waste acid is neutralized.

Transfer Facility means any transportation-related facility, including loading docks, parking areas, storage areas, and other similar areas, where shipments of hazardous waste are held during the normal course of transportation.

Transport Vehicle means a motor vehicle or rail car used for the transportation of cargo by any mode. Each cargo-carrying body (trailer, railroad freight car, etc.) is a separate transport vehicle.

Transportation means the movement of hazardous waste by air, rail, highway, or water.

Transporter means a person engaged in the off-site transportation of hazardous waste by air, rail, highway, or water.

Treatability Study means a study in which a hazardous waste is subjected to a treatment process to determine: (1) whether the waste is amenable to the treatment process, (2) what pretreatment (if any) is required, (3) the optimal process conditions needed to achieve the desired treatment, (4) the efficiency of a treatment process for a specific waste or wastes, or (5) the characteristics and volumes of residuals from a particular treatment process. Also included in this definition for the purpose of the 261.4(e) and (f) exemptions are liner compatibility, corrosion, and other material compatibility studies and toxicological and health effects studies. A "treatability study" is not a means to commercially treat or dispose of hazardous waste.

Treatment means any method, technique, or process, including neutralization, designed to change the physical, chemical, or biological character or composition of any hazardous waste so as to neutralize such waste, or so as to recover energy or materials resources from the waste, or so as to render such waste nonhazardous or less hazardous; safer to transport, store, or dispose of; or amenable for recovery, amenable for storage, or reduced in volume.

Treatment Zone means a soil area of the unsaturated zone of a land treatment unit within which hazardous constituents are degraded, transformed, or immobilized.

Underground Injection means the subsurface emplacement of fluids through a bored, drilled, or driven well; or through a dug well, where the depth of the dug well is greater than the largest surface dimension. (See also "injection well.")

Underground Tank means a device meeting the definition of tank in Section 260.10 whose entire surface area is totally below the surface of and covered by the ground.

Unfit-for-Use Tank System means a tank system that has been determined through an integrity assessment or other inspection to be no longer capable of storing or treating hazardous waste without posing a threat of release of hazardous waste into the environment.

Unsaturated Zone or *Zone of Aeration* means the zone between the land surface and the water table.

United States means the 50 states, the District of Columbia, the Commonwealth of Puerto Rico, the U.S. Virgin Islands, Guam, American Samoa, and the Commonwealth of the Northern Mariana Islands.

Uppermost Aquifer means the geologic formation nearest the natural ground surface that is an aquifer, as well as lower aquifers that are hydraulically interconnected with this aquifer within the facility's property boundary.

Vessel includes every description of watercraft used or capable of being used as a means of transportation on the water.

Wastewater Treatment Unit means a device that:

(1) Is part of a wastewater treatment facility that is subject to regulation under either Section 402 or Section 307(b) of the Clean Water Act

(2) Receives and treats or stores an influent wastewater that is a hazardous waste as defined in Section 261.3 of 40 CFR, or generates and accumulates a wastewater treatment sludge that is a hazardous waste as defined in Section 261.3 of 40 CFR, or treats or stores a wastewater treatment sludge that is a hazardous waste as defined in Section 261.3 of 40 CFR

(3) Meets the definition of tank in Section 260.10 of 40 CFR.

Water (Bulk Shipment) means the bulk transportation of hazardous waste that is loaded or carried on board a vessel without containers or labels.

Well means any shaft or pit dug or bored into the earth, generally of a cylindrical form, and often walled with bricks or tubing to prevent the earth from caving in.

Zone of Engineering Control means an area under the control of the owner or operator that, upon detection of a hazardous waste release, can be readily cleaned up prior to the release of hazardous waste or hazardous constituents to groundwater or surface water.

HMTA DEFINITIONS

Atmospheric Gases means gases that are commercially derived through an air separation process. For purposes of this subchapter, "atmospheric gases" means argon, krypton, neon, nitrogen, oxygen, and xenon.

Bottle means a container having a neck of relatively smaller cross section than the body and an opening capable of holding a closure for retention of the contents.

Break-Bulk means packages of hazardous materials that are handled individually, palletized, or unitized for purposes of transportation, as opposed to bulk and containerized freight.

BTU means British thermal unit.

Bulk Packaging means a packaging, other than a vessel or a barge, including a transport vehicle or freight container, in which hazardous materials are loaded with no intermediate form of containment and that which has (1) an internal volume greater than 450 liters (118.9 gallons) as a receptacle for a liquid; (2) a capacity greater than 400 kilograms (881.8 pounds) as a receptacle for a solid; or (3) a water capacity greater than 1,000 pounds (453.6 kilograms) as a receptacle for a gas as defined in 49 CFR 173.300.

Cargo tank means any tank permanently attached to or forming a part of any motor vehicle or any bulk liquid or compressed gas packaging not permanently attached to any motor vehicle that by reason of its size, construction, or attachment to a motor vehicle is loaded or unloaded without being removed from the motor vehicle. Any packaging fabricated under specifications for cylinders is not a cargo tank.

Cargo Vessel means: (1) any vessel other than a passenger vessel; and (2) any ferry being operated under authority of a change of character certificate issued by a Coast Guard Officer-in-Charge, Marine Inspection.

Carrier means a person engaged in the transportation of passengers or property by: (1) land or water, as a common, contract, or private carrier, or (2) civil aircraft.

CC means closed-cup.

Character of Vessel means the type of service in which the vessel is engaged at the time of carriage of a hazardous material.

Consumer Commodity means a material that is packaged and distributed in a form intended or suitable for sale through retail sales agencies or instrumentalities for consumption by individuals for purposes of personal care or household use. This term also includes drugs and medicines.

Container Ship means a cargo vessel designed and constructed to transport, within specifically designed cells, portable tanks and freight containers that are lifted on and off with their contents intact.

Cylinder means a pressure vessel designed for pressures higher than 40 psi and having a circular cross section. It does not include a portable tank, multiunit tank, car tank, cargo tank, or tank car.

Designated Facility means a hazardous waste treatment, storage, or disposal facility that has been designated on the manifest by the generator.

Flash Point means the minimum temperature at which a substance gives off flammable vapors that in contact with spark or flame, will ignite. For liquids, see 49 CFR 173.115, and for solids, see 49 CFR 173.150.

Freight Container means a reusable container having a volume of 64 cubic feet or more, designed and constructed to permit being lifted with its contents intact and intended primarily for containment of packages (in unit form) during transportation.

Fuel Tank means a tank other than a cargo tank, used to transport flammable or combustible liquid, or compressed gas for the purpose of supplying fuel for propulsion of the transport vehicle to which it is attached or for the operation of other equipment on the transport vehicle.

Gross Weight means the weight of a packaging plus the weight of its contents.

Hazardous Material means a substance or material, including a hazardous substance, that has been determined by the Secretary of Transportation to be capable of posing an unreasonable risk to health, safety, and property when transported in commerce and has been so designated.

Hazardous Substance for the purposes of this subchapter, means a material, including its mixtures and solutions, that:

(1) Is listed in the Appendix to 49 CFR 172.101 of this subchapter;

(2) Is in a quantity, in one package, that equals or exceeds the reportable quantity (RQ) listed in the Appendix to 49 CFR 172.101 of this subchapter; and

(3) When in a mixture or solution, is in a concentration by weight that equals or exceeds the concentration corresponding to the RQ of the material, as shown in the following table:

	Concentration by weight	
RQ pounds (kilograms)	Percent	ppm
5,000 (2,270)	10	100,000
1,000 (454)	2	20,000
100 (45.4)	0.2	2,000
10 (4.54)	0.02	200
1 (0.454)	0.002	20

This definition does not apply to petroleum products that are lubricants or fuels (see 40 CFR 300.6).

Hazardous Waste, for the purposes of this chapter, means any material that is subject to the Hazardous Waste Manifest Requirements of the U.S. Environmental Protection Agency specified in 40 CFR Part 262.

Intermodal Container means a freight container designed and constructed to permit it to be used interchangeably in two or more modes of transport.

Intermodal Portable Tank or ***IM Portable Tank*** means a specific class of portable tank designed primarily for international use.

Limited Quantity, when specified as such in a section applicable to a particular material, with the exception of Poison B materials, means the maximum amount of a hazardous material for which there is a specific labeling and packaging exception.

Liquid means a material that has a vertical flow of over 2 inches (50 mm) within a 3-minute period or a material having 1 gram (1 g) or more liquid separation, when determined in accordance with the procedures specified in ASTM D 4359-84, "Standard Test Method for Determining whether a Material is a Liquid or Solid," 1984 edition.

Marking means applying the descriptive name, instructions, cautions, weight, or specification marks or combination thereof required by this subchapter to be placed upon outside containers of hazardous materials.

Mixture means a material composed of more than one chemical compound or element.

Mode means any of the following transportation methods: rail, highway, air, or water.

Motor Vehicle includes a vehicle, machine, tractor, trailer, or semitrailer, or any combination thereof, propelled or drawn by mechanical power and used upon the highways in the transportation of passengers or property. It does not include a vehicle, locomotive, or car operated exclusively on a rail or rails, or a trolley bus operated by electric power derived from a fixed overhead wire, furnishing local passenger transportation similar to street-railway service.

Name of contents means the proper shipping name as specified in 49 CFR 172.101 or 172.102 (when authorized).

Navigable Waters means, for the purposes of this subchapter, waters of the United States, including the territorial seas.

Net Weight means a measure of weight referring only to the contents of a package and does not include the weight of any packaging material.

Nonbulk Packaging means a packaging that has (1) an internal volume of 450 liters (118.9 gallons) or less as a receptacle for a liquid; (2) a capacity of 400 kilograms (881.8 pounds) or less as a receptacle for a solid; or (3) a water capacity of 1,000 pounds (453.6 kilograms) or less as a receptacle for a gas as defined in 49 CFR 173.300.

NOS means not otherwise specified.

NRC (Nonreusable Container) means a container whose reuse is restricted in accordance with the provisions of 49 CFR 173.28.

Operator means a person who controls the use of an aircraft, vessel, or vehicle.

Organic Peroxide See 49 CFR 173.151.

ORM means other regulated materials.

Outside Container means the outermost enclosure used in transporting a hazardous material other than a freight container.

Overpack, except when referenced to a packaging specified in 49 CFR Part 178, means an enclosure that is used by a single consignor to provide protection or convenience in handling of a package or to consolidate two or more packages. *Overpack* does not include a freight container.

Packaging means the assembly of one or more containers and any other components necessary to assure compliance with the minimum packaging requirements of this subchapter and includes containers (other than freight containers or overpacks), portable tanks, cargo tanks, tank cars, and multitank car tanks. For radioactive materials, see 49 CFR 173.403.

Placarded Car means a rail car that is placarded in accordance with the requirements of 49 CFR Part 172, except those cars displaying only the FUMIGATION placards as required by 172.510.

Portable Tank means a bulk packaging (except a cylinder having a water capacity of 1,000 pounds or less) designed primarily to be loaded onto or on or temporarily attached to a transport vehicle or ship and equipped with skids, mountings, or accessories to facilitate handling of the tank by mechanical means. It does not include a cargo tank, tank car, multiunit tank car tank, or trailer carrying 3AX, 3AAX, or 3T cylinders.

Proper Shipping Name means the name of the hazardous material shown in roman print (not italics) in 49 CFR 172.101.

Residue means the hazardous material remaining in a packaging, including a tank car, after its contents have been unleaded to the maximum extent practicable and before the packaging is either refilled or cleaned of hazardous material and purged to remove any hazardous vapors.

RSPA means the Research and Special Programs Administration, U.S. Department of Transportation, Washington, DC 20590.

STC (Single-Trip Container) means a container that may not be refilled and reshipped after having been previously emptied, except as provided in 49 CFR 173.28.

Solid means a material that has a vertical flow of 2 inches (50 mm) or less within a 3-minute period or a separation of 1 gram (l g) or less of liquid when determined in accordance with the procedures specified in ASTM D 4359-84 "Standard Test Method for Determining Whether a Material is a Liquid or Solid," 1984 edition.

Solution means any homogeneous liquid mixture of two or more chemical compounds or elements that will not undergo any segregation under conditions normal to transportation.

Spontaneously Combustible Material (Solid) means a solid substance (including sludges and pastes) that may undergo spontaneous heating or self-ignition under conditions normally incident to transportation or may upon contact with the atmosphere undergo an increase in temperature and ignite.

Strong Outside Container means the outermost enclosure that provides protection against the unintentional release of its contents under conditions normally incident to transportation.

Technical Name means a recognized chemical name currently used in scientific and technical handbooks, journals, and texts. Generic descriptions authorized for use as technical names are, organic phosphate compound, organic phosphorus compound mixture, organic phosphorus compound mixture, methyl parathion, and parathion.

Transport Vehicle means a cargo-carrying vehicle, such as an automobile, van, tractor, truck, semitrailer, tank car, or rail car, used for the transportation of cargo by any mode. Each cargo-carrying body (trailer, rail car, etc.) is a separate transport vehicle.

Viscous Liquid means a liquid material that has a measured viscosity in excess of 2,500 centistokes at 25 degrees C (77 degrees F) when determined in accordance with the procedures specified in ASTM Method D 445-72 "Kinematic Viscosity of Transparent and Opaque Liquids (and the Calculation of Dynamic Viscosity)" or ASTM Method D 1200-70 "Viscosity of Paints, Varnishes, and Lacquers by Ford Viscosity Cup."

Volatility refers to the relative rate of evaporation of materials to assume the vapor state.

Water Reactive Material (Solid) means any solid substance (including sludges and pastes) that, by interaction with water, is likely to become spontaneously flammable or to give off flammable or toxic gases in dangerous quantities.

Appendix B

Acronyms

AA	Assistant Administrator
ACL	Alternate Concentration Level
ADI	Acceptable Daily Intake
AEA	Atomic Energy Act
ALJ	Administrative Law Judge
ANPR	Advanced Notice of Proposed Rulemaking
AO	Administrative Order
ATA	American Trucking Association
ATSDR	Agency for Toxic Substances and Disease Registry
AX	Administrator's Office
BDAT	Best Demonstrated Available Technology
BOE	Bureau of Explosives
BTU	British Thermal Unit
CA	Corrective Action
CA	Cooperative Agreement
CAA	Clean Air Act
CAO	Corrective Action Order
CAP	Corrective Action Plan
CAR	Corrective Action Report
CAS	Chemical Abstract Service
CASRN	Chemical Abstract Service Registry Number
CATS	Corrective Action Tracking System
CBI	Confidential Business Information
CEA	Cooperative Enforcement Agreement

CEI	Compliance Evaluation Inspection
CERCLA	Comprehensive Environmental Response, Compensation, and Liability Act
CERCLIS	Comprehensive Environmental Response, Compensation, and Liability Information System
CERI	Center for Environmental Research Information
CFR	Code of Federal Regulations
CGL	Comprehensive General Liability
CHRIS	Chemical Hazards Response Information System
CM	Corrective Measures
CME	Comprehensive Monitoring Evaluation
CMI	Corrective Measures Implementation
CMS	Corrective Measures Study
COG	Compliance Order Guidance
CPSC	Consumer Product Safety Commission
DCM	Dangerous Cargo Manifest
DOC	Department of Commerce
DOD	Department of Defense
DOE	Department of Energy
DOL	Department of Labor
DOT	Department of Transportation
DRE	Destruction and Removal Efficiency
EA	Enforcement Agreement
EA	Environmental Assessment
EIL	Environmental Impairment Liability
EIR	Exposure Information Report
EIS	Environmental Impact Statement
EODAA	Explosives and Other Dangerous Articles Act
EPA	Environmental Protection Agency
EP TOX	Extraction Procedure for Toxicity Characteristic
ES	Enforcement Strategy
FDA	Food and Drug Administration
FML	Flexible Membrane Liner
FMP	Facility Management Plan
FOIA	Freedom of Information Act
FR	Federal Register
FRA	Federal Railroad Administration
FY	Fiscal Year
GA	Grant Agreement
GPO	Government Printing Office
GWM	Groundwater Monitoring
GWPS	Groundwater Protection Standard

HA	Health Advisory
HC	Hazardous Constituents
HEEP	Health & Environmental Effects Profile
HLW	High Level Waste
HMR	RCRA/Superfund Hotline Monthly Report
HMIR	Hazardous Materials Incident Report
HMT	Hazardous Materials Table
HMTA	Hazardous Materials Transportation Act
HSL	Hazardous Substance List
HSWA	Hazardous and Solid Waste Amendments of 1984
HW	Hazardous Waste
IM	Intermodal
IRP	Installation Restoration Program (DOD)
ISCL	Interim Status Compliance Letter
LCRS	Leachate Control and Removal System
LLW	Low-Level Radioactive Waste
LNG	Liquified Natural Gas
LOIS	Loss of Interim Status
LWT	Legal Weight Truck
MCL	Maximum Contaminant Level
MCLG	Maximum Contaminant Level Goal
MOA	Memorandum of Agreement
MOD	Memorandum of Decision
MOU	Memorandum of Understanding
MPRSA	Marine Protection, Research, and Sanctuaries Act
MQG	Medium-Quantity Generator
MSDS	Material Safety Data Sheet
MTB	Materials Transportation Board
MTR	Minimum Technological Requirements
MTU	Mobile Treatment Unit
NEIC	National Enforcement Investigations Center
NEPA	National Environmental Policy Act
NMFC	National Motor Freight Council
NOD	Notice of Decision
NOD	Notice of Deficiency
NON	Notice of Noncompliance
NOS	Not Otherwise Specified
NOV	Notice of Violation
NPDES	National Pollutant Discharge Elimination System
NPS	National Permit Strategy
NRC	National Response Center
NRC	Nonreuseable Container
NRC	Nuclear Regulatory Commission

OECM	Office of Enforcement and Compliance Monitoring
OERR	Office of Emergency and Remedial Response
OHMT	Office of Hazardous Materials Transportation
OMB	Office of Management and Budget
O/O	Owner or Operator
ORC	Office of Regional Counsel
ORM	Other Regulated Material
OSHA	Occupational Safety and Health Administration
OSW	Office of Solid Waste
OTS	Office of Toxic Substances
OUST	Office of Underground Storage Tanks
OWPE	Office of Waste Programs Enforcement
OWT	Overweight Truck
PAGM	Permit Applicants Guidance Manual
PAT	Permit Assistance Team
PATRAM	Packaging and Transportation of Radioactive Materials
PCB	Polychlorinated Biphenyl
PCP	Pentachlorophenol
PFLT	Paint Filter Liquids Test
PHC	Principal Hazardous Constituent
PIG	Program Implementation Guidance
PIP	Public Involvement Plan
PL	Public Law
POC	Point of Compliance
POE	Point of Exposure
POHC	Principal Organic Hazardous Constituent
POTW	Publicly Owned Treatment Works
PR	Preliminary Review
QA	Quality Assurance
QC	Quality Control
RA	Regional Administrator
RAATS	RCRA Administrative Action Tracking System
RAM	Radioactive Material
RCRA	Resource Conservation and Recovery Act
RD&D	Research, Demonstration, and Development
RFA	RCRA Facility Assessment
RFI	RCRA Facility Inspection
RI	Remedial Investigation
RIA	Regulatory Impact Analysis
RIL	Regulatory Interpretation Letter
RIM	Regulatory Information Memorandum
RIP	RCRA Implementation Plan
RMCL	Recommended Maximum Contaminant Level
RSPA	Research and Special Programs Administration

RTECS	Registry of Toxic Effects of Chemical Substances
RU	Regulated Unit
SAB	Science Advisory Board
SARA	Superfund Amendments and Reauthorization Act
SDWA	Safe Drinking Water Act
SI	Site Identification
SIC	Standard Industrial Classification
SNC	Significant Noncomplier
SOSG	Standard Operating Safety Guidance
SPA	State Program Advisory
SPCC	Spill Prevention, Control, and Countermeasure Plan
SQG	Small-Quantity Generator
STC	Single Trip Container
STCC	Standard Transportation Commodity Code
SV	Sampling Visit
SWDA	Solid Waste Disposal Act
SWMU	Solid Waste Management Unit
T/A	Timely and Appropriate
TCLP	Toxicity Characteristic Leaching Procedure
TDGR	Transportation of Dangerous Goods Regulations
TEGD	Technical Enforcement Guidance Document
TOC	Total Organic Carbon
TOT	Time of Travel
TOX	Total Organic Halogen
TSCA	Toxic Substances Control Act
TSD	Treatment, Storage, and/or Disposal
TSS	Total Suspended Solids
UIC	Underground Injection Control
UN/NA	United Nations/North America
U.S.C.	United States Code
USDW	Underground Source of Drinking Water
UST	Underground Storage Tank
UZM	Unsaturated Zone Monitoring
VHS	Vertical and Horizontal Spread
VOC	Volatile Organic Compound
VSI	Visual Site Inspection
WAP	Waste Analysis Plan
WIC	Washington Information Center
ZRL	Zero Risk Level

Appendix C

RCRA-Listed Hazardous Wastes

EPA waste number	Hazardous waste	Hazard code[1]
	HAZARDOUS WASTES FROM NONSPECIFIC SOURCES	
F001	The following spent halogenated solvents used in degreasing: tetrachloroethylene, trichloroethylene, methylene chloride, 1,1,1-trichloroethane, carbon tetrachloride, chlorinated fluorocarbons, all spent solvent mixtures/blends used in degreasing containing, before use, a total of 10 percent or more (by volume) of one or more of the above halogenated solvents or those solvents listed in F002, F004, and F005; and still bottoms from the recovery of these spent solvents and spent solvent mixtures	(T)
F002	The following spent halogenated solvents: tetrachloroethylene, methylene chloride, trichloroethylene, 1,1,1-trichloroethane, chlorobenzene, 1,1,2-trichloro-1,2,2-trifluoroethane, *o*-dichlorobenzene, and trichlorofluoromethane; all spent solvent mixtures/blends containing, before use, a total of 10 percent or more of the above halogenated solvents or those listed in F001, F004, or F005; and still bottoms from the recovery of these spent solvents and spent solvent mixtures	(T)
F003	The following spent nonhalogenated solvents: xylene, acetone, ethyl acetate, ethyl benzene, ethyl ether, methyl isobutyl ketone, *n*-butyl alcohol, cyclohexanone, methanol; all spent solvent mixtures/blends containing, before use, one or more of the above	(I)

[1]Hazard codes are: C = corrosive, H = acutely hazardous, I = ignitable, R = reactive, and T = toxic.

EPA waste number	Hazardous waste	Hazard code[1]
	nonhalogenated solvents, and a total of 10 percent or more (by volume) of one or more of those solvents listed in F001, F002, F004, and F005; and still bottoms from the recovery of these spent solvents and spent solvent mixtures	
F004	The following spent nonhalogenated solvents: cresols and cresylic acid, nitrobenzene; all spent solvent mixtures/blends containing, before use, a total of 10 percent or more (by volume) of one or more of the above nonhalogenated solvents or those solvents listed in F001, F002, and F005; and the still bottoms from the recovery of these spent solvents and spent solvent mixtures	(T)
F005	The following spent nonhalogenated solvents: toluene, methyl ethyl ketone, carbon disulfide, isobutanol, pyridine; all spent solvent mixtures/blends containing, before use, a total of 10 percent or more (by volume) of one or more of the above nonhalogenated solvents or those listed in F001, F002, and F004; and the still bottoms from the recovery of these spent solvents and spent solvent mixtures	(I,T)
F006	Wastewater treatment sludges from electroplating operations except from the following processes: (1) sulfuric acid anodizing of aluminum; (2) tin plating on carbon steel; (3) zinc plating (segregated basis) on carbon steel; (4) aluminum orzinc-aluminum plating on carbon steel; (5) cleaning/stripping associated with tin, zinc, and aluminum plating on carbon steel; and (6) chemical etching and milling of aluminum	(T)
F019	Wastewater treatment sludges from the chemical conversion coating of aluminum	(T)
F007	Spent cyanide plating bath solutions from electroplating operations (except for precious metals electroplating spent cyanide plating bath solutions)	(R,T)
F008	Plating bath sludges from the bottom of plating baths from electroplating operations for which cyanides are used in the process (except for precious metals electroplating plating bath sludges)	(R,T)
F009	Spent stripping and cleaning bath solutions from electroplating operations for which cyanides are used in the process (except for precious metals electroplating spent stripping and cleaning bath solutions)	(R,T)
F010	Quenching bath sludges from oil baths from metal heat treating operations for which cyanides are used in the process (except for precious metals heat-treating quenching bath sludges)	(R,T)
F011	Spent cyanide solutions from salt bath pot cleaning from metal heat-treating operations (except for precious metals heat-treating spent cyanide solutions from salt bath pot cleaning)	(R,T)

[1]Hazard codes are: C = corrosive, H = acutely hazardous, I = ignitable, R = reactive, and T = toxic.

EPA waste number	Hazardous waste	Hazard code[1]
F012	Quenching wastewater treatment sludges from metal heat treating operations for which cyanides are used in the process (except for precious metals heat treating quenching wastewater treatment sludges)	(T)
F024	Wastes including but not limited to distillation residues, heavy ends, tars, and reactor cleanout wastes from the production of chlorinated aliphatic hydrocarbons, having carbon content from one to five, utilizing free radical catalyzed processes (does not include light ends, spent filters and filter aids, spent desiccants, wastewater, wastewater treatment sludges, spent catalysts, and wastes listed in 261.32)	(T)
F020	Wastes (except wastewater and spent carbon from hydrogen chloride purification) from the production or manufacturing use (as a reactant, chemical intermediate, or component in a formulating process) of tri- or tetrachlorophenol or of intermediates used to produce their pesticide derivatives (does not include wastes from the production of hexachlorophene from highly purified 2,4,5-trichlorophenol)	(H)
F021	Wastes (except wastewater and spent carbon from hydrogen chloride purification) from the production or manufacturing use (as a reactant, chemical intermediate, or component in a formulating process) of pentachlorophenol or of intermediates used to produce its derivatives	(H)
F022	Wastes (except wastewater and spent carbon from hydrogen chloride purification) from the manufacturing use (as a reactant, chemical intermediate, or component in a formulating process) of tetra-, penta-, or hexachlorobenzenes under alkaline conditions	(H)
F023	Wastes (except wastewater and spent carbon from hydrogen chloride purification) from the production of materials on equipment previously used for the production or manufacturing use (as a reactant, chemical intermediate, or component in a formulating process) of tri- and tetrachlorophenols (does not include wastes from equipment used only for the production or use of hexachlorophene from highly purified 2,4,5-trichlorophenol)	(H)
F026	Wastes (except wastewater and spent carbon from hydrogen chloride purification) from the production of materials on equipment previously used for the manufacturing use (as a reactant, chemical intermediate, or component in a formulating process) of tetra-, penta-, or hexachlorobenzene under alkaline conditions	(H)
F027	Discarded unused formulations containing tri-, tetra-, or pentachlorophenol or discarded unused formulations containing compounds derived from these chlorophenols (does not include	(H)

[1]Hazard codes are: C = corrosive, H = acutely hazardous, I = ignitable, R = reactive, and T = toxic.

EPA waste number	Hazardous waste	Hazard code[1]
	formulations containing hexachlorophene synthesized from prepurified 2,4,5-trichlorophenol as the sole component	
F028	Residues resulting from the incineration or thermal treatment of soil contaminated with EPA hazardous wastes numbered F020, F021, F022, F023, F026, and F027	(T)

HAZARDOUS WASTES FROM SPECIFIC SOURCES

Wood Preservatives

EPA waste number	Hazardous waste	Hazard code[1]
K001	Bottom sediment sludge from the treatment of wastewaters from wood preserving processes that use creosote and/or pentachlorophenol	(T)

Inorganic Pigments

EPA waste number	Hazardous waste	Hazard code[1]
K002	Wastewater treatment sludge from the production of chrome yellow and orange pigments	(T)
K003	Wastewater treatment sludge from the production of molybdate orange pigments	(T)
K004	Wastewater treatment sludge from the production of zinc yellow pigments	(T)
K005	Wastewater treatment sludge from the production of chrome green pigments	(T)
K006	Wastewater treatment sludge from the production of chrome oxide green pigments (anhydrous and hydrated)	(T)
K007	Wastewater treatment sludge from the production of iron blue pigments	(T)
K008	Oven residue from the production of chrome oxide green pigments	(T)

Organic Chemicals

EPA waste number	Hazardous waste	Hazard code[1]
K009	Distillation bottoms from the production of acetaldehyde from ethylene	(T)
K010	Distillation side cuts from the production of acetaldehyde from ethylene	(T)
K011	Bottom stream from the wastewater stripper in the production of acrylonitrile	(R,T)
K013	Bottom stream from the acetonitrile column in the production of acrylonitrile	(R,T)
K014	Bottoms from the acetonitrile purification column in the production of acrylonitrile	(T)

[1]Hazard codes are: C = corrosive, H = acutely hazardous, I = ignitable, R = reactive, and T = toxic.

EPA waste number	Hazardous waste	Hazard code[1]
K015	Still bottoms from the distillation of benzyl chloride	(T)
K016	Heavy ends or distillation residues from the production of carbon tetrachloride	(T)
K017	Heavy ends (still bottoms) from the purification column in the production of epichlorohydrin	(T)
K018	Heavy ends from the fractionation column in ethyl chloride production	(T)
K019	Heavy ends from the distillation of ethylene dichloride in tethylene dichloride production	(T)
K020	Heavy ends from the distillation of vinyl chloride in vinyl chloride monomer production	(T)
K021	Aqueous spent antimony catalyst waste from fluoromethanes production	(T)
K022	Distillation bottom tars from the production of phenol/acetone from cumene	(T)
K023	Distillation light ends from the production of phthalic anhydride from naphthalene	(T)
K024	Distillation bottoms from the production of phthalic anhydride from naphthalene	(T)
K093	Distillation light ends from the production of phthalic anhydride from *o*-xylene	(T)
K094	Distillation bottoms from the production of phthalic anhydride from *o*-xylene	(T)
K025	Distillation bottoms from the production of nitrobenzene by the nitration of benzene	(T)
K026	Stripping still tails from the production of methyl ethyl pyridines	(T)
K027	Centrifuge and distillation residues from toluene diisocyanate production	(R,T)
K028	Spent catalyst from the hydrochlorinator reactor in the production of 1,1,1-trichloroethane	(T)
K029	Waste from the product steam stripper in the production of 1,1,1-trichloroethane	(T)
K095	Distillation bottoms from the production of 1,1,1-trichloro ethane	(T)
K096	Heavy ends from the heavy ends column from the production of 1,1,1-trichloroethane	(T)
K030	Column bottoms or heavy ends from the combined production of trichloroethylene and perchloroethylene	(T)
K083	Distillation bottoms from aniline production	(T)
K103	Process residues from aniline extraction from the production of aniline	(T)
K104	Combined wastewater streams generated from nitrobenzene/aniline production	(T)

[1]Hazard codes are: C = corrosive, H = acutely hazardous, I = ignitable, R = reactive, and T = toxic.

EPA waste number	Hazardous waste	Hazard code[1]
K085	Distillation or fractionation column bottoms from the production of chlorobenzenes	(T)
K105	Separated aqueous stream from the reactor product washing step in the production of chlorobenzenes	(T)
K111	Product washwaters from the production of dinitrotoluene via nitration of toluene	(C,T)
K112	Reaction by-product water from the drying column in the production of toluenediamine via hydrogenation of dinitrotoluene	(T)
K113	Condensed liquid light ends from the purification of toluenediamine in the production of toluenediamine via hydrogenation of dinitrotoluene	(T)
K114	Vicinals from the purification of toluenediamine in the production of toluenediamine via hydrogenation of dinitrotoluene	(T)
K115	Heavy ends from the purification of toluenediamine in the production of toluenediamine via hydrogenation of dinitrotoluene	(T)
K116	Organic condensate from the solvent recovery column in the production of toluene diisocyanate via phosgenation of toluenediamine	(T)
K117	Wastewater from the reactor vent gas scrubber in the production of ethylene dibromide via bromination of ethene	(T)
K118	Spent adsorbent solids from purification of ethylenedibromide via bromination of ethene	(T)
K136	Still bottoms from the purification of ethylene dibromide in the production of ethylene dibromide via bromination of ethene	(T)
	Inorganic Chemicals	
K071	Brine purification muds from the mercury cell process in chlorine production for which separately prepurified brine is not used	(T)
K073	Chlorinated hydrocarbon waste from the purification step of the diaphragm cell process using graphite anodes in chlorine production	(T)
K106	Wastewater treatment sludge from the mercury cell process in chlorine production	(T)
	Pesticides	
K031	By-product salts generated in the production of MSMA and cacodylic acid	(T)
K032	Wastewater treatment sludge from the production of chlordane	(T)
K033	Wastewater and scrub water from the chlorination of cyclopentadiene in the production of chlordane	(T)

[1]Hazard codes are: C = corrosive, H = acutely hazardous, I = ignitable, R = reactive, and T = toxic.

EPA waste number	Hazardous waste	Hazard code[1]
K034	Filter solids from the filtration of hexachlorocyclopentadiene in the production of chlordane	(T)
K097	Vacuum stripper discharge from the chlordane chlorinator in the production of chlordane	(T)
K035	Wastewater treatment sludges generated in the production of creosote	(T)
K036	Still bottoms from toluene reclamation distillation in the production of disulfoton	(T)
K037	Wastewater treatment sludges from the production of disulfoton	(T)
K038	Wastewater from the washing and stripping of phorate production	(T)
K039	Filter cake from the distillation of diethylphosphorodithioic acid in the production of phorate	(T)
K040	Wastewater treatment sludge from the production of phorate	(T)
K041	Wastewater treatment sludge from the production of toxaphene	(T)
K098	Untreated process wastewater from the production of toxaphene	(T)
K042	Heavy ends or distillation residues from the distillation of tetrachlorobenzene in the production of 2,4,5-T	(T)
K043	2,6-Dichlorophenol waste from the production of 2,4-D	(T)
K099	Untreated wastewater from the production of 2,4-D	(T)
	Explosives	
K044	Wastewater treatment sludges from the manufacturing and processing of explosives	(R)
K045	Spent carbon from the treatment of wastewater containing explosives	(R)
K046	Wastewater treatment sludges from the manufacturing, formulation, and loading of lead-based initiating compounds	(R)
K047	Pink/red water from TNT operations	(R)
	Petroleum Refining	
K048	Dissolved air floatation (DAF) float from the petroleum refining industry	(T)
K049	Slop oil emulsion solids from the petroleum refining industry	(T)
K050	Heat exchanger bundle cleaning sludge from the petroleum refining industry	(T)
K051	API separator sludge from the petroleum refining industry	(T)
K052	Tank bottoms (leaded) from the petroleum refining industry	(T)
	Iron and Steel	
K061	Emission control dust/sludge from the primary production of steel in electric furnaces	(T)

[1]Hazard codes are: C = corrosive, H = acutely hazardous, I = ignitable, R = reactive, and T = toxic.

EPA waste number	Hazardous waste	Hazard code[1]
K062	Spent pickle liquor generated by steel finishing operations of facilities within iron and steel industry SIC codes 331 and 332	(C,T)
K064	Acid plant blowdown slurry/sludge resulting from the thickening of blowdown slurry from primary copper production	(T)
K065	Surface impoundment solids contained in and dredged from surface impoundments at primary lead smelting facilities	(T)
K066	Sludge from treatment of process wastewater and/or acid plant blowdown from primary zinc production	(T)
K088	Spent potliners from primary aluminum reduction	(T)
K090	Emission control dust or sludge from ferrochromium silicon production	(T)
K091	Emission control dust or sludge from ferrochromium production	(T)
	Secondary Lead	
K069	Emission control dust/sludge from secondary lead smelting	(T)
K100	Waste leaching solution from acid leaching of emission control dust/sludge from secondary lead smelting	(T)
	Veterinary Pharmaceuticals	
K084	Wastewater treatment sludges generated during the production of veterinary pharmaceuticals from arsenic or organoarsenic compounds	(T)
K101	Distillation tar residues from the distillation of aniline-based compounds in the production of veterinary pharmaceuticals from arsenic or organoarsenic compounds	(T)
K102	Residue from the use of activated carbon for decolorization in the production of veterinary pharmaceuticals from arsenic or organoarsenic compounds	(T)
	Ink Formulation	
K086	Solvent washes and sludges, caustic washes and sludges, or water washes and sludges from cleaning tubs and equipment used in the formulation of ink from pigments, driers, soaps, and stabilizers containing chromium and lead	(T)
	Coking	
K060	Ammonia still lime sludge from coking operations	(T)
K087	Decanter tank tar sludge from coking operations	(T)

[1]Hazard codes are: C = corrosive, H = acutely hazardous, I = ignitable, R = reactive, and T = toxic.

EPA waste number	Hazardous waste	Hazard code[1]

COMMERCIAL CHEMICAL PRODUCTS

The following P code wastes are considered acutely (H) hazardous

P023 Acetaldehyde, chloro-
P002 Acetamide, N-(aminothioxomethyl)-
P057 Acetamide, 2-fluoro-
P058 Acetic acid, fluoro-, sodium salt
P066 Acetimidic acid, N-[(methylcarbamoyl)oxy]thio-, methyl ester
P001 3-(alpha-acetonylbenzyl)-4-hydroxycoumarin and salts, when present at concentrations greater than 0.3 percent
P002 1-Acetyl-2-thiourea
P003 Acrolein
P070 Aldicarb
P004 Aldrin
P005 Allyl alcohol
P006 Aluminum phosphide
P007 5-(Aminomethyl)-3-isoxazolol
P008 4-aAminopyridine
P009 Ammonium picrate (R)
P119 Ammonium vanadate
P010 Arsenic acid
P012 Arsenic (III) oxide
P011 Arsenic (V) oxide
P011 Arsenic pentoxide
P012 Arsenic trioxide
P038 Arsine, diethyl
P054 Aziridine

P013 Barium cyanide
P024 Benzenamine, 4-chloro-
P077 Benzenamine, 4-nitro-
P028 Benzene, (chloromethyl)-
P042 1,2-Benzenediol, 4-[(1-hydroxy-2-(methyl-amino)ethyl)]-
P014 Benzenethiol
P028 Benzyl chloride
P015 Beryllium dust
P016 Bis(chloromethyl) ether
P017 Bromoacetone
P018 Brucine

P021 Calcium cyanide
P123 Camphene, octachloro-

[1]Hazard codes are: C = corrosive, H = acutely hazardous, I = ignitable, R = reactive, and T = toxic.

EPA waste number	Hazardous waste	Hazard code[1]
P103	Carbamimidoselenoic acid	
P022	Carbon bisulfide	
P022	Carbon disulfide	
P095	Carbonyl chloride	
P033	Chlorine cyanide	
P023	Chloroacetaldehyde	
P024	*p*-Chloroaniline	
P026	1-(*o*-Chlorophenyl)thiourea	
P027	3-Chloropropionitrile	
P029	Coppercyanides	
P030	Cyanides (soluble cyanide salts), not elsewhere specified	
P031	Cyanogen	
P033	Cyanogen chloride	
P036	Dichlorophenylarsine	
P037	Dieldrin	
P038	Diethylarsine	
P039	O,O-Diethyl S-[2-(ethylthio)ethyl] phosphorodithioate	
P041	Diethyl-*p*-nitrophenyl phosphate	
P040	O,O-Diethyl O-pyrazinyl phosphorothioate	
P043	Diisopropyl fluorophosphate	
P044	Dimethoate	
P045	3,3-Dimethyl-1-(methylthio)-2-butanone,O-[(methylamino) carbonyl] oxime	
P071	O,O-Dimethyl O-*p*-nitrophenyl phosphorothioate	
P082	Dimethylnitrosamine	
P046	alpha,alpha-Dimethylphenethylamine	
P047	4,6-Dinitro-*o*-cresol and salts	
P034	4,6-Dinitro-*o*-cyclohexylphenol	
P048	2,4-Dinitrophenol	
P020	Dinoseb	
P085	Diphosphoramide, octamethyl	
P039	Disulfoton	
P049	2,4-Dithiobiuret	
P109	Dithiopyrophosphoric acid, tetraethyl ester	
P050	Endosulfan	
P088	Endothall	
P051	Endrin	
P042	Epinephrine	
P046	Ethanamine, 1,1-dimethyl-2-phenyl-	

[1]Hazard codes are: C = corrosive, H = acutely hazardous, I = ignitable, R = reactive, and T = toxic.

EPA waste number	Hazardous waste	Hazard code[1]
P084	Ethenamine, N-methyl-N-nitroso-	
P101	Ethyl cyanide	
P054	Ethylenimine	
P097	Famphur	
P056	Fluorine	
P057	Fluoroacetamide	
P058	Fluoroacetic acid, sodium salt	
P065	Fulminic acid, mercury(II) salt (R,T)	
P059	Heptachlor	
P051	1,2,3,4,10,10-Hexachloro-6,7-epoxy-1,4,4a,5,6,7,8,8a-octahydro-endo,endo-1,4:5,8-dimethanonaphthalene	
P037	1,2,3,4,10,10,-Hexachloro-6,7-epoxy-1,4,4a,5,6,7,8,8a-octahydro- endo, exo-1,4:5,8-dimethanonaphthalene	
P060	1,2,3,4,10,10-Hexachloro-1,4,4a,5,8,8a-hexahydro-1,4:5,8-endo, endo-dimethanonaphthalene	
P004	1,2,3,4,10,10-Hexachloro-1,4,4a,5,8,8a-hexahydro-1,4:5,8-endo, exo-dimethanonaphthalene	
P060	Hexachlorohexahydro-exo, exo-dimethanonaphthalene	
P062	Hexaethyl tetraphosphate	
P116	Hydrazinecarbothioamide	
P068	Hydrazine, methyl-	
P063	Hydrocyanic acid	
P063	Hydrogen cyanide	
P096	Hydrogen phosphide	
P064	Isocyanic acid, methyl ester	
P007	3(2H)-isoxazolone, 5-(aminomethyl)-	
P092	Mercury, (acetato-0)phenyl-	
P065	Mercury fulminate (R,T)	
P016	Methane, oxybis(chloro)-	
P112	Methane, tetranitro- (R)	
P118	Methanethiol, trichloro-	
P059	4,7-Methano-1H-indene, 1,4,5,6,7,8,8-heptachloro-3a,4,7,7a-tetrahydro-	
P066	Methomyl	
P067	2-Methylaziridine	
P068	Methyl hydrazine	
P064	Methyl isocyanate	
P069	2-Methyllactonitrile	
P071	Methyl parathion	

[1]Hazard codes are: C = corrosive, H = acutely hazardous, I = ignitable, R = reactive, and T = toxic.

EPA waste number	Hazardous waste	Hazard code[1]
P072	alpha-Naphthylthiourea	
P073	Nickel carbonyl	
P074	Nickle cyanide	
P074	Nickle(II) cyanide	
P073	Nickle tetracarbonyl	
P075	Nicotine and salts	
P076	Nitric oxide	
P077	*p*-Nitroaniline	
P078	Nitrogen dioxide	
P076	Nitrogen(II) oxide	
P078	Nitrogen(IV) oxide	
P081	Nitroglycerine (R)	
P082	N-Nitrosodimethylamine	
P084	N-Nitrosomethylvinylamine	
P050	5-Norbornene-2,3-dimethanol,1,4,5,6,7,7-hexachloro, cyclic sulfite	
P085	Octamethylpyrophosphoramide	
P087	Osmium oxide	
P087	Osmium tetroxide	
P088	7-Oxabicyclo-[2.2.1] heptane-2,3-dicarboxylic acid	
P089	Parathion	
P034	Phenol, 2-cyclohexyl-4,6-dinitro-	
P048	Phenol, 2,4-dinitro-	
P047	Phenol, 2,4-dinitro-6-methyl-	
P020	Phenol, 2,4-dinitro-6-(1-methylpropyl)-	
P009	Phenol, 2,4,6-trinitro-, ammonium salt (R)	
P036	Phenyl dichloroarsine	
P092	Phenylmercuric acetate	
P093	N-Phenylthiourea	
P094	Phorate	
P095	Phosgene	
P096	Phosphine	
P041	Phosphoric acid, diethyl *p*-nitrophenyl ester	
P044	Phosphorodithioic acid, O,O-dimethyl S-[2-(methylamino)-2-oxyethyl] ester	
P043	Phosphorofluoric acid, bis(1-methylethyl)ester	
P094	Phosphorothioic acid, O,O-diethyl S-(ethylthio) methyl ester	
P089	Phosphorothioic acid, O,O-diethyl O(*p*-nitrophenyl) ester	
P040	Phosphorothioic acid, O,O-diethyl O-pyrazinyl ester	

[1]Hazard codes are: C = corrosive, H = acutely hazardous, I = ignitable, R = reactive, and T = toxic.

EPA waste number	Hazardous waste	Hazard code[1]
P097	Phosphorothioic acid, O,O-dimethyl O-[*p*(dimethylamino)-sulfonyl) phenyl] ester	
P110	Plumbane, tetraethyl-	
P098	Potassium cyanide	
P099	Potassium silver cyanide	
P070	Propanal, 2-methyl-2-(methylthio)-O-[(methylamino) carbonyl]-oxime	
P101	Propanenitrile	
P027	Propanenitrile, 3-chloro	
P069	Propanenitrile, 2-hydroxy-2-methyl-	
P081	1,2,3-Propanetriol, trinitrate- (R)	
P017	2-Propanone, 1-bromo-	
P102	Propargyl alcohol	
P003	2-propenal	
P005	2-propen-1-ol	
P067	1,2-Propylenimine	
P102	2-Propyn-1-ol	
P008	4-Pyridinamine	
P075	Pyridine, (S)-3-(1-methyl-2-pyrrolidinyl)-, and salts	
P111	Pyrophosphoric acid, tetraethyl ester	
P103	Selenourea	
P104	Silver cyanide	
P105	Sodium azide	
P106	Sodium cyanide	
P107	Strontium sulfide	
P108	Strychnidin-10-one, and salts	
P018	Strychnidin-10-one, 2,3-dimethoxy-	
P108	Strychnine and salts	
P115	Sulfuric acid, thallium(I) salts	
P109	Tetraethyldithiopyrophosphate	
P110	Tetraethyl lead	
P111	Tetraethylpyrophosphate	
P112	Tetranitromethane (R)	
P062	Tetraphosphoric acid, hexaethyl ester	
P113	Thallic oxide	
P113	Thallium(III) oxide	
P114	Thallium (I) selenite	
P115	Thallium(I) sulfate	
P045	Thiofanax	

[1]Hazard codes are: C = corrosive, H = acutely hazardous, I = ignitable, R = reactive, and T = toxic.

EPA waste number	Hazardous waste	Hazard code[1]
P049	Thiomidodicarbonic diamide	
P014	Thiophenol	
P116	Thiosemicarbazide	
P026	Thiourea, (2-chlorophenyl)-	
P072	Thiourea, 1-naphthalenyl-	
P093	Thiourea, phenyl	
P123	Toxaphene	
P118	Trichloromethanethiol	
P119	Vanadic acid, ammonium salt	
P120	Vanadium pentoxide	
P120	Vanadium(V) oxide	
P001	Warfarin, when present at concentrations greater than 0.3 percent	
P121	Zinc cyanide	
P122	Zinc phosphide, when present at concentrations greater than 10 percent	

The following U code wastes are nonacutely hazardous

EPA waste number	Hazardous waste	Hazard code[1]
U001	Acetaldehyde (I)	
U034	Acetaldehyde, trichloro-	
U187	Acetamide, N-(4-ethoxyphenyl)-	
U005	Acetamide, N-9H-fluoren-2-yl-	
U112	Acetic acid, ethyl ester (I)	
U144	Acetic acid, lead salt	
U214	Acetic acid, thallium(I) salt	
U002	Acetone (I)	
U003	Acetonitrile (I,T)	
U248	3-(alpha-Acetonylbenzyl)-4-hydroxycoumarin and salts, when present at concentations of 0.3 percent or less	
U004	Acetophenone	
U005	2-Acetylaminofluorene	
U006	Acetyl chloride (C,R,T,)	
U007	Acrylamide	
U008	Acrylic acid (I)	
U009	Acrylonitrile	
U150	Alanine, 3-[*p*-bis(2-chloroethyl)amino] phenyl-, L-	
U328	2-Amino-I-methylbenzene	
U353	4-Amino-I-methylbenzene	
U011	Amitrole	

[1]Hazard codes are: C = corrosive, H = acutely hazardous, I = ignitable, R = reactive, and T = toxic.

EPA waste number	Hazardous waste	Hazard code[1]
U012	Aniline (I,T)	
U014	Auramine	
U015	Azaserine	
U010	Azirino (2′,3′,3′,4)pyrrolo (1,2-a)indole-4,7-dione, 6-amino-8-[((aminocarbonyl)oxy)methyl)-1,1a,2,8,8a,8b-hexahydro-8a-methoxy-5-methyl-	
U157	Benz(j)aceanthrylene, 1,2-dihydro-3-methyl-	
U016	Benz(c)acridine	
U016	3,4-Benzacridine	
U017	Benzal chloride	
U018	Benz(a)anthracene	
U018	1,2-Benzanthracene	
U094	1,2-Benzanthracene, 7,12-dimethyl-	
U012	Benzenamine (I,T)	
U014	Benzenamine, 4,4′-carbonimidoylbis(N,N-dimethyl)-	
U049	Benzenamine, 4-chloro-2-methyl-	
U093	Benzenamine, N,N′-dimethyl-4-phenylazo-	
U158	Benzenamine, 4,4′-methylenebis(2-chloro)-	
U222	Benzenamine, 2-methyl-,hydrochloride	
U181	Benzenamine, 2-methyl-,5-nitro	
U019	Benzene (I,T)	
U038	Benzeneacetic acid, 4-chloro-alpha-(4-chloro-phenyl)-alpha-hydroxy,ethyl ester	
U030	Benzene, 1-bromo-4-phenoxy-	
U037	Benzene, chloro	
U190	1,2-Benzenedicarboxylic acid anhydride	
U028	1,2-Benzenedicarboxylic acid [bis(2-ethyl-hexyl)] ester	
U069	1,2-Benzenedicarboxylic acid, dibutyl ester	
U088	1,2-Benzenedicarboxylic acid, diethyl ester	
U102	1,2-Benzenedicarboxylic acid, dimethyl ester	
U107	1,2-Benzenedicarboxylic acid, di-*n*-octyl ester	
U070	Benzene, 1,2-dichloro-	
U071	Benzene, 1,3-dichloro-	
U072	Benzene, 1,4-dichloro-	
U017	Benzene, (dichloromethyl)-	
U223	Benzene, 1,3-diisocyanatomethyl- (R,T)	
U239	Benzene, dimethyl- (I,T)	
U201	1,3-Benzenediol	
U127	Benzene, hexachloro-	
U056	Benzene, hexahydro- (I)	
U188	Benzene, hydroxy-	

[1]Hazard codes are: C = corrosive, H = acutely hazardous, I = ignitable, R = reactive, and T = toxic.

EPA waste number	Hazardous waste	Hazard code[1]
U220	Benzene, methyl-	
U105	Benzene, 1-methyl-1,2,4-dinitro-	
U106	Benzene, 1-methyl-2,6-dinitro-	
U203	Benzene, 1,2-methylenedioxy-4-allyl-	
U141	Benzene, 1,2-methylenedioxy-4-propenyl-	
U090	Benzene, 1,2-methylenedioxy-4-propyl-	
U055	Benzene, (1-methylethyl) (I)	
U169	Benzene, nitro- (I,T)	
U183	Benzene, pentachloro-	
U185	Benzene, pentachloro-nitro-	
U020	Benzenesulfonic acid chloride (C,R)	
U020	Benzenesulfonyl chloride (C,R)	
U207	Benzene, 1,2,4,5-tetrachloro-	
U023	Benzene, (trichloromethyl)- (C,R,T)	
U234	Benzene, 1,3,5-trinitro (R,T)	
U021	Benzidine	
U202	1,2-Benzisothiazolin-3-one,1,1-dioxide	
U120	Benzo(j,k)fluorene	
U022	Benzo(a)pyrene	
U022	3,4-Benzopyrene	
U197	*p*-Benzoquinone	
U023	Benzotrichloride (C,R,T)	
U050	1,2-Benzphenanthrene	
U085	2,2′-Bioxirane (I,T)	
U021	(1,1′-Biphenyl)-4,4′-diamine	
U073	(1,1′-Biphenyl)-4,4′-diamine, 3,3′-dichloro-	
U091	(1,1′-Biphenyl)-4,4′-diamine, 3,3′-dimethoxy-	
U095	(1,1′-Biphenyl)-4,4′-diamine, 3,3′-dimethyl-	
U024	Bis(2-chloroethoxy) methane	
U027	Bis(2-chloroisopropyl) ether	
U244	Bis(dimethylthiocarbamoyl) disulfide	
U028	Bis(2-ethyhexyl)phthalate (DEHP)	
U246	Bromine cyanide	
U225	Bromoform	
U030	4-Bromophenyl phenyl ether	
U128	1,3-Butadiene, 1,1,2,3,4,4-hexachloro	
U172	1-Butanamine, N-butyl-N-nitroso-	
U035	Butanoic acid, 4-[bis(2-chloroethyl)amino]benzene-	
U031	1-Butanol (I)	
U159	2-Butanone (I,T)	
U160	2-Butanone peroxide (R,T)	
U053	2-Butenal	

[1]Hazard codes are: C = corrosive, H = acutely hazardous, I = ignitable, R = reactive, and T = toxic.

EPA waste number	Hazardous waste	Hazard code[1]
U074	2-Butene, 1,4-dichloro- (I,T)	
U031	*n*-Butyl alcohol (I)	
U136	Cacodylic acid	
U032	Calcium chromate	
U238	Carbamic acid, ethyl ester	
U178	Carbamic acid, methylnitroso-, ethyl ester	
U176	Carbamide, N-ethyl-N-nitroso-	
U177	Carbamide, N-methyl-N-nitroso-	
U219	Carbamide, thio-	
U097	Carbamoyl chloride, dimethyl-	
U215	Carbonic acid, dithallium(I)salt	
U156	Carbonochloridic acid, methyl ester (I,T)	
U033	Carbon oxyfluoride (R,T)	
U211	Carbon tetrachloride	
U033	Carbonyl fluoride (R,T)	
U034	Chloral	
U035	Chlorambucil	
U036	Chlordane, technical	
U026	Chlornaphazine	
U037	Chlorobenzene	
U039	4-Chloro-*m*-cresol	
U041	1-Chloro-2,3-epoxypropane	
U042	2-Chloroethyl vinyl ether	
U044	Chloroform	
U046	Chloromethyl methyl ether	
U047	beta-Chloronaphthalene	
U048	*o*-Chlorophenol	
U049	4-Chloro-*o*-toluidine, hydrochloride	
U032	Chromic acid, calcium salt	
U050	Chrysene	
U051	Creosote	
U052	Cresols	
U052	Cresylic acid	
U053	Crotonaldehyde	
U055	Cumene (I)	
U246	Cyanogen bromide	
U197	1,4-Cyclohexadienedione	
U056	Cyclohexane (I)	
U057	Cyclohexanone (I)	
U130	1,3-Cyclopentadiene, 1,2,3,4,5,5-hexa-chloro-	
U058	Cyclophosphamide	

[1]Hazard codes are: C = corrosive, H = acutely hazardous, I = ignitable, R = reactive, and T = toxic.

EPA waste number	Hazardous waste	Hazard code[1]
U240	2,4-D, salts and esters	
U059	Daunomycin	
U060	DDD	
U061	DDT	
U142	Decachloro octahydro-1,3,4-metheno-2H-cyclobuta(c,d)pentalen-2-one	
U062	Diallate	
U133	Diamine (R,T)	
U221	Diaminotoluene	
U063	Dibenz(a,h)anthracene	
U063	1,2:5,6-Dibenzanthracene	
U064	1,2:7,8-Dibenzopyrene	
U064	Dibenz(a,i)pyrene	
U066	1,2-Dibromo-3-chloropropane	
U069	Dibutyl phthalate	
U062	S-(2,3-Dichloroallyl)diisopropylthiocarbamate	
U070	*o*-Dichlorobenzene	
U071	*m*-Dichlorobenzene	
U072	*p*-Dichlorobenzene	
U073	3,3′-Dichlorobenzidine	
U074	1,4-Dichloro-2-butene (I,T)	
U075	Dichlorodifluoromethane	
U192	3,5-Dichloro-N-(1,1-dimethyl-2-propynyl)benzamide	
U060	Dichloro diphenyl dichloroethane	
U061	Dichloro diphenyl trichloroethane	
U078	1,1-Dichloroethylene	
U079	1,2-Dichloroethylene	
U025	Dichloroethyl ether	
U081	2,4-Dichlorophenol	
U082	2,6-Dichlorophenol	
U240	2,4-Dichlorophenoxyacetic acid, salts and esters	
U083	1,2-Dichloropropane	
U084	1,3-Dichloropropene	
U085	1,2:3,4-Diepoxybutane (I,T)	
U108	1,4-Diethylene dioxide	
U086	N,N-Diethylhydrazine	
U087	O,O-Diethyl-S-methyl-dithiophosphate	
U088	Diethyl phthalate	
U089	Diethylstilbestrol	
U148	1,2-Dihydro-3,6-pyradizinedione	
U090	Dihydrosafrole	
U091	3,3′-Dimethoxybenzidine	

[1]Hazard codes are: C = corrosive, H = acutely hazardous, I = ignitable, R = reactive, and T = toxic.

EPA waste number	Hazardous waste	Hazard code[1]
U092	Dimethylamine (I)	
U093	Dimethylaminoazobenzene	
U094	7,12-Dimethylbenz(a)anthracene	
U095	3,3′-Dimethylbenzidine	
U096	alpha,alpha-Dimethylbenzylhydroperoxide (R)	
U097	Dimethylcarbamoyl chloride	
U098	1,1-Dimethylhydrazine	
U099	1,2-Dimethylhydrazine	
U101	2,4-Dimethylphenol	
U102	Dimethyl phthalate	
U103	Dimethyl sulfate	
U105	2,4-Dinitrotoluene	
U106	2,6-Dinitrotoluene	
U107	Di-*n*-octyl phthalate	
U108	1,4-Dioxane	
U109	1,2-Dipheylhydrazine	
U110	Dipropylamine (I)	
U111	Di-N-propylnitrosamine	
U001	Ethanal (I)	
U174	Ethanamine, N-ethyl-N-nitroso-	
U067	Ethane, 1,2-dibromo-	
U076	Ethane, 1,1-dichloro-	
U077	Ethane, 1,2-dichloro-	
U114	1,2-Ethanediylbiscarbamodithioic acid	
U131	Ethane, 1,1,1,2,2,2-hexachloro-	
U024	Ethane, 1,1′-[methylenebis(oxy)]bis(2-chloro)-	
U003	Ethanenitrile (I,T)	
U117	Ethane, 1,1′-oxybis- (I)	
U025	Ethane, 1,1′-oxybis(2-chloro)-	
U184	Ethane pentachloro-	
U208	Ethane, 1,1,1,2-tetrachloro-	
U209	Ethane, 1,1,2,2-tetrachloro-	
U218	Ethanethioamide	
U247	Ethane, 1,1,1-trichloro-2,2-bis(p-methoxyphenyl)	
U227	Ethane, 1,2,1-trichloro-	
U043	Ethene, chloro-	
U042	Ethene, 2-chloroethoxy-	
U078	Ethene, 1,1-dichloro-	
U079	Ethene, *trans*-1,2-dicloro-	
U210	Ethene, 1,1,2,2-tetrachloro-	
U173	Ethanol, 2,2′-(nitrosoimino)bis-	

[1]Hazard codes are: C = corrosive, H = acutely hazardous, I = ignitable, R = reactive, and T = toxic.

EPA waste number	Hazardous waste	Hazard code[1]
U004	Ethanone, 1-phenyl-	
U006	Ethanoyl chloride (C,R,T)	
U112	Ethyl acetate (I)	
U113	Ethyl acrylate (I)	
U238	Ethyl carbamate (urethan)	
U038	Ethyl 4,4′-dichlorobenzilate	
U359	Ethylene glycol monoethyl ether	
U114	Ethylenebis(dithiocarbamic acid)	
U067	Ethylene dibromide	
U077	Ethylene dichloride	
U115	Ethylene oxide (I,T)	
U116	Ethylene thiourea	
U117	Ethyl ether	
U076	Ethylidene dichloride	
U118	Ethylmethacrylate	
U119	Ethyl methanesulfonate	
U139	Ferric dextran	
U120	Fluoranthene	
U122	Formaldehyde	
U123	Formic acid (C,T)	
U124	Furan (I)	
U125	2-Furancarboxaldehyde (I)	
U147	2,5-Furandione	
U213	Furan, tetrahydro- (I)	
U125	Furfural (I)	
U124	Furfuran (I)	
U206	D-Glucopyranose,2-deoxy-2(3-methyl-3-nitro-soureido)-	
U126	Glycidylaldehyde	
U163	Guanidine, N-nitroso-N-methyl-N′nitro-	
U127	Hexachlorobenzene	
U128	Hexachlorobutadiene	
U129	Hexachlorocyclohexane(gamma isomer)	
U130	Hexachlorocyclopentadiene	
U131	Hexachloroethane	
U132	Hexachlorophene	
U243	Hexachloropropene	
U133	Hydrazine (R,T)	
U086	Hydrazine, 1,2-diethyl-	
U098	Hydrazine, 1,1-dimethyl-	

[1]Hazard codes are: C = corrosive, H = acutely hazardous, I = ignitable, R = reactive, and T = toxic.

EPA waste number	Hazardous waste	Hazard code[1]
U099	Hydrazine, 1,2-dimethyl-	
U109	Hydrazine, 1,2-diphenyl-	
U134	Hydrofluoric acid (C,T)	
U134	Hydrogen fluoride (C,T)	
U135	Hydrogen sulfide	
U096	Hydroperoxide, 1-methyl-1-phenylethyl- (R)	
U136	Hydroxydimethylarsine oxide	
U116	2-Imidazolidinethione	
U137	Indeno(1,2,3-cd)pyrene	
U140	Isobutyl alcohol (I,T)	
U141	Isosafrole	
U142	Kepone	
U143	Lasiocarpine	
U144	Lead acetate	
U145	Lead phosphate	
U146	Lead subacetate	
U129	Lindane	
U147	Maleic anhydride	
U148	Maleic hydrazide	
U149	Malononitrile	
U150	Melphalan	
U151	Mercury	
U152	Methacrylonitrile (I,T)	
U092	Methanamine, N-methyl- (I)	
U029	Methane, bromo-	
U045	Methane, chloro- (I,T)	
U046	Methane, chloromethoxy-	
U068	Methane, dibromo-	
U080	Methane, dichloro-	
U075	Methane, dichlorodifluoro-	
U138	Methane, iodo-	
U119	Methanesulfonic acid, ethyl ester	
U211	Methane, tetrachloro-	
U121	Methane, trichlorofluoro-	
U153	Methanethiol (I,T)	
U225	Methane, tribromo-	
U044	Methane, trichloro-	
U121	Methane, trichlorofluoro-	
U123	Methanoic acid (C,T)	
U036	4,7-Methanoindan, 1,2,4,5,6,7,8,8-octachloro-3a,4,7,7a-tetrahydro-	

[1]Hazard codes are: C = corrosive, H = acutely hazardous, I = ignitable, R = reactive, and T = toxic.

EPA waste number	Hazardous waste	Hazard code[1]
U154	Methanol (I)	
U155	Methapyrilene	
U247	Methoxychlor	
U154	Methyl alcohol (I)	
U029	Methyl bromide	
U186	1-Methylbutadiene (I)	
U045	Methyl chloride (I,T)	
U156	Methyl chlorocarbonate (I,T)	
U226	Methyl chloroform	
U157	3-Methylcholanthrene	
U158	4,4′-Methylenebis(2-chloroaniline)	
U132	2,2′-Methylenebis(3,4,6-trichlorophenol)	
U068	Methylene bromide	
U080	Methylene chloride	
U122	Methylene oxide	
U159	Methyl ethyl ketone (I,T)	
U160	Methyl ethyl ketone peroxide (R,T)	
U138	Methyl iodide	
U161	Methyl isobutyl ketone (I)	
U162	Methyl methacrylate (I,T)	
U163	N-Methyl-N′-nitro-N-nitrosoguanidine	
U161	4-Methyl-2-pentanone (I)	
U164	Methylthiouracil	
U010	Mitomycin C	
U059	5,12-Naphthacenedione,(8S-*cis*)-8-acetyl-10-[(3-amino-2,3,6-trideoxy-alpha-L-lyxo-hexopyranosyl)oxyl)-7,8,9,10-tetrahydro-6,8,11-trihydroxy-1-methyoxy-	
U165	Naphthalene	
U047	Naphthalene,2-chloro-	
U166	1,4-Naphthalenedione	
U236	2,7-Naphthalenedisulfonic acid,3,3′-[(3,3′-dimethyl-(1,1′biphenyl)-4,4′diyl)]-bis(azo)bis(5-amino-4-hydroxy)-, tetrasodium salt	
U166	1,4,Naphthaquinone	
U167	1-Naphthylamine	
U168	2-Naphthylamine	
U167	alpha-Naphthylamine	
U168	beta-Naphthylamine	
U026	2-Naphthylamine, N,N′-bis(2-chloromethyl)-	
U169	Nitrobenzene (I,T)	
U170	*p*-Nitrophenol	
U171	2-Nitropropane (I)	

[1]Hazard codes are: C = corrosive, H = acutely hazardous, I = ignitable, R = reactive, and T = toxic.

EPA waste number	Hazardous waste	Hazard code[1]
U172	N-Nitrosodi-*n*-butylamine	
U173	N-Nitrosodiethanolamine	
U174	N-Nitrosodiethylamine	
U111	N-Nitroso-N-propylamine	
U176	N-Nitroso-N-ethylurea	
U177	N-Nitroso-N-methylurea	
U178	N-Nitroso-N-methylurethane	
U179	N-Nitrosopiperidine	
U180	N-Nitrosopyrrolidine	
U181	5-Nitro-*o*-toluidine	
U193	1,2-Oxathiolane,2,2-dioxide	
U058	2H-1,3,2-Oxazaphosphorine,2-[bis(2-chloroethyl)amino] tetrahydro-,oxide 2-	
U115	Oxirane (I,T)	
U041	Oxirane, 2-(chloromethyl)-	
U182	Paraldehyde	
U183	Pentachlorobenzene	
U184	Pentachloroethane	
U185	Pentachloronitrobenzene	
U186	1,3-Pentadiene (I)	
U187	Phenacetin	
U188	Phenol	
U048	Phenol, 2-chloro-	
U039	Phenol, 4-chloro-3-methyl	
U081	Phenol, 2,4-dichloro-	
U082	Phenol, 2,6-dichloro-	
U101	Phenol, 2,4-dimethyl-	
U170	Phenol, 4-nitro-	
U137	1,10-(1,2-phenylene)pyrene	
U145	Phosphoric acid, lead salt	
U087	Phosphorodithioic acid 0,0-diethyl-,S-methylester	
U189	Phosphorous sulfide (R)	
U190	Phthalic anhydride	
U191	2-Picoline	
U192	Pronamide	
U194	1-Propanamine (I,T)	
U110	1-Propanamine, N-propyl- (I)	
U066	Propane, 1,2-dibromo-3-chloro-	
U149	Propanedinitrile	

[1]Hazard codes are: C = corrosive, H = acutely hazardous, I = ignitable, R = reactive, and T = toxic.

EPA waste number	Hazardous waste	Hazard code[1]
U171	Propane, 2-nitro- (I)	
U027	Propane, 2,2′-oxybis(2-chloro)-	
U193	1,3-Propane sultone	
U235	1-Propanol, 2,3-dibromo-,phosphate(3:1)	
U126	1-Propanol, 2,3-epoxy-	
U140	1-Propanol, 2-methyl- (I,T)	
U002	2-Propanone (I)	
U007	2-Propenamide	
U084	Propene, 1,3-dichloro-	
U243	1-Propene, 1,1,2,3,3,3-hexachloro-	
U009	2-Propenenitrile	
U152	2-Propenenitrile, 2-methyl- (I,T)	
U008	2-Propenoic acid (I)	
U113	2-Propenoic acid, ethyl ester (I)	
U118	2-Propenoic acid, 2-methyl-, ethyl ester	
U162	2-Propenoic acid, 2-methyl, methyl ester (I,T)	
U194	*n*-Propylamine (I,T)	
U083	Propylene dichloride	
U196	Pyridine	
U155	Pyridine, 2-[(2-(dimethylamino)-2-thenylamino)]	
U179	Pyridine, hexahydro-N-nitroso-	
U191	Pyridine, 2-methyl-	
U164	4(1H)-Pyrimidinone, 2,3-dihydro-6-methyl-2-thioxo-	
U180	Pyrrole, tetrahydro-N-nitroso-	
U200	Reserpine	
U201	Resorcinol	
U202	Saccharin and salts	
U203	Safrole	
U204	Selenious acid	
U204	Selenium dioxide	
U205	Selenium disulfide (R,T)	
U015	L-Serine, diazoacetate (ester)	
U089	4,4′-Stilbenediol,alpha,alpha′-diethyl-	
U206	Streptozotocin	
U135	Sulfur hydride	
U103	Sulfuric acid, dimethyl ester	
U189	Sulfur phosphide (R)	
U205	Sulfur selenide (R,T)	
U207	1,2,4,5-Tetrachlorobenzene	
U208	1,1,1,2-Tetrachloroethane	

[1]Hazard codes are: C = corrosive, H = acutely hazardous, I = ignitable, R = reactive, and T = toxic.

EPA waste number	Hazardous waste	Hazard code[1]
U209	1,1,2,2-Tetrachloroethane	
U210	Tetrachloroethylene	
U213	Tetrahydrofuran (I)	
U214	Thallium(I)acetate	
U215	Thallium(I)carbonate	
U216	Thallium(I)chloride	
U217	Thallium(I)nitrate	
U218	Thioacetamide	
U153	Thiomethanol (I,T)	
U219	Thiourea	
U244	Thiram	
U220	Toluene	
U221	Toluenediamine	
U223	Toluenediisocyanate (R,T)	
U328	*o*-Toluidine	
U222	O-Toluidine hydrochloride	
U353	*p*-Toluidine	
U011	1H-1,2,4-Triazol-3-amine	
U226	1,1,1-Trichloroethane	
U227	1,1,2-Trichloroethane	
U228	Trichloroethene	
U228	Trichloroethylene	
U121	Trichloromonofluoromethane	
U234	sym-Trinitrobenzene (R,T)	
U182	1,3,5-Trioxane, 2,4,5-trimethyl-	
U235	Tris(2,3-dibromopropyl)phosphate	
U236	Trypan blue	
U237	Uracil, 5[bis(2-chloromethyl)amino]-	
U237	Uracil mustard	
U043	Vinyl chloride	
U248	Warfarin, when present at concentrations of 0.3 percent or less	
U239	Xylene (I)	
U200	Yohimban-16-carboxylic acid, 11,17-dimethoxy-18-[(3,4,5 trimethoxy-benzoyl)oxy]-methyl ester	
U249	Zinc phosphide, when present at concentrations of 10 percent or less	

[1] Hazard codes are: C = corrosive, H = acutely hazardous, I = ignitable, R = reactive, and T = toxic.

Appendix D

The Hazardous Constituents (The Appendix VIII Constituents)

Common Name
Acetonitrile
Acetophenone
2-Acetylaminofluorene
Acetyl chloride
1-Acetyl-2-thiourea
Acrolein
Acrylamide
Acrylonitrile
Aflatoxins
Aldicarb
Aldrin
Allyl alcohol
Aluminum phosphide
4-Aminobiphenyl
5-(Aminomethyl)-3-isoxazolol
4-Aminopyridine
Amitrole
Ammonium vanadate
Aniline
Antimony and compounds, N.O.S.
Aramite
Arsenic and compounds, N.O.S.
Arsenic acid
Arsenic pentoxide
Arsenic trioxide
Auramine
Azaserine
Barium and compounds, N.O.S.
Barium cyanide
Benz[c]acridine
Benz[a]anthracene
Benzal chloride
Benzene
Benzenearsonic acid
Benzidine
Benzo[b]fluoranthene
Benzo[j]fluoranthene
Benzo[a]pyrene
p-Benzoquinone
Benzotrichloride
Benzyl chloride
Beryllium and compounds, N.O.S.
Bromoacetone
Bromoform
4-Bromophenyl phenyl ether

Common Name
Brucine
Butyl benzyl phthalate
Cacodylic acid
Cadmium and compounds, N.O.S.
Calcium chromate
Calcium cyanide
Carbon disulfide
Carbon oxyfluoride
Carbon tetrachloride
Chloral
Chlorambucil
Chlordane (alpha and gamma isomers)
Chlorinated benzenes, N.O.S.
Chlorinated ethane, N.O.S.
Chlorinated fluorocarbons, N.O.S.
Chlorinated naphthalene, N.O.S.
Chlorinated phenol, N.O.S.
Chlornaphazine
Chloroacetaldehyde
Chloroalkyl ethers, N.O.S.
p-Chloroaniline
Chlorobenzene
Chlorobenzilate
p-Chloro-*m*-cresol
1-Chloro-2,3-epoxypropane
2-Chloroethyl vinyl ether
Chloroform
Chloromethyl methyl ether
beta-Chloronaphthalene
o-Chlorophenol
1-(*o*-Chlorophenyl) thiourea
Chloroprene
3-Chloropropionitrile
Chromium and compounds, N.O.S.
Chrysene
Citrus red no. 2
Coal tar
Copper cyanide
Creosote
Cresol (cresylic acid)
Crotonaldehyde
Cyanides (soluble salts and complexes), N.O.S.

Common Name
Cyanogen
Cyanogen bromide
Cyanogen chloride
Cycasin
2-Cyclohexyl-4,6-dinitrophenol
Cyclophosphamide
2,4-D, salts and esters
Daunomycin
DDD
DDE
Bis(2-chloromethoxy) ethane
Bis(2-chloroethyl) ether
Bis(chloromethyl) ether
Bis(2-ethylhexyl) phthalate
DDT
Diallate
Dibenz[a,h]acridine
Dibenz[a,j]acridine
Dibenz[a,h]anthracene
7H-Dibenzo[c,g]carbazole
Dibenzo[a,e]pyrene
Dibenzo[a,h]pyrene
Dibenzo[a,i]pyrene
1,2-Dibromo-3-chloropropane
Dibutyl phthalate
o-Dichlorobenzene
m-Dichlorobenzene
p-Dichlorobenzene
Dichlorobenzene, N.O.S.
3,3′-Dichlorobenzidine
1,4-Dichloro-2-butene
Dichlorodifluoromethane
Dichloroethylene, N.O.S.
1,1-Dichloroethylene
1,2-Dichloroethylene
2,4-Dichlorophenol
2,6-Dichlorophenol
Dichlorophenylarsine
Dichloropropane, N.O.S.
Dichloropropanol, N.O.S.
Dichloropropene, N.O.S.
1,3-Dichloropropene
Dieldrin

Common Name
1,2:3,4-Diepoxybutane
Diethylarsine
1,4 Diethyleneoxide
N,N′-Diethylhydrazine
Diethylhexyl phtalate
O,O-Diethyl S-methyl dithiophosphate
Diethyl-*p*-nitrophenyl phosphate
Diethyl phthalate
O,O-Diethyl O-pyrazinyl phosphorothioate
Diethylstilbesterol
Dihydrosafrole
3,4-Dihydroxy-alpha-(methylamino) methyl benzyl alcohol
Diisopropylfluorophosphate(DFP)
Dimethoate
3,3′-Dimethylbenzidine
p-Dimethylaminoazobenzene
7,12-Dimethylbenz[a]anthracene
3,3pr-Dimethylbenzidine
Dimethylcarbamoyl chloride
1,1-Dimethylhydrazine
1,2-Dimethylhydrazine
alpha, alpha-Dimethylphenethylamine
2,4-Dimethylphenol
Dimethyl phthalate
Dimethyl sulfate
Dinitrobenzene, N.O.S.
4,6-Dinitro-*o*-cresol and salts
2,4-Dinitrophenol
2,4-Dinitrotoluene
2,6-Dinitrotoluene
Dinoseb
Di-*n*-octylphthalate
Diphenylamine
1,2-Diphenylhydrazine
Di-*n*-propylnitrosamine
Disulfoton
Dithiobiuret
Endosulfan
Endothal
Endrin and metabolites

Common Name
Ethyl carbamate (urethane)
Ethyl cyanide
Ethylenebisdithiocarbamic acid, salts and esters
Ethylene dibromide
Ethylene dichloride
Ethylene glycol monoethyl ether
Ethyleneimine
Ethylene oxide
Ethylenethiourea
Ethylidene dichloride
Ethyl methacrylate
Ethylmethane sulfonate
Famphur
Fluoranthene
Fluorine
Fluoroacetamide
Fluoroacetic acid, sodium salt
Formaldehyde
Glycidylaldehyde
Halomethanes, N.O.S.
Heptachlor
Heptachlor epoxide (alpha, beta, and gamma isomers)
Hexachlorobenzene
Hexachlorobutadiene
Hexachlorocyclopentadiene
Hexachlorodibenzo-*p*-dioxins
Hexachlorodibenzofurans
Hexachloroethane
Hexachlorophene
Hexachloropropene
Hexaethyl tetraphosphate
Hydrazine
Hydrogen cyanide
Hydrogen fluoride
Hydrogen sulfide
Indeno[1,2,3-cd]pyrene
Isobutyl alcohol
Isodrin
Isosafrole

Common Name
Kepone
Lasiocarpine
Lead and compounds, N.O.S.
Lead acetate
Lead phosphate
Lead subacetate
Lindane
Maleic anhydride
Maleic hydrazide
Malononitrile
Melphalan
Mercury and compounds, N.O.S.
Mercury fulminate
Methacrylonitrile
Methapyrilene
Methomyl
Methoxychlor
Methyl bromide
Methyl chloride
Methyl chlorocarbonate
Methyl chloroform
3-Methylcholanthrene
4,4′-Methylenebis(2-chloroaniline)
Methylene bromide
Methylene chloride
Methyl ethyl ketone (MEK)
Methyl ethyl ketone peroxide
Methyl hydrazine
Methyl iodide
Methyl isocyanate
2-Methyllactonitrile
Methyl methacrylate
Methyl methanesulfonate
Methyl parathion
Methylithiouracil
Mitomycin C
MNNG
Mustard gas
Naphthalene
1,4-Naphthoquinone
alpha-Naphthylamine
beta-Naphthylamine

Common Name
alpha-Naphthylthiourea
Nickel and compounds, N.O.S.
Nickel carbonyl
Nickel cyanide
Nicotine and salts
Nitric oxide
p-Nitroaniline
Nitrobenzene
Nitrogen dioxide
Nitrogen mustard and hydrochloride salt
Nitrogen mustard N-oxide and hydrochloride salt
Nitroglycerin
p-Nitrophenol
2-Nitropropane
4-Nitroquinoline-1 oxide
Nitrosamines, N.O.S.
N-Nitrosodi-*n*-butylamine
N-Nitrosodiethanolamine
N-Nitrosodiethylamine
N-Nitroso-N-ethylurea
N-Nitrosomethylethylamine
N-Nitroso-N-methylurea
N-Nitroso-N-methylurethane
N-Nitrosomethylvinylamine
N-Nitrosomorpholine
N-Nitrosonornicotine
N-Nitrosopiperidine
N-Nitrosopyrrolidine
N-Nitrososarcosine
5-Nitro-*o*-toluidine
Octamethylpyrophosphoramide
Osmium tetroxide
Paraldehyde
Parathion
Pentachlorobenzene
Pentachlorodibenzo-*p*-dioxins
Pentachlorodibenzofurans
Pentachloroethane
Pentachloronitrobenzene (PCNB)
Pentachlorophenol
Phenacetin

Common Name
Phenol
Phenylenediamine
Phenylmercury acetate
Phenylthiourea
Phosgene
Phosphine
Phorate
Phthalic acid esters, N.O.S.
Phthalic anhydride
2-Picoline
Polychlorinated biphenyls, N.O.S.
Potassium cyanide
Potassium silver cyanide
Pronamide
1,3-Propane sultone
n-Propylamine
Propargyl alcohol
Propylene dichloride
1,2-Propylenimine
Propylthiouracil
Pyridine
Reserpine
Resorcinol
Saccharin and salts
Safrole
Selenium and compounds, N.O.S.
Selenium dioxide
Selenium sulfide
Selenourea
Silver and compounds, N.O.S.
Silver cyanide
Silvex (2,4,5-TP)
Sodium cyanide
Streptozotocin
Strontium sulfide
Strychnine and salts
TCDD
1,2,4,5-Tetrachlorobenzene
Tetrachlorodibenzo-*p*-dioxins
Tetrachlorodibenzofurans

Common Name
Tetrachloroethane, N.O.S.
1,1,1,2-Tetrachloroethane
1,1,2,2-Tetrachloroethane
Tetrachloroethylene
2,3,4,6-Tetrachlorophenol
Tetraethyldithiopyrophosphate
Tetraethyl lead
Tetraethyl pyrophosphate
Tetranitromethane
Thallium and compounds, N.O.S.
Thallic oxide
Thallium(I) acetate
Thallium(I) carbonate
Thallium(I) chloride
Thallium(I) nitrate
Thallium selenite
Thallium(I) sulfate
Thioacetamide
Thiofanox
Thiomethanol
Thiophenol
Thiosemicarbazide
Thiourea
Thiram
Toluene
Toluenediamine
2,4-Toluenediamine
2,6-Toluenediamine
3,4-Toluenediamine
Toluene diisocyanate
p-Toluidine
o-Toluidine hydrochloride
Toxaphene
1,2,4-Trichlorobenzene
1,1,2-Trichloroethane
Trichloroethylene
Trichloromethanethiol
Trichloromonofluoromethane
2,4,5-Trichlorophenol
2,4,6-Trichlorophenol
2,4,5-T
Trichloropropane, N.O.S.
1,2,3-Trichloropropane
O,O,O-Triethylphosphorothioate

Common Name
sym-Trinitrobenzene
Tris(1-aziridinyl)phosphine sulfide
Tris(2,3-dibromopropyl)phosphate
Trypan blue
Uracil mustard

Common Name
Vanadium pentoxide
Vinyl chloride
Warfarin and salts
Zinc cyanide
Zinc phosphide

Appendix E

Groundwater Monitoring Constituents (The Appendix IX Constituents)

Common Name
Acenaphthene
Acenaphthylene
Acetone
Acetophenone
Acetonitrile (methyl cyanide)
2-Acetylaminofluorene (2-AAF)
Acrolein
Acrylonitrile
Aldrin
Allyl chloride
4-Aminobiphenyl
Aniline
Anthracene
Antimony
Aramite
Arsenic
Barium
Benzene
Benzo[a]anthracene (benzathracene)
Benzo[b]fluoranthene

Common Name
Benzo[k]fluoranthene
Benzo[ghi]perylene
Benzo[a]pyrene
Benzyl alcohol
Beryllium
alpha-BHC
beta-BHC
delta-BHC
gamma-BHC; Lindane
Bis(2-chloroethoxy)methane
Bis(2-chloroethyl)ether
Bis(2-chloro-1-methylethyl) ether; 2,2′-Dichlorodiisopropyl ether
Bis(2-ethylhexyl)phthalate
Bromodichloromethane
Bromoform (tribromomethane)
4-Bromophenyl phenyl ether
Butyl benzyl phthalate (benzyl butyl phthalate)
Cadmium
Carbon disulfide

Common Name
Carbon tetrachloride
Chlordane
p-Chloroaniline
Chlorobenzene
Chlorobenzilate
p-Chloro-*m*-cresol
Chloroethane (ethyl chloride)
Chloroform
2-Chloronaphthalene
2-Chlorophenol
4-Chlorophenyl phenyl ether
Chloroprene
Chromium
Chrysene
Cobalt
Copper
m-Cresol
o-Cresol
p-Cresol
Cyanide
2,4-D (2,4-dichlorophenoxyacetic acid)
4,4′-DDD
4,4′-DDE
4,4′-DDT
Diallate
Dibenz[a,h]anthracene
Dibenzofuran
Dibromochloromethane; chlorodibromomethane
1,2-Dibromo-3-chloropropane (DBCP)
1,2-Dibromethane (ethylene dibromide)
Di-*n*-butyl phthalate
o-Dichlorobenzene
m-Dichlorobenzene
p-Dichlorobenzene
3,3′-Dichlorobenzidine
trans-1,4-Dichloro-2-butene
Dichlorodifluoromethane
1,1-Dichloroethane
1,2-Dichloroethane (ethylene dichloride)

Common Name
1,1-Dichloroethylene (vinylidene chloride)
trans-1,2-Dichloroethylene
2,4-Dichlorophenol
2,6-Dichlorophenol
1,2-Dichloropropane
cis-1,3-Dichloropropene
trans-1,3-Dichloropropene
Dieldrin
Diethyl phthalate
O,O-Diethyl O-2 pyrazinyl phosphorothioate (thionazin)
Dimethoate
p-(Dimethylamino)azobenzene
7,12-dimethylbenz[a]anthracene
3,3′-Dimethylbenzidine
alpha, alpha-Dimethylphenethylamine
2,4-Dimethylphenol
Dimethyl phthalate
m-Dinitrobenzene
4,6-Dinitro-*o*-cresol
2,4-Dinitrophenol
2,4-Dinitrotoluene
2,6-Dinitrotoluene
Dinoseb (DNBP or 2-sec-Butyl-4,6-dinitrophenol)
Di-*n*-octyl phthalate
1,4-Dioxane
Diphenylamine
Disulfoton
Endosulfan I
Endosulfan II
Endosulfan sulfate
Endrin
Endrin aldehyde
Ethylbenzene
Ethyl methacrylate
Ethyl methanesulfonate
Famphur
Fluoranthene
Fluorene

Common Name
Heptachlor
Heptachlor epoxide
Hexachlorobenzene
Hexachlorobutadiene
Hexachlorocyclopentadiene
Hexachloroethane
Hexachlorophene
Hexachloropropene
2-Hexanone
Indeno(1,2,3-cd)pyrene
Isobutyl alcohol
Isodrin
Isophorone
Isosafrole
Kepone
Lead
Mercury
Methacrylonitrile
Methapyrilene
Methoxychlor
Methyl bromide (bromomethane)
Methyl chloride (chloromethane)
3-Methylcholanthrene
Methylene bromide (dibromomethane)
Methylene chloride (dichloromethane)
Methyl ethyl ketone (MEK)
Methyl iodide (iodomethane)
Methyl methacrylate
Methyl methanesulfonate
2-Methylnaphthalene
Methyl parathion (parathion methyl)
4-Methyl-2-pentanone (methyl isobutyl ketone)
Naphthalene
1,4-Naphthoquinone
1-Naphthylamine
2-Naphthylamine
Nickel
o-Nitroaniline

Common Name
m-Nitroaniline
p-Nitroaniline
Nitrobenzene
o-Nitrophenol
p-Nitrophenol
4-Nitroquinoline 1-oxide
N-Nitrosodi-*n*-butylamine
N-Nitrosodiethylamine
N-Nitrosodimethylamine
N-Nitrosodiphenylamine
N-Nitrosodipropylamine (di-*n*-propylnitrosamine)
N-Nitrosomethylethylamine
N-Nitrosomorpholine
N-Nitrosopiperidine
N-Nitrosopyrrolidine
5-Nitro-*o*-toluidine
Parathion
Polychlorinated biphenyls (PCBs)
Polychlorinated dibenzo-*p*-dioxins (PCDDs)
Polychlorinated dibenzofurans (PCDFs)
Pentachlorobenzene
Pentachlorethane
Pentachloronitrobenzene
Pentachlorophenol
Pentachloronitrobenzene
Pentachlorophenol
Phenacetin
Phenanthrene
Phenol
p-Phenylenediamine
Phorate
2-Picoline
Pronamide
Propionitrile (ethyl cyanide)
Pyrene
Pyridine
Safrole
Selenium
Silver

Common Name
Silvex (2,4,5-TP)
Styrene
Sulfide
2,4,5-T (2,4,5-Trichlorophenoxyacetic acid)
2,3,7,8-TCDD (2,3,7,8-Tetrachlorodibenzo-*p*-dioxin)
1,2,4,5-Tetrachlorobenzene
1,1,1,2-Tetrachloroethane
1,1,2,2-Tetrachloroethane
Tetrachloroethylene (perchloroethylene or tetrachloroethene)
2,3,4,6-Tetrachlorophenol
Tetraethyl dithiopyrophosphate (Sulfotepp)
Thallium
Tin
Toluene
o-Toluidine

Common Name
Toxaphene
1,2,4-Trichlorobenzene
1,1,1-Trichloroethane (methylchloroform)
1,1,2-Trichloroethane
Trichloroethylene (trichloroethene)
Trichlorofluoromethane
2,4,5-Trichlorophenol
2,4,6-Trichlorophenol
1,2,3-Trichloropropane
O,O,O-Triethyl phosphorothioate
sym-Trinitrobenzene
Vanadium
Vinyl acetate
Vinyl chloride
Xylene (total)
Zinc

Appendix F

Classifications of Permit Modifications

Modifications	Class
A. *General Permit Provisions*	
1: Administrative and informational changes	1
2: Correction of typographical errors	1
3: Equipment replacement or upgrading with functionally equivalent components (e.g., pipes, valves, pumps, conveyors, controls)	1
4: Changes in the frequency of or procedures for monitoring reporting, sampling, or maintenance activities by the permittee:	
(a) To provide for more frequent monitoring, reporting, sampling or maintenance	1
(b) Other changes	2
5: Schedule of compliance:	
(a) Changes in interim compliance dates, with prior approval of the director	*1
(b) Extension of final compliance date	3
6: Changes in expiration date of permit to allow earlier permit termination, with prior approval of the director	*1
7: Changes in ownership or operational control of a facility, provided the procedures of 270.40(b) are followed	*1

Note: When a permit modification (such as introduction of a new unit) requires a change in facility plans or other general facility standards, that change shall be reviewed under the same procedures as the permit modification.
Note: See 270.42(g) for modification procedures to be used for the management of newly listed or identified wastes.
* Class 1 modifications requiring prior EPA approval.
Source: 40 CFR, Appendix I to Section 270.42.

Modifications	Class
B. *General Facility Standards*	
1: Changes to waste sampling or analysis methods:	
(a) To conform with agency guidance or regulations	1
(b) Other changes	2
2: Changes to analytical quality assurance/control plan:	
(a) To conform with agency guidance or regulations	1
(b) Other changes	2
3: Changes in procedures for maintaining the operating record	1
4: Changes in frequency or content of inspection schedules	2
5: Changes in the training plan:	
(a) That affect the type or decrease the amount of training given to employees	2
(b) Other changes	1
6: Contingency plan:	
(a) Changes in emergency procedures (i.e., spill or release response procedures)	2
(b) Replacement with functionally equivalent equipment, upgrade, or relocate emergency equipment listed	1
(c) Removal of equipment from emergency equipment list	2
(d) Changes in name, address, or phone number of coordinators or other persons or agencies identified in the plan	1
C. *Groundwater Protection*	
1: Changes to wells:	
(a) Changes in the number, location, depth, or design of upgradient or downgradient wells of permitted groundwater monitoring system	2
(b) Replacement of an existing well that has been damaged or rendered inoperable, without change to location, design, or depth of the well	1
2: Changes in groundwater sampling or analysis procedures or monitoring schedule, with prior approval of the director	*1
3: Changes in statistical procedure for determining whether a statistically significant change in groundwater quality between upgradient and downgradient wells has occurred, with prior approval of the director	*1
4: Changes in point of compliance	2
5: Changes in indicator parameters, hazardous constituents, or concentration limits (including ACLs):	
(a) As specified in the groundwater protection standard	3
(b) As specified in the detection monitoring program	2

* Class I modifications requiring prior EPA approval.

Modifications	Class
6: Changes to a detection monitoring program as required by 264.98(j), unless otherwise specified in this appendix	2
7: Compliance monitoring program:	
(a) Addition of compliance monitoring program as required by 264.98(h)(4) and 264.99	3
(b) Changes to a compliance monitoring program as required 264.99(k), unless otherwise specified in this appendix	2
8: Corrective action program:	
(a) Addition of a corrective action program as required by 264.99(i)(2) and 264.100	3
(b) Changes to a corrective action program as required by 264.100(h), unless otherwise specified in this appendix	2
D. *Closure*	
1: Changes to the closure plan:	
(a) Changes in estimate of maximum extent of operations or maximum inventory of waste on-site at any time during the active life of the facility, with prior approval of the director	*1
(b) Changes in the closure schedule for any unit, changes in the final closure schedule for the facility, or extension of the closure period, with prior approval of the director	*1
(c) Changes in the expected year of final closure, where other permit conditions are not changed, with prior approval of the director	*1
(d) Changes in procedures for decontamination of facility equipment or structures, with prior approval of the director	*1
(e) Changes in approved closure plan resulting from unexpected events occurring during partial or final closure, unless otherwise specified in this appendix	2
2: Creation of a new landfill unit as part of closure	3
3: Addition of the following new units to be used temporarily for closure activities:	
(a) Surface impoundments	3
(b) Incinerators	3
(c) Waste piles that do not comply with 264.250(c)	3
(d) Waste piles that comply with 264.250(c)	2
(e) Tanks or containers (other than specified below)	2
(f) Tanks used for neutralization, dewatering, phase separation, or component separation, with prior approval of the director	*1

* Class I modifications requiring prior EPA approval.

Modifications	Class
E. *Post-Closure*	
1: Changes in name, address, or phone number of contact in post-closure plan	1
2: Extension of post-closure care period	2
3: Reduction in the post-closure care period	3
4: Changes to the expected year of final closure, where other permit conditions are not changed	1
5: Changes in post-closure plan necessitated by events occurring during the active life of the facility, including partial and final closure	2
F. *Containers*	
1: Modification or addition of container units:	
(a) Resulting in greater than 25 percent increase in the facility's container storage capacity	3
(b) Resulting in up to 25 percent increase in the facility's container storage capacity	2
2: (a) Modification of a container unit without increasing the capacity of the unit	2
(b) Addition of a roof to a container unit without alteration of the containment system	1
3: Storage of different wastes in containers:	
(a) That require additional or different management practices from those authorized in the permit	3
(b) That do not require additional or different management practices from those authorized in the permit	2
4: Other changes in container management practices (e.g., aisle space; types of containers; segregation)	2
G. *Tanks*	
1: (a) Modification or addition of tank units resulting in greater than 25 percent increase in the facility's tank capacity, except as provided in G(1)(d) of this appendix	3
(b) Modification or addition of tank units resulting in up to 25 percent increase in the facility's tank capacity, except as provided in G(1)(d) of this appendix	*2
(c) Addition of a new tank that will operate for more than 90 days using any of the following physical or chemical treatment technologies: neutralization, dewatering, phase separation, or component separation	*1

* Class I modifications requiring prior EPA approval.

Modifications	Class
(d) After prior approval of the director, addition of a new tank that will operate for up to 90 days using any of the following physical or chemical treatment technologies: neutralization, dewatering, phase separation, or component separation	1
2: Modification of a tank unit or secondary containment system without increasing the capacity of the unit	2
3: Replacement of a tank with a tank that meets the same design standards and has a capacity within + 1–10 percent of the replaced tank provided	1
—The capacity difference is no more than 1,500 gallons	
—The facility's permitted tank capacity is not increased	
—The replacement tank meets the same conditions in the permit	
4: Modification of a tank management practice	2
5: Management of different wastes in tanks:	
(a) That require additional or different management practices, tank design, different fire protection specifications, or significantly different tank treatment process from that authorized in the permit	3
(b) That do not require additional or different management practices, tank design, different fire protection specifications, or significantly different tank treatment process than authorized in the permit	2
H. *Surface Impoundments*	
1: Modification or addition of surface impoundment units that result in increasing the facility's surface impoundment storage or treatment capacity	3
2: Replacement of a surface impoundment unit	3
3: Modification of a surface impoundment unit without increasing the facility's surface impoundment storage or treatment capacity	3
4: Modification of a surface impoundment management practice	2
5: Treatment, storage, or disposal of different wastes in surface impoundments:	
(a) That require additional or different management practices or different designs of the liner or leak detection system from those authorized in the permit	3
(b) That do not require additional or different management practices or different designs of the liner or leak detection system from those authorized in the permit	2

* Class I modifications requiring prior EPA approval.

Modifications	Class
I. *Enclosed Waste Piles* For all waste piles except those complying with 264.250(c), modifications are treated the same as for a landfill. The following modifications are applicable only to waste piles complying with 264.250(c).	
1: Modification or addition of waste pile units:	
(a) Resulting in greater than 25 percent increase in the facility's waste pile storage or treatment capacity	3
(b) Resulting in up to 25 percent increase in the facility's waste pile storage or treatment capacity	2
2: Modification of waste pile unit without increasing the capacity of the unit	2
3: Replacement of a waste pile unit with another waste pile unit of the same design and capacity and meeting all waste pile conditions in the permit	1
4: Modification of a waste pile management practice	2
5: Storage or treatment of different wastes in waste piles:	
(a) That require additional or different management practices or different design of the unit	3
(b) That do not require additional or different management practices or different design of the unit	2
J. *Landfills and Unenclosed Waste Piles*	
1: Modification or addition of landfill units that result in increasing the facility's disposal capacity	3
2: Replacement of a landfill	3
3: Addition or modification of a liner, leachate collection system, leachate detection system, run-off control, or final cover system	2
4: Modification of a landfill unit without changing a liner, leachate collection system, leachate detection system, run-off control, or final cover system	2
5: Modification of a landfill management practice	2
6: Landfill different wastes:	
(a) That require additional or different management practices, different design of the liner, leachate collection system, or leachate detection system	3
(b) That do not require additional or different management practices, different design of the liner, leachate collection system, or leachate detection system	2

* Class I modifications requiring prior EPA approval.

Modifications	Class
K. *Land Treatment*	
1: Lateral expansion of or other modification of a land treatment unit to increase a real extent	3
2: Modification of run-on control system	2
3: Modify run-off control system	3
4: Other modifications of land treatment unit component specifications or standards required in permit	2
5: Management of different wastes in land treatment units:	
(a) That require a change in permit operating conditions or unit design specifications	2
(b) That do not require a change in permit operating conditions or unit design specifications	2
6: Modification of a land treatment unit management practice to:	
(a) Increase rate or change method of waste application	3
(b) Decrease rate of waste application	1
7: Modification of a land treatment unit management practice to change measures of pH or moisture content, or to enhance microbial or chemical reactions	2
8: Modification of a land treatment unit management practice to grow food-chain crops, to add to or replace existing permitted crops with different food-chain crops, or to modify operating plans for distribution of animal feeds resulting from such crops	3
9: Modification of operating practice due to detection of releases from the land treatment unit pursuant to 264.278(g)(2)	3
10: Changes in the unsaturated zone monitoring system, resulting in a change to the location, depth, number of sampling points, or replace unsaturated zone monitoring devices or components of devices with devices or components that have specifications different from permit requirements	3
11: Changes in the unsaturated zone monitoring system that do not result in a change to the location, depth, number of sampling points, or that replace unsaturated zone monitoring devices or components of devices with devices or components having specifications different from permit requirements	
12: Changes in background values for hazardous constituents in soil and soil-pore liquid	2
13: Changes in sampling, analysis, or statistical procedure	2
14: Changes in land treatment demonstration program prior to or during the demonstration	2

* Class I modifications requiring prior EPA approval.

Modifications	Class
15: Changes in any condition specified in the permit for a land treatment unit to reflect results of the land treatment demonstration, provided performance standards are met and the director's prior approval has been received	*1
16: Changes to allow a second land treatment demonstration to be conducted when the results of the first demonstration have not shown the conditions under which the wastes can be treated completely, provided the conditions for the second demonstration are substantially the same as the conditions for the first demonstration	*1
17: Changes to allow a second land treatment demonstration to be conducted when the results of the first demonstration have not shown the conditions under which the wastes can be treated completely, where the conditions for the second demonstration are not substantially the same as the conditions for the first demonstration	3
18: Changes in vegetative cover requirements for closure	2
L. *Incinerators*	
1: Changes to increase by more than 25 percent any of the following limits authorized in the permit: a thermal feed rate limit, or an organic chlorine feed rate limit. The director will require a new trial burn to substantiate compliance with the regulatory performance standards unless this demonstration can be made through other means,	3
2: Changes to increase by up to 25 percent any of the following limits authorized in the permit: a thermal feed rate limit, a waste feed limit, or an organic chlorine feed rate limit. The director will require a new trial burn to substantiate compliance with the regulatory performance standards unless this demonstration can be made through other means.	2
3: Modification of an incinerator unit by changing the internal size or geometry of the primary or secondary combustion units, by adding a primary or secondary combustion unit, by substantially changing the design of any component used to remove HCL or particulate from the combustion gases, or by changing other features of the incinerator that could affect its capability to meet the regulatory performance standards. The director may require a new trial burn to demonstrate compliance with the regulatory performance standards.	2
4: Modification of an incinerator unit in a manner that would not likely affect the capability of the unit to meet the regulatory performance standards, but that would change the operating conditions or moni-	

* Class I modifications requiring prior EPA approval.

Modifications	Class
toring requirements specified in the permit. The director may require a new trial burn to demonstrate compliance with the regulatory performance standards.	2
5: Operating requirements:	
(a) Modification of the limits specified in the permit for minimum combustion gas temperature, minimum combustion gas residence time, or oxygen concentration in the secondary combustion chamber. The director will require a new trial burn to substantiate compliance with the regulatory performance standards unless this demonstration can be made through other means	
(b) Modification of any stack gas emission limits specified in the permit, or modification of any conditions in the permit concerning emergency shutdown or automatic waste-feed-cutoff procedures or controls	3
(c) Modification of any other operating condition or any inspection or recordkeeping requirement specified in the permit	3
6: Incineration of different wastes:	
(a) If the waste contains a POHC that is more difficult to incinerate than authorized by the permit or if incineration of the waste requires compliance with regulatory performance standards unless this demonstration can be made through other means	3
(b) If the waste does not contain a POHC that is more difficult to incinerate than authorized by the permit and if incineration of the waste does not require compliance with different regulatory performance standards than specified in the permit	
7: Shakedown and trial burn:	
(a) Modification of the trial burn plan or any of the permit conditions applicable during the shakedown period for determining operational readiness after construction, the trial burn period, or the period immediately following the trial burn	2
(b) Authorization of up to an additional 720 hours of waste incineration during the shakedown period for determining operational readiness after construction, with the prior approval of the director	*1
(c) Changes in the operating requirements set in the permit for conducting a trial burn, provided the change is minor and has received the prior approval of the director	*1

* Class I modifications requiring prior EPA approval.

Modifications	Class
(d) Changes in the operating requirements set in the permit to reflect the results of the trial burn, provided the change is minor and has received the prior approval of the director	*1
8: Substitution of an alternate type of fuel that is not specified in the permit	1

* Class I modifications requiring prior EPA approval.

Appendix G

Information Resources

GUIDANCE AND REGULATORY INFORMATION SOURCES

No one source of information will assist you with all your regulatory needs. As a professional, you also cannot rely solely on the regulatory agencies for information. Many sources provide information concerning regulations, guidance, and policy. Using all the sources listed below or selecting those that best suit your needs will ensure that your knowledge is the most complete, accurate, and current possible. These sources and summaries of the services provided are as follows.

The Federal Register

The *Federal Register* is the most complete and helpful source of regulatory information. Proposed and final regulations, as well as selected notices, publications, interpretations, and other related federal government information, is published in the *Federal Register*. Each regulation (final and proposed) published contains a preamble, which is essentially a discussion of EPA's rationale for the regulation and background information and interpretive guidance. In addition, each April and October, EPA publishes its semiannual regulatory agenda, listing all projected significant regulations, the status of previously projected regulations, and a technical contact for each.

The *Federal Register*, published daily, is available for $340 per year from:

Superintendent of Documents
U.S. Government Printing Office
Washington, DC 20402
202-783-3238

Code of Federal Regulations

Annually, the *Federal Register* is codified into the *Code of Federal Regulations* (CFR). The CFR contains promulgated final regulations only, not preambles or proposed regulations. This document, available from the Government Printing Office, is updated in July of each year, but is not available until October or November. Title 40 of the CFR, contains EPA's regulations, and Title 49 contains DOT's regulations.

Hazardous Waste Collection Database

Although there is no comprehensive collection of EPA documents concerning hazardous waste, the Hazardous Waste Collection Database is the most complete. It contains the OSWER Directive System as well as nongovernment publications concerning hazardous waste and Superfund.

The database and hard copies of the database are available for viewing at EPA's headquarters library and all EPA regional libraries. The National Technical Information Service (NTIS) has the data base on diskette for sale (order no. PB-87-945001, $125). If you have any questions concerning the collection, contact:

Hazards Waste Collection Coordinator
U.S. EPA
Headquarters Library
202-260-5934

Written Regulatory Interpretation

Anyone can obtain an interpretation of a regulation or policy by writing to EPA and requesting one. However, it takes at least 8 weeks to obtain a response. To expedite a response, a letter should be routed through one of your Congressmen. A letter from a member of Congress requesting a regulatory interpretation has a set time for a response, usually 15 days.

Freedom of Information

The Freedom of Information Act (FOIA), enacted in 1966, established an effective statutory right of access to federal government records. Under the provisions of FOIA, all records under the custody and control of federal executive branch agencies (with specific exceptions) are covered. FOIA requests are required to be responded to within 10 working days. Additionally, the requester can be charged for reasonable research and copying fees.

You can use FOIA to access EPA and DOT policies, nonconfidential enforcement records, biennial report statistics, and many other types of information. For further information concerning the limitations and procedures of the Freedom of Information Act, contact GPO for the publication, *Your Right to Federal Records, Questions and Answers on the Freedom of Information Act and the Privacy Act* (order no. 052-071-00752-1, $1.75).

EPA Telephone Book

You should obtain a copy of the EPA headquarters telephone directory. The directory contains helpful information including organization charts, telephone numbers, and a listing of EPA personnel. For information on the availability of phone books, call 202-260-2626.

Hotlines and Clearinghouses

RCRA/Superfund Hotline

The RCRA/Superfund Hotline has traditionally been the best source for obtaining regulatory information and documents concerning the Resource Conservation and Recovery Act (RCRA) and Superfund. The hotline provides some good information, such as regulatory status, general answers, information concerning *Federal Register* notices, and information on public hearings.

Each month, the hotline produces the *RCRA/Superfund Hotline Monthly Report*. This document, available only through an FOIA request, contains informative EPA interpretations and clarification of policy and guidance.

The hotline does send out a free document, *OSW Documents in Demand* (EPA/530-SW-87-0168). This semiannual catalog identifies the Office of Solid Waste's most frequently requested documents and their availability.

RCRA/Superfund Hotline - 8:30 to 7:30 EST
800-424-9346
703-920-9810

TSCA Hotline

The TSCA Hotline provides regulatory information and publications concerning the Toxic Substances Control Act (TSCA). Of special interest are regulations and publications concerning PCBs and asbestos. The hotline does not provide a list of publications, but will suggest and provide documents. In addition, the TSCA Assistance Office publishes a free bimonthly publication concerning TSCA regulations, *The Chemicals-in-Progress Bulletin*. You can be placed on the mailing list by contacting the hotline.

TSCA Hotline - 8:30 to 5:00 EST
202-554-1404

Small Business Ombudsman/Asbestos Hotline

The Small Business Ombudsman Hotline acts as a liaison between small business and EPA and serves as EPA's Asbestos Hotline. The hotline answers questions covering many EPA subjects, although reportedly more than 50 percent of the callers request hazardous waste information. It is a good general information service that will also direct callers to an appropriate contact in EPA if they have a technical question. In addition, it publishes a quarterly bulletin, *Update on Recent Asbestos and Small Business Activities*. You can be placed on the mailing list by contacting the hotline.

Small Business/Asbestos Hotline - 8:00 to 4:30 EST
800-368-5888
703-557-1938

DOT Hazardous Materials Information Hotline

The Regulations Development Branch under DOT's Special Research and Administration Programs answers technical questions concerning the transportation of hazardous wastes. In addition, it provides lists of available publications and courses. Additionally, DOT publishes an informative quarterly newsletter, *The Hazardous Materials Newsletter*. This newsletter provides information on upcoming and recent rules, enforcement notes, interpretations, questions and answers, and training news.

Regulatory questions 202-366-4488
Publications 202-366-2301

Miscellaneous Information Sources

Safe Drinking Water Act	800-426-4791 202-260-5533
SARA Title III	800-535-0202
Pesticides (regulatory)	703-557-7760
Pesticides (nonregulatory)	317-494-6614
Air (regulatory)	919-629-5651
Air (nonregulatory)	919-541-0850
Water	202-260-5400
OSHA	202-523-7031
Acid Rain	202-260-7407

Marine Protection	202-260-7166
Radiation Program	703-557-9710
NOAA	202-377-4190
NIOSH	202-472-7234
EPA Personnel Locator	202-260-2090
EPA Freedom of Information	202-260-4048
EPA Library (main)	202-260-5922
EPA Library (pesticides and toxic substances)	202-260-3568

PUBLICATION SOURCES

EPA Office of Research and Development

EPA's Office of Research and Development (ORD), located in Cincinnati, Ohio, conducts an agencywide, integrated program of research and development relevant to pollution sources and control, fate and transport processes, health and ecological effects, measurement and monitoring, and risk assessment. ORD provides technical documents; two of particular interest are *The EPA/ORD Technical Assistance Directory* (CERI-87-51) and *The Hazardous Waste Research Locator* (EPA 600/9-87/007).

ORD Publications - 513-569-7562

Government Printing Office

The Government Printing Office (GPO) has many publications concerning pollution control, including both EPA and other federal agency publications. Because GPO has thousands of documents, it does not send a list of publications. However, you can call or write to GPO and order up to six selected "subject bibliographies" free of charge, for example:

Air Pollution SB# 46
Environmental Protection SB# 88
Waste Management SB# 95
Water Pollution SB# 50

U.S. GPO - 202-783-3238

The National Technical Information Service

NTIS is a depository for federal government publications. In recent years, EPA has sent a majority of its publications to NTIS for distribution. A complete listing of all NTIS documents is contained in *U.S. Government Reports and Announcement Documents*, which is available at any U.S. Government Depository Library. (For a listing of all such libraries, contact GPO for a free directory.) NTIS publishes a listing of all EPA documents available from NTIS in *Recent EPA Publications Bibliography* (order no. PB-86-904 203, $22).

National Technical Information Service
U.S. Department of Commerce
5285 Port Royal Road
Springfield, VA 22161
703-487-4650

The General Accounting Office

The General Accounting Office is an independent government agency that provides oversight functions on behalf of Congress. Periodically, Congress requests that GAO conduct an investigation of a particular EPA or DOT program. Upon the completion of the investigation, GAO prepares and distributes its report. These reports provide in-depth information on the program of interest.

You can contact GAO and obtain a list of publications issued for the previous year. In addition, it will also conduct custom searches by subject matter.

The General Accounting Office - 202-275-6241

The Office of Technology Assessment

The Office of Technology Assessment (OTA) provides Congress with analyses and recommendations on particular government programs as requested. Many reports on selected EPA programs offer excellent summaries and insights. OTA does provide a list of publications.

The Office of Technology Assessment - 202-224-8996

Miscellaneous Publication Sources

Air	919-541-2777
Pesticides	703-557-4460
EPA's National Enforcement Investigations Center	303-236-5122
Drinking Water	202-260-5533
Water Planning & Standards	202-260-7115
Radiation	703-557-9351

REGULATORY AGENCIES

Federal Agencies

EPA Headquarters,
401 M Street, SW,
Washington, DC 20460

DOT Headquarters,
400 7th Street, SW,
Washington, DC 20590

EPA Regional Offices

Region I	JFK Federal Building Room 2203 Boston, MA 02203 617-565-3698
Region II	26 Federal Plaza New York, NY 10278 212-264-9628
Region III	841 Chestnut Street Philadelphia, PA 19107 215-597-9492
Region IV	345 Courtland Street, NE Atlanta, GA 30365 404-347-2234
Region V	230 South Dearborn Street Chicago, IL 60604 312-353-7579
Region VI	1445 Ross Avenue Suite 1200 Dallas, Texas 75270 214-655-6700
Region VII	726 Minnesota Avenue Kansas City, KS 66101 913-236-2930
Region VIII	1 Denver Place 999 18th Street Suite 500 Denver, CO 80202 303-293-7540
Region IX	75 Hawthorne St. San Francisco, CA 94105 415-744-1500
Region X	1200 Sixth Avenue Seattle, WA 98101 206-442-2782

States in EPA's Regions

Region I	Connecticut, Maine, Massachusetts, New Hampshire, Rhode Island, and Vermont
Region II	New Jersey, New York, and Puerto Rico
Region III	Delaware, District of Columbia, Maryland, Pennsylvania, Virginia, and West Virginia
Region IV	Alabama, Florida, Georgia, Kentucky, Mississippi, North Carolina, South Carolina, and Tennessee
Region V	Illinois, Indiana, Michigan, Minnesota, Ohio, and Wisconsin
Region VI	Arkansas, Louisiana, New Mexico, Oklahoma, and Texas

Region VII	Iowa, Kansas, Missouri, and Nebraska
Region VIII	Colorado, Montana, North Dakota, South Dakota, Utah, and Wyoming
Region IX	Arizona, California, Guam, Hawaii, and Nevada
Region X	Alaska, Idaho, Oregon, and Washington

State Contacts

Alabama	205-271-7730
Alaska	907-465-2671
Arizona	602-257-2270
Arkansas	502-562-7444 ext 504
California	916-445-4171
Colorado	303-320-8333 ext 4364
Connecticut	203-566-5712
Delaware	302-739-3689
Florida	904-488-0300
Georgia	404-656-2833
Hawaii	808-548-8548
Idaho	208-334-5879
Illinois	217-333-8941
Indiana	317-232-3292
Iowa	515-281-8693
Kansas	913-296-1590 ext 290
Kentucky	504-564-3350 ext 214
Louisiana	504-342-1354
Maine	207-289-2651
Maryland	301-631-3304
Massachusetts	617-292-5961
Michigan	517-373-2730
Minnesota	612-631-3774
Mississippi	601-961-4733
Missouri	314-751-3176
Montana	406-444-2821
Nebraska	402-471-2186
Nevada	702-885-4670
New Hampshire	603-271-2900
New Jersey	609-633-1408
New Mexico	505-827-0020
New York	518-457-3446
North Carolina	919-733-2178
North Dakota	701-224-2366
Ohio	614-466-7220
Oklahoma	405-271-5338
Oregon	503-229-5696
Pennsylvania	717-787-9870
Puerto Rico	809-722-1175
Rhode Island	401-277-2797
South Carolina	803-734-5200
South Dakota	605-773-3153
Tennessee	615-741-3424
Texas	512-458-7111
Utah	801-538-6101
Vermont	802-244-8755
Virginia	804-225-2667
Washington	206-459-6316
West Virginia	304-348-2107
Wisconsin	608-266-1327
Wyoming	307-777-7752

INDEX